T0186640

INTERFACIAL APPLICATIONS IN ENVIRONMENTAL ENGINEERING

SURFACTANT SCIENCE SERIES

ADDITIONAL VOLUMES IN PREPARATION

INTERFACIAL APPLICATIONS IN ENVIRONMENTAL ENGINEERING

edited by

Mark A. Keane

University of Kentucky
Lexington, Kentucky, U.S.A.

MARCEL DEKKER, INC.

NEW YORK • BASEL

ISBN: 0-8247-0866-0

This book is printed on acid-free paper.

Headquarters
Marcel Dekker, Inc.
270 Madison Avenue, New York, NY 10016
tel: 212-696-9000; fax: 212-685-4540

Eastern Hemisphere Distribution
Marcel Dekker AG
Hutgasse 4, Postfach 812, CH-4001 Basel, Switzerland
tel: 41-61-260-6300; fax: 41-61-260-6333

World Wide Web
http://www.dekker.com

The publisher offers discounts on this book when ordered in bulk quantities. For more
information, write to Special Sales/Professional Marketing at the headquarters address
above.

Current printing (last digit):
10 9 8 7 6 5 4 3 2 1

PRINTED IN THE UNITED STATES OF AMERICA

Preface

The contents of this book are based loosely on presentations at a special symposium, "Application of Interface Science to Environmental Pollution Control," held as part of the ACS National Meeting in Chicago, August 26–30, 2001. This symposium offered an opportunity for researchers from a range of disciplines to discuss the role of interface science in environmental remediation. The development of an archival book based on this meeting is a timely contribution to a burgeoning area of research that is now attracting the attention of a diverse research community. The topics covered include fundamental studies of general interest and/or overviews of strategies for pollution abatement—in short, any research that can lead to improvements in or protection of the quality of our air, water, and land.

The content is broad and encompasses subjects ranging from physical separations (e.g., adsorption, absorption, and ion exchange) to chemical reactions (e.g., catalytic oxidation and reduction, photocatalysis, and sensing). The book is structured to focus on the relevance of interface science to four topics critical to any study of environmental remediation: (1) NOx/SOx abatement, (2) water treatment, (3) application of catalysis to organic pollutant remediation, (4) waste minimization/recycle. Each contribution has either a theoretical significance or practical utility or both. *Interfacial Applications in Environmental Engineering* is an invaluable resource for chemists, chemical engineers, environmental scientists/engineers, environmental regulators, and the industrial sector. Moreover, it can serve as a comprehensive reference source to supplement educational coursework and both fundamental and applied research.

The contributions to this book come from a combination of scientists and engineers based in the United States, Canada, the United Kingdom, France, Spain, China, Japan, and Argentina. They serve to illustrate the global importance of interfacial science as applied to environmental protection. I must express my gratitude to all the authors, who have contributed their time and effort and willingness to share their research results in this collaborative effort. Special thanks go to Dr. Arthur Hubbard for his unswerving encouragement in getting this book project off the ground and his insightful advice in seeing it develop into an archival contribution to our understanding of what is and will undoubtedly continue to be an important dimension to the study of interface science.

Mark A. Keane

Contents

Part IV. Waste Minimization: Recycle of Waste Plastics

Contributors

Salmiaton Ali Environmental Technology Centre, Department of Chemical Engineering, University of Manchester Institute of Science and Technology, Manchester, United Kingdom

Isaac Asencio Department of Chemical Engineering, University of Castilla–La Mancha, Ciudad Real, Spain

David Atwood Department of Chemistry, University of Kentucky, Lexington, Kentucky, U.S.A.

Dorothée Berthomieu Laboratoire de Matériaux Catalytiques et Catalyse en Chimie Organique, ENSCM–CNRS, Montpellier, France

Heather A. Bullen Department of Chemistry, Michigan State University, East Lansing, Michigan, U.S.A.

Steven S. C. Chuang Department of Chemical Engineering, The University of Akron, Akron, Ohio, U.S.A.

Bernard Coq Laboratoire de Matériaux Catalytiques et Catalyse en Chimie Organique, ENSCM–CNRS, Montpellier, France

Gérard Delahay Laboratoire de Matériaux Catalytiques et Catalyse en Chimie Organique, ENSCM–CNRS, Montpellier, France

Fernando Dorado Department of Chemical Engineering, University of Castilla–La Mancha, Ciudad Real, Spain

Peter K. Dorhout Department of Chemistry, Colorado State University, Fort Collins, Colorado, U.S.A.

Lisa Dysleski Department of Chemistry, Colorado State University, Fort Collins, Colorado, U.S.A.

Sarah E. Frank Department of Chemistry, Colorado State University, Fort Collins, Colorado, U.S.A.

Arthur Garforth Environmental Technology Centre, Department of Chemical Engineering, University of Manchester Institute of Science and Technology, Manchester, United Kingdom

Simon J. Garrett Department of Chemistry, Michigan State University, East Lansing, Michigan, U.S.A.

Annick Goursot Laboratoire de Matériaux Catalytiques et Catalyse en Chimie Organique, ENSCM–CNRS, Montpellier, France

Trevor R. Griffiths Department of Chemistry, The University of Leeds, Leeds, United Kingdom

David H. Harris Engelhard Corporation, Iselin, New Jersey, U.S.A.

Richard H. Harris Department of Chemistry, Cardiff University, Cardiff, United Kingdom

Ian D. Hudson BNFL, Seascale, United Kingdom

Graham J. Hutchings Department of Chemistry, Cardiff University, Cardiff, United Kingdom

Xihai Kang Department of Chemical Engineering, The University of Akron, Akron, Ohio, U.S.A.

Mark A. Keane Department of Chemical and Materials Engineering, University of Kentucky, Lexington, Kentucky, U.S.A.

C. Y. Li College of Chemistry and Chemical Engineering, University of Petroleum, Dongying, Shandong Province, People's Republic of China

Xi Lin Department of Chemistry, Massachusetts Institute of Technology, Cambridge, Massachusetts, U.S.A.

Nicholas T. Loux National Exposure Research Laboratory, U.S. Environmental Protection Agency, Athens, Georgia, U.S.A.

Matthew Matlock Department of Chemistry, University of Kentucky, Lexington, Kentucky, U.S.A.

Norma S. Nudelman Department of Organic Chemistry, University of Buenos Aires, Buenos Aires, Argentina

Colin Park Synetix, Billingham, United Kingdom

Stella Maris Ríos Department of Chemistry, National University of Patagonia, Comodoro Rivadavia, Argentina

Amaya Romero Department of Chemical Engineering, University of Castilla–La Mancha, Ciudad Real, Spain

Yusaku Sakata Department of Applied Chemistry, Okayama University, Tsushima Naka, Japan

Paula Sánchez Department of Chemical Engineering, University of Castilla–La Mancha, Ciudad Real, Spain

H. H. Shan College of Chemistry and Chemical Engineering, University of Petroleum, Dongying, Shandong Province, People's Republic of China

Ron A. Shigeishi Department of Chemistry, Carleton University, Ottawa, Ontario, Canada

Steven H. Strauss Department of Chemistry, Colorado State University, Fort Collins, Colorado, U.S.A.

Stuart H. Taylor Department of Chemistry, Cardiff University, Cardiff, United Kingdom

Bernhardt L. Trout Department of Chemical Engineering, Massachusetts Institute of Technology, Cambridge, Massachusetts, U.S.A.

Azhar Uddin* Department of Applied Chemistry, Okayama University, Tsushima Naka, Japan

Jose L. Valverde Department of Chemical Engineering, University of Castilla–La Mancha, Ciudad Real, Spain

C. H. Yang College of Chemistry and Chemical Engineering, University of Petroleum, Dongying, Shandong Province, People's Republic of China

Q. M. Yuan College of Chemistry and Chemical Engineering, University of Petroleum, Dongying, Shandong Province, People's Republic of China

B. Y. Zhao College of Chemistry and Chemical Engineering, University of Petroleum, Dongying, Shandong Province, People's Republic of China

J. F. Zhang College of Chemistry and Chemical Engineering, University of Petroleum, Dongying, Shandong Province, People's Republic of China

J. S. Zheng College of Chemistry and Chemical Engineering, University of Petroleum, Dongying, Shandong Province, People's Republic of China

* *Current affiliation*: The University of Newcastle, Callghan, New South Wales, Australia.

Environmental Engineering at the Interface: An Overview

I. ENVIRONMENTAL POLLUTION

It is fair to state that public awareness and concern about the condition of the local and global environment have grown dramatically over the past decade. Such developments have resulted in the appearance of "environmental issues" on various political agenda. Environmental pollution can be anthropogenic and/or geochemical in nature but the principal source of appreciable pollution by organic and inorganic species is the waste generated from an array of commercial/industrial processes [1–5]. The entry of pollutants into the environment is linked directly to such effects as global warming, climate change, and loss of biodiversity. As a direct consequence, stringent legislation has been introduced to limit those emissions from commercial operations that lead to contamination of water/land/air [6,7]. The legislation imposed by the regulatory bodies is certain to become increasingly more restrictive, and the censure of defaulters is now receiving high priority in Europe and the United States. The latter has lent an added degree of urgency to the development of effective control strategies. In addition to the legislative demands, the economic pressures faced by the commercial sector in the 21st century include loss of potentially valuable resources through waste, escalating disposal charges, and increasing raw material/energy costs. Effective waste management must address [8,9] waste avoidance, waste reuse, waste recovery, and, as the least progressive option, waste treatment. The ultimate goal must now be the achievement of zero waste, the development of novel, low-energy, cleaner manufacturing technologies that support pollution avoidance/prevention at source, i.e., what has become known as "green" processing. Indeed, the groundswell of public opinion has been so great that manufacturers and advertisers have targeted green consumerism and the notion of "green marketing" has now taken hold [10]. Nonetheless, in times of recession, the "dark green" altruism whereby consumers will choose to purchase a more expensive but more environmentally friendly product is unlikely. A shift in attitude from "green at any price" to "greener than before" gained ground as the economic recession began to bite in the 1990s. A "sustainable development" rather than no development is generally viewed as a viable option, with an emphasis on conserving natural resources through better management.

Urban air pollution became a decided concern during the period of rapid industrialization in Europe and North America that began in the late 18th century [11]. The British Parliament passed the Alkali Act in 1862 and the Rivers Pollution Act in 1876 to combat excessive emissions that were clearly responsible for adverse public health impacts. From the outset, a conflict was extant between a curtailment of manufacturing, with a consequent reduction in employment/prosperity, and human health concerns. Environmental protection has always been bedeviled by compromise of this nature. Until recently, chemical industries, in the main, ran output-oriented processes in which raw material processing generated the target product and "unavoidable" waste. There is now a concerted move toward a holistic approach, a comprehensive examination of every aspect of an industrial process, from raw material input/process operability to the final output. This has led to the "life cycle" concept involving a full assessment of the environmental burdens associated with a product, process, or activity. A comprehensive life cycle assessment (LCA) collects, analyzes, and assesses the associated environmental impacts "from cradle to grave" [12]. The LCA scope can be narrow or broad and is an invaluable and progressive production tool to facilitate pollution prevention and possible energy savings.

In many instances, pollution has been considered inevitable and necessary in large technologically advanced communities. Nevertheless, environmental regulatory bodies designate emission limits and quality limits and strongly encourage the application of green commercial technology. The main factors contributing to environmental deterioration are population growth, affluence, and technology. Poor air quality in large cities, caused by excessive motor traffic, remains a grave cause for concern, although this has been somewhat alleviated through the phasing out of Pb additives in gasoline and the use of catalytic converters. Energy generation can be considered the most ubiquitous cause of pollution, with the established appreciable environmental damage associated with coal mining, petroleum extraction/refining, and fossil fuel combustion [3]. Biomass represents a renewable energy source but has the decided drawback of polycyclic aromatic hydrocarbon (PAH) production, while an indiscriminate harvesting of wood and combustible vegetation can result in irretrievable land degradation. The past five decades have seen ever-increasing chemical synthesis activity, and it is estimated that some 10,000 chemicals are in current commercial use. An explicit link between exposure of these chemicals and human health complaints (respiratory and neurological) is now forthcoming; this book focuses on the treatment of some well-established environmental toxins. The levels of indoor air pollution (in the home and workplace) can exceed those recorded in an urban outdoor environment [3,13], a fact that is particularly worrying given the prolonged exposure times and the emergence of "sick office" syndrome. Typical indoor pollutants comprise a complex mixture of volatile organic compounds (VOCs), nitrogen oxides (NOx), and carbon oxides (CO/CO_2).

The study of the causes, effects, and control of pollution remains a fast-moving field of research, characterized by changes of emphasis and often of perception. Authoritative scientific data with a solid interpretative basis are essential to ensure significant progress in terms of environmental pollution control. It is now accepted that a progressive approach to chemical processing must embrace waste reduction, chemical reuse/recycling, and energy recovery [14]. These issues are addressed from a number of perspectives in this book, which focuses on four topics that underpin the role that interfacial science must play in environmental protection: (1) NOx/SOx abatement, (2) water treatment: heavy metal and organic removal, (3) catalytic approaches to organic pollutant remediation, (4) waste minimization: recycle of waste plastics. A brief treatment of the generic aspects of these four topics follows.

II. NOx/SOx ABATEMENT

The growth in environmentalism has seen the introduction of the concept of "environmental quality," which is typically applied to the air we breathe and the water we consume. In terms of NOx/SOx, if the air contains more than 0.1 parts per million (ppm) NO_2 or SO_2, persons with respiratory complaints may experience breathing difficulties; if it contains more than 2.5 ppm NO_2 or 5 ppm SO_2, healthy persons can also be affected [15]. Policymakers have acknowledged the potential dangers posed by excessive NOx/SOx release, and the Kyoto Protocol sets out measures to reduce such emissions by the year 2008 to levels below those recorded in 1990 [6]. NOx/SOx release is explicitly linked to consumption of fossil fuel, i.e., coal, oil, and natural gas. Even allowing for steady improvements in energy efficiency, future generations will use massive quantities of energy. If current trends prevail and this demand is met by burning fossil fuels, the environmental implications are grave. Energy technologies drawing on renewable energy serve to minimize the negative environmental impacts associated with the fossil fuel cycle. Such technologies, which are either reasonably well established or in the formative stage, convert sunlight, wind, flowing water, the heat of the earth and oceans, certain plants, and other resources into useful energy. The use of renewables can still impact the environment, but the effect is far smaller than that of the present dependence on deployment of nonrenewable resources. Because vehicle exhaust contains appreciable levels of toxic emissions, much can be done to alleviate the environmental burden through economies of motor fuel consumption and engine/combustion modifications. Fuel cell developments suggest that these devices will make a valuable contribution to future power generation [16]. Fuel cells that operate on pure hydrogen as fuel produce only water as byproduct, thus eliminating all emissions associated with standard methods of electricity production. Hydrogen production/storage remains something of an obstacle in fuel cell commercialization. Fuel cells have yet to make a serious impression on the energy market, and mass market zero-emission automobiles are far from realization.

The most abundant nitrogen oxide in the environment is nitrous oxide (N_2O), which is relatively unreactive and not regarded as a primary pollutant. Nitric oxide (NO) and nitrogen dioxide (NO_2) comprise the predominant atmospheric burden and are denoted by the collective term NOx [17,18]. NOx is produced mainly in high-temperature combustion processes involving atmospheric nitrogen (or as a fossil fuel/biomass component) and oxygen and is associated with power stations, refineries, transport, agriculture, and domestic applications. In addition to contributing, as a heat-trapping pollutant, to the greenhouse effect, NOx directly impacts on the environment in three ways [3]: depletion of the ozone layer, production of acid rain, and general air pollution. Of the two oxides of sulfur, SO_2 and SO_3 (collectively SOx), the former is far more abundant in the atmosphere [19]. Sulfur dioxide reacts on the surface of a variety of airborne solid particles, is soluble in water, and can be oxidized within airborne water droplets. Natural sources of sulfur dioxide include releases from volcanoes, oceans, biological decay, and forest fires. The most important man-made SO_2 sources are fossil fuel combustion, smelting, manufacture of sulfuric acid, conversion of wood pulp to paper, incineration of refuse, and production of elemental sulfur. Coal and oil burning are the predominant sources of atmospheric SOx, which can contribute to respiratory illness, alterations in pulmonary defenses, and aggravation of existing cardiovascular disease. In the atmosphere, SOx mixes with water vapor, producing sulfuric acid, which can be transported over hundreds of kilometers and deposited as acid rain [20]. Sulfur dioxide and the sulfuric acid that it generates have four established adverse effects: (1) toxicity to humans, (2) acidification of lakes and surface waters, (3) damage to trees and crops, and (4) damage to buildings.

Control of NOx/SOx emissions can follow two strategies [21]: a direct curtailment of NOx/SOx formation (primary measures), and a secondary, downstream treatment (end-of-pipe solutions). Effective emissions reduction requires controls on both stationary and mobile sources. One viable approach in reducing NOx production focuses on fuel denitrogenation, in which the nitrogen component is removed from liquid fuels by intimate mixing with hydrogen at elevated temperatures to produce ammonia and cleaner fuel. This technology can reduce the nitrogen contained in both naturally occurring and synthetic fuels. In any fuel combustion application, combustion control can focus on [22]: (1) reduction of the peak temperature in the combustion zone, (2) lowering gas residence time in the high-temperature zone, and (3) reduction of oxygen concentration in the combustion zone. Process modifications can include staged combustion, flue-gas recirculation, and water/steam injection [23]. Flue-gas treatment is highly effective in reducing NOx emissions and can call on selective catalytic and noncatalytic reduction. An effective lowering of SOx production typically involves flue-gas desulfurization (reaction of SO_2 with lime) or fluidized-bed combustion [24].

The environmental damage caused by NOx release is addressed in Chapters 1–3 of this book. Delahay et al. (Chapter 1) consider and assess the options available to limit NOx production and discuss in some detail the role of zeolite-based catalysts, drawing on quantum chemical calculations to gain an insight into the architecture of the surface active sites. Kang and Chuang (Chapter 2) focus on selective catalytic reduction (SCR) using Cu-ZSM-5, employing in situ FTIR to probe the nature of the surface reaction as a means of enhancing N_2 production and limiting CO_2 formation. Valverde et al. continue this theme in Chapter 3 and consider SCR of NO by propene promoted by Cu-ZSM-5 and Ti-based pillared clays, in which the redox cycle associated with the supported Cu cations is shown to be critical in governing SCR efficiency. The implications of SOx release and possible remediation actions are addressed in Chapters 4 and 5. Lin and Trout (Chapter 4) provide a comprehensive review of the chemistry of sulfur oxides on transition metals, in which the emphasis is on controlling automobile emissions. Li et al., in Chapter 5, examine catalytic strategies for improved gasoline desulfurization.

III. WATER TREATMENT: HEAVY METAL AND ORGANIC REMOVAL

Water is perhaps the most fundamental of resources; without it, as the cliché has it, life could not exist on land. Water pollution has been defined as "the introduction by man into the environment of substances or energy liable to cause hazards to human health, harm to living resources and ecological systems, damage to structure or amenity, or interference with legitimate uses of the environment" [17,25]. Water quality is typically assessed on the basis of three easily measured parameters [26]: pH, conductivity, and color. The study of water contamination can be conveniently divided into two groups of pollutants—organic and inorganic (heavy metals)—an approach taken in this book. It has to be borne in mind that, with the exception of synthetic elements/nuclides, all pollutant metals are naturally present in the aquatic environment, where the concentrations are the result of intricate biogeochemical cycles operating over time scales of thousands to millions of years. Heavy metals are widely used in electronics and "high-tech" applications and tend to reach the environment from an array of anthropogenic sources. Some of the "oldest" cases of environmental pollution are due to mining and smelting of Cu, Hg, and Pb. The fate and overall impact of any pollutant metal that enters the aquatic environment are difficult, if not impossible, to assess given the prevailing complex interrelated bioprocesses/cycles. The extent of organic pollution can be quantified in terms of biological/biochemical oxygen demand (BOD) or chemical oxygen demand (COD) and total organic carbon [27]. Sources of organic pollution include an array of commercial chemical plants, sewage treatment works, breweries, dairies, food processing plants, and, with the intensification of livestock rearing, agricultural effluent. A concentrated discharge of organic material into natural waterways is broken down by microorganisms that utilize oxygen, to the detriment of the stream biota [17]. The ecological impact is dependent on the nature and concentration of the organic discharge and the rate of transport/dispersion, which is controlled [28] by advection (mass movement) and mixing or diffusion (without net movement of water). Water movement/turbulence affects solid particulate suspension that can occlude light, thereby eliminating photosynthetic organisms [25]. The release of nutrients during the breakdown of organic matter and discharge of phosphates (in particular) stimulate the growth of aquatic plants, a process termed *eutrophication* that results in a decline in aquatic species diversity [29]. The introduction of such toxic pollutants as heavy metals, pesticides, herbicides, PCBs, phenols, acids, and alkalis can have acute or cumulative toxic effects. The World Health Organization has set guideline values for acceptable levels of heavy metals/organics in drinking water [30].

The chemistry of heavy metals in natural water is extremely complex because of the virtual cocktail of organic and inorganic components that participate in a range of (possibly redox) steps (notably complexation and adsorption) responsible for metal speciation [17,31]. Effective water treatment strategies to remove excessive organic/inorganic contaminants can draw on these naturally occurring processes. In general, industrial wastewaters are more readily and most economically treated in admixture with domestic wastewaters rather than in isolation. Water treatment methodologies can be classified as biological, chemical, and physical. Biological treatment can be divided into aerobic and anaerobic and further subdivided into dispersed growth and fixed film, in which tolerance level is a critical issue [32]. In the treatment of toxic waste, a microbial population must be developed that is acclimatized to the presence of the toxin and, in the case of degradable toxins, a sufficient concentration of organism capable of metabolizing the toxin must be in place. The established physical methods that serve to separate, concentrate, and recover (potentially valuable material) include solvent extraction, reverse osmosis, ion exchange, and adsorption. Chemical treatment of heavy metals typically involves pH adjustment to facilitate sedimentation. Organic contamination can be tackled by chemical means, usually involving some form of catalyzed oxidation, in which case care must be taken to avoid any toxic byproducts. These issues are evaluated and discussed in Chapters 6–11.

The fundamental and applied aspects of heavy metal removal from water forms the basis of Chapters 6–8. Keane (Chapter 6) examines the role of synthetic zeolite ion exchange materials in batch and continuous heavy metal remediation and considers the feasibility of metal recovery/zeolite reuse. Dysleski and co-workers describe (in Chapter 7) the action of a new class of stable spinel-like materials that are effective in the ion exchange of Hg^{2+} and Pb^{2+} from aqueous waste. In Chapter 8, Matlock and Atwood discuss the various chemical methods to chemically bind heavy metals and facilitate precipitation. Pollution by organic compounds and possible remediation strategies are examined in Chapters 9 and 10. Nudelman and Ríos (Chapter 9) consider the impact of oil residues on the environment and propose that an adsorption on natural solids is a viable clean-up methodology. Park and Keane, in Chapter 10, focus on the problem of phenolic waste contamination and consider the feasibility of employing novel carbon na-nofibers as effective adsorbents. Loux (Chapter 11) tackles the complexities associated with adsorption on environmental surfaces and addresses the strengths and limitations of the existing models.

IV. CATALYTIC APPROACHES TO ORGANIC POLLUTANT REMEDIATION

Two terms and their acronyms are widely used in environmental remediation circles to categorize organic-based pollutants [33,34], i.e., volatile organic compounds (VOCs) and persistent organic pollutants (POPs). The VOCs encompass a broad range of substances that are easily vaporized and that contain carbon and different proportions of other elements, such as hydrogen, oxygen, fluorine, chlorine, bromine, sulfur, and nitrogen. A significant number of the VOCs are commonly employed as solvents (paint thinner, lacquer thinner, degreasers, and dry cleaning fluids). The POPs are chemical substances that persist in the environment, bioaccumulate through the food web, and pose a risk of adverse effects to human health and the environment. The POPs regarded as high-priority pollutants include a range of halogenated compounds: pesticides such as DDT, industrial chemicals such as polychlorinated biphenyls (PCBs), and unwanted industrial byproducts such as dioxins and furans. The POPs and VOCs are of anthropogenic origin, associated with industrial processes, product use and applications, waste disposal, leaks and spills, combustion of fuels, and waste incineration. With the evidence of long-range transport of these substances to regions where they have never been used or produced, it is now clear that POPs and VOCs pose a serious immediate threat to the global environment.

Catalysis, particularly heterogeneous catalysis, has always had an environmental dimension, for the deployment of catalysts ensures lower operating temperatures and/or pressures, with a resultant reduction in fuel usage/waste production. The emergence of "environmental catalysis" as a discipline has focused on the development of catalysts to either decompose environmentally unacceptable compounds or provide alternative catalytic syntheses of important compounds without the formation of environmentally unacceptable byproducts. Catalysts now play key roles in the production of clean fuels, the conversion of waste and green raw materials into energy, clean combustion, including control of NOx/SOx and soot production, reduction of greenhouse gases, and water treatment. The challenges associated with the growing demand for clean technology and zero-waste processes can be met through novel catalytic strategies that alleviate the dependence on industrial solvents and the need for solvent vaporization. The role of environmental catalysis in organic pollution control is addressed in Chapters 12–15 and it is evaluated against applicable noncatalytic approaches. Taylor and co-workers record in Chapter 12 that uranium oxide–based catalysts are highly effective in the oxidative destruction of benzene and propane as model VOCs, in which water can serve as both a promoter and an inhibitor. Keane (Chapter 13) presents catalytic hydrodehalogenation over supported nickel as a viable low-energy, nondestructive means of transforming highly recalcitrant haloarene gas streams into reusable raw material. In Chapter 14, Bullen and Garrett investigate the fundamental issues that underpin the photocatalytic properties of TiO_2 and highlight some interesting applications in environmental remediation. Trevor Griffiths' Chapter 15 completes this section with a demonstration of a novel application of interfacial chemistry in fuel combustion that calls on the catalytic action of a Pt/Rh aerosol.

V. WASTE MINIMIZATION: RECYCLE OF WASTE PLASTICS

The concept of "waste minimization" encompasses the reduction of waste at its source, combined with environmentally sound recycling. Even when hazardous wastes are stringently regulated and managed, they can pose environmental concerns, and accidents during handling/transportation can result in significant releases to the environment. "Waste," in this context, represents material that was not used for its intended purpose or unwanted material produced as a consequence of a poorly controlled process. Waste minimization fits within the ethos of the "waste management hierarchy," which sets out a preferred sequence of waste management options [38,39]. The first and most preferred option is source reduction; the next preferred option is recycling—the reclamation of useful constituents of a waste for reuse or the use/reuse of a waste as a substitute for a commercial feedstock. Although it is impossible to have an entirely "clean" manufacturing process, any associated toxic waste can be reduced significantly through better process control or avoided entirely by an alternative process.

The final two chapters address a specific aspect of waste minimization, one that is growing in ever-increasing importance: recycle of waste plastics. Liquefaction of waste plastics into fuel oil by thermal or catalytic degradation is emerging as a progressive means of waste reuse as a potential energy source. The amount of waste plastics is increasing annually worldwide, and disposal by landfilling and incineration can no longer be regarded as viable options, due to limited landfill space and the possibility of appreciable environmental toxin production during incineration. In Chapter 16, Ali et al. provide a general overview of catalytic polymer recycling and assess the viability of employing "fresh" and "used" catalysts as recycle agents, with a consideration of economic factors. Uddin and Sakata, in the final chapter, consider the recycling of halogen-containing polymers, notably PVC. A thermal degradation of PVC-based waste will generate a range of chlorine-containing organic compounds that cannot be used as fuel. To circumvent this problem, Uddin and Sakata have undertaken a comprehensive study of catalytic dehalogenation to selectively remove the halogen component and facilitate the production of a waste-plastic-derived oil.

VI. SUMMATION

While global environmental systems are extremely resilient, there is a limit to the pollution burden that can be sustained. An unabated entry of heavy metals, toxic organics, and NOx/SOx into the environment will undoubtedly result in dramatic adverse effects on human health, agricultural productivity, and natural ecosystems. Interfacial science has a role to play in this abatement. It is hoped that the original research and remediation evaluations presented in this book serve to illustrate the extent of this role.

REFERENCES

1. RE Hester, RM Harrison. Volatile Organic Compounds in the Environment, Issues in Environmental Science and Technology. Cambridge, UK: Royal Society of Chemistry, 1995.
2. Toxics Release Inventory, Public Data Release. Washington DC: USEPA, Office of Pollution Prevention and Toxics, 1991.
3. BJ Alloway, DC Ayres. Chemical Principles of Environmental Pollution. Oxford: Chapman & Hall, 1993.
4. DT Allen, DR Shonnard. Green Engineering: Environmentally Conscious Design of Chemical Processes. Upper Saddle River, NJ: Prentice Hall, 2002.
5. JW Davis. Fast Track to Waste-Free Manufacturing: Straight Talk from a Plant Manager (Manufacturing and Production). Portland, OR: Productivity Press, 1999.
6. C Rolf. Kyoto Protocol to the United Nations Framework Convention on Climate

    Change: A Guide to the Protocol and Analysis of Its Effectiveness. Vancouver: West
    Coast Environmental Law Association, 1998.

 7. JF McEldowney, S McEldowney. Environment and the Law. Essex, UK: Longman,
    1996.

 8. MD Lagrega, PD Buckingham, JC Evans. Hazardous Waste Management. 2nd ed.
    New York: McGraw-Hill, 2000.

 9. HM Freeman, ed. Standard Handbook of Hazardous Waste Treatment and Disposal.
    2nd ed. New York: McGraw-Hill, 1997.

10. J Ottman. Green Marketing: Challenges and Opportunities for the New Marketing
    Age. New York: NTC, 1994.

11. K Thomas. Man and the Natural World. London: Penguin Books, 1983.

12. R Welford, A Gouldson. Environmental Management and Business Strategy. Lon-
    don: Pitman, 1993.

13. JD Spengler, JF McCarthy, JM Samet. Indoor Air Quality Handbook. New York:
    McGraw-Hill, 1988.

14. D Ellis. Environments at Risk: Case Histories of Impact Assessment. Berlin:
    Springer Verlag, 1989.

15. CE Kupchella, MC Hyland. Environmental Science. 2nd ed. Needham Heights, MA:
    Allyn & Bacon, 1986.

16. L Carrette, KA Friedrich, U Stimming. Fuel cells: principles, types, fuels and appli-
    cations. Chem Phys Chem 1:162–193, 2000.

17. RM Harrison, ed. Pollution: Causes, Effects and Control. 2nd ed. Cambridge, UK:
    Royal Society of Chemistry, 1990.

18. T Schneider, L Grant. Air Pollution by Nitrogen Oxides: Studies in Environmental
    Science. Amsterdam: Elsevier Science, 1982.

19. D Van Velzen. Sulphur Dioxide and Nitrogen Oxides in Industrial Waste Gases:
    Emission, Legislation, and Abatement. Dordrecht, Netherlands: Kluwer, 1991.

20. AH Legge, SV Krupa, eds. Acidic Deposition: Sulfur and Nitrogen Oxides. Chelsea,
    MI: Lewis, 1990.

21. A. Tomita, ed. Emissions Reduction: NOx/SOx Suppression. Amsterdam: Elsevier
    Science, 2001.

22. CD Cooper, FC Alley. Air Pollution Control: A Design Approach. Prospect Heights,
    IL: Waveland Press, 1986.

23. AJ Bounicore, WT Davis, eds. Air Pollution Engineering Manual. New York: Van
    Nostrand Reinhold, 1992.

24. M Allaby. Basics of Environmental Science. London: Routledge, 1996.

25. MW Holdgate. A Perspective of Environmental Pollution. Cambridge, UK: Cam-
    bridge University Press, 1979.

26. S Watts, L Halliwell, eds. Essential Environmental Science. London: Routledge,
    1996.

27. VD Adams. Water and Wastewater Examination Manual. Chelsea, MI: Lewis, 1990.

28. RE Hester, ed. Understanding our Environment. Cambridge, UK: Royal Society of
    Chemistry, 1986.

29. JE Middlebrooks. Modelling the Eutrophication Process. Ann Arbor, MI: Ann Arbor
    Press, 1974.

30. Guidelines for Drinking-Water Quality: Health Criteria & Other Supporting Information. World Health Organization, 1996.
31. CJM Kramer, JC Duinker, eds. Complexation of Trace Metals in Natural Waters. The Hague: Nijhoff/Junk, 1984.
32. NP Cheremisinoff. Biotechnology for Waste and Wastewater Treatment. Park Ridge, NJ: Noyes, 1996.
33. HF Rafson, ed. Odor and VOC Control. New York: McGraw-Hill, 1998.
34. S Harrad. Persistent Organic Pollutants: Environmental Behaviour and Pathways for Human Exposure. Dordrecht, Netherlands: Kluwer, 2001.
35. JN Armor. Environmental Catalysis. Washington, DC: American Chemical Society, 1994.
36. FJJG Janssen, RA van Santen, eds. Environmental Catalysis. London: Imperial College Press, 1999.
37. G Ertl, H Knozinger, J Weitkamp, eds. Environmental Catalysis. New York: Wiley, 1999.
38. HM Freeman. Hazardous Waste Minimization. New York: McGraw-Hill, 1990.
39. F Domenic. The Role of Waste Minimization. Washington, DC: National Governors' Association, 1989.

Mark A. Keane

INTERFACIAL APPLICATIONS IN ENVIRONMENTAL ENGINEERING

1

Zeolite-Based Catalysts for the Abatement of NOx and N₂O Emissions from Man-Made Activities

GÉRARD DELAHAY, DOROTHÉE BERTHOMIEU, ANNICK GOURSOT, and BERNARD COQ Laboratoire de Matériaux Catalytiques et Catalyse en Chimie Organique, ENSCM–CNRS, Montpellier, France

I. INTRODUCTION

It is a truism to assert that man-made activities have an impact on the environment. But the negative effect of this impact has been growing exponentially from the beginning of the industrial era. That concerns the development of harmful products, of dangerous and/or energy-inefficient processes, and unsafe waste streams. Many of these environmentally damaging associated issues are of chemical origin, so one should therefore state "What a chemist knows how to make, he has to know how to unmake." To that end, catalysis is of vital importance to promote greener and/or energy-saving processes and cleaner fuels and to reduce pollutants emissions in gaseous and liquid streams. The main pollutants in gaseous emissions concern: volatile organic compounds (VOCs), greenhouse gases, NOx, and SOx. We will present only the abatement of NOx and N_2O emissions, which can potentially be treated by zeolites.

For 8000 years, the temperature of earth's atmosphere has stayed constant, but a sudden rise has been occurring since the last century ($+1°C$) with the concurrent increase of CO_2 concentration from 280 ppm (in 1860) to 350 ppm at present. This is due to global warming from extra emissions of greenhouse gases from anthropogenic activities: CO_2, CH_4, N_2O, O_3, CFCs. The contributions to global warming effect, which integrates the emission flows and the global warming potential, are ca 81% for CO_2, 7% for CH_4, and 9% for N_2O. Policymakers acknowledged the potential dangers of these emissions and implemented the Kyoto Protocol in 1997 to reduce emissions by the year 2008 by 7% (U.S.), 8% (E.U.), or 6% (Japan) below 1990 levels. The gases of main concern were CO_2, N_2O, and

CH_4, but only emissions of N_2O can be controlled by technologies using zeolites. Besides its contribution to global warming, N_2O contributes to the depletion of the stratospheric ozone layer. N_2O emissions amount to $4-13 \times 10^6$ ton-N year^{-1} from agriculture ($\approx 43\%$), biomass burning ($\approx 18\%$), vehicles ($\approx 13\%$), and power plants and industrial processes ($\approx 23\%$) [1]. These last are the easiest to control.

NOx (NO + NO_2) emissions are responsible for acid rain (deforestation), photochemical smog (health disease), and intensification of ground-level ozone. Total emissions of NOx amount to ca 45×10^6 ton-N year^{-1}, with 75% from anthropogenic activities. On a global scale, the major sources are the combustion of fossil fuels ($\approx 48\%$), biomass burning ($\approx 16\%$), decomposition in soils ($\approx 13\%$), and lightning ($\approx 10\%$). NOx emitted from man-made activities are distributed among mobile sources ($\approx 70\%$) and stationary sources ($\approx 30\%$), with 20% from industrial processes and 80% from large combustion plants. Obviously, emissions from stationary sources are the easiest to control, and technologies with zeolites are currently in use.

Some issues regarding technological achievements, with or without zeolites, by catalysis for pollution abatement were reviewed recently [2,3], as was a survey of the best available technologies for reducing NOx and N_2O emissions from industrial activities [4].

II. CONTROL OF N₂O EMISSIONS

Emission levels for N_2O are expected to become regulated in the near future, and are already imposed by taxes in France. There is therefore a strong incentive toward emission control. In principle, two methods are available, i.e., reducing N_2O formation (primary measures) and after-treatment (end-of-pipe solutions). Regarding the end-of-pipe solutions, there are two very different situations, depending on the N_2O concentration in the tail gas. The treatment of highly concentrated N_2O streams (20–40%) coming from adipic acid plants, glyoxal plants, etc. is nearing completion. Two alternatives exist regarding catalytic routes. One is based on the catalytic decomposition under adiabatic conditions; due to the high exothermicity of the process, the temperature ranges between 770 and 1070 K, and the catalytic materials are composed of promoted mixed oxides for their thermal stability. The second route, patented in 1988 [5,6], which does not seem industrially implemented to date, is based on the valorization of N_2O as a strong and selective oxidant of benzene to phenol. The preferred catalyst is Fe-ZSM-5, in which active iron species will be composed of binuclear oxocations as extra-framework cationic species [7]. Good yield and near 100% selectivity to phenol were claimed when the reaction was carried out at ca 623 K. It was recently reported that the generation of active iron species is well achieved by steaming at 873 K of framework-incorporated [Fe]ZSM-5 [8]. The steaming causes the breaking of Fe—O—Si bonds and leads to well-dispersed extraframework iron species. Details about some postulated structures of these species will be given later. The same iron-exchanged zeolites are also active in the selective oxidation by N$_2$O of methane to methanol [7,9].

In contrast, the treatment of streams from nitric acid plants and power plants with low N$_2$O concentrations remains a challenge. In this respect, catalysis is of vital importance in N$_2$O removal technologies, which are based on the catalytic decomposition of N$_2$O to N$_2$ and O$_2$ without or with the help of a reductant. The catalytic decomposition of N$_2$O (2N$_2$O $\rightarrow$ 2N$_2$ + O$_2$) can be described, in its simplest form, by Eqs. (1)–(3):

$$N_2O + * \rightarrow N_2 + O* \tag{1}$$

$$N_2O + O* \rightarrow N_2 + O_2 + * \tag{2}$$

$$2O* \leftrightarrow O_2 + 2* \tag{3}$$

In the presence of a reductant, e.g., CO, C$_3$H$_6$, or NH$_3$, the surface oxygen O* can also be removed according to:

$$CO + O* \rightarrow CO_2 + * \tag{4}$$

$$C_3H_6 + 9O* \rightarrow 3CO_2 + 3H_2O + 9* \tag{5}$$

$$2NH_3 + 3O* \rightarrow N_2 + 3H_2O + 3* \tag{6}$$

Among a huge number of catalytic formulations that have been evaluted for the reaction, transition-metal-ion-exchanged zeolites (TMI-zeolite) have shown high activities (see Refs. 10 and 11 for a review). Moreover, they are not, or weakly, inhibited by the presence of excess O$_2$ [Eq. (3) backwards], as compared to transition metal oxide catalysts. This is the remarkable case of MOR, ZSM-5, FER, BEA exchanged with Co or Fe, which do not suffer any inhibition by O$_2$ and exhibit the highest activity. These materials have received particular attention, and it was shown that the preparation protocol employed for Fe-ZSM-5 has a very great influence on the catalytic performances with respect to activity and stability [12–15]. Depending on whether the catalyst has been prepared by aqueous ion exchange, solid-state exchange, chemical vapor deposition, incorporation of Fe in the framework, or dryness impregnation, various Fe species have been identified. Even though it is difficult to compare the properties of catalysts prepared by various groups, it would seem that the most efficient materials, regarding Fe-ZSM-5, will be prepared from hydrothermal treatment at 800–900 K of [Fe]ZSM-5, with Fe isomorphously substituted in the framework [8,15]. Upon steaming, Fe—O—Si bonds are broken, which leads to the generation of the active iron species. A wide variety of species have been identified, or postulated, in the final catalyst. There is no general agreement as to the exact nature of the active species, except on one point: The large iron oxide aggregates exhibit very poor activity for N$_2$O conversion.

From spectroscopic studies (ESR, Mössbauer, EXAFS, etc.), thermal analysis (TPR, TPO, etc.), and quantum chemical modeling, several models have been designed for the architecture of the highly reactive oxo-cation formed by interaction of N_2O with Fe-ZSM-5. Panov et al. [7], prompted by a biological model, designed a binuclear Fe complex with various peroxide bridges. Because of the sharp increase of rate with Fe loading in Fe-ZSM-5, prepared from CVD or solid-state exchange, El-Malki et al. [14] also concluded that binuclear sites and nanoclusters are more active for N_2O decomposition than mononuclear sites. Without any more precision about the nuclearity of the species, Ribera et al. [8] proposed that high-spin rhombic iron, with signal at $g = 6$ in ESR spectra, are active species for N_2O activation after steaming of [Fe]ZSM-5. From ab initio quantum chemical calculations, Arbuznikov and Zhidomirov [16] confirmed the probability of binuclear oxo-cations with peroxide bridges and hydroxyl groups. Lázár et al. [17] also proposed an oxygen bridge between two Fe atoms, one of the two belonging to the zeolite framework. In contrast, in the reduction of N_2O + O_2 by NH_3 Fe-BEA exhibited a strong increase of TOF, a mole of N_2O transformed per mole of Fe, at low Fe content [18]. It was proposed that the treatment of Fe^{2+}-BEA by N_2O, obtained from reduction of exchanged and calcined samples, led to the formation of both mononuclear and binuclear Fe oxo-cations. The former are more easily reduced by H_2 and CO, and their formations are preferred at low Fe content [19]. From DFT computations, Yoshizawa et al. [20] proposed that active sites in Fe-ZSM-5 should have relevance to mononuclear iron-oxo species of the type $(FeO)^+$. It was also proposed that the reactivity of Fe-TON in N_2O decomposition should be attributed mainly to Fe species of framework origin [21]. Grubert et al. [22] have prepared materials Fe-ZSM-5 and Fe-MCM-41, the former containing a majority of iron-oxo nanoclusters of typical-size Fe_4O_4, while iron is present mainly in the form of isolated cation in the latter. They concluded that both species are active sites for N_2O reaction. It is clear that the debate about the exact nature of these sites still remains open in Fe-zeolite catalysts. Moreover, compared to the rigidity of the MFI framework, the BEA is very flexible, which allows (1) reversible changes from tetrahedral to octahedral coordination of Al (and Fe maybe) and (2) the presence of defects, dangling bonds, and partially coordinated aluminic fragments. That could explain the specificity of Fe-BEA, as compared to Fe-ZSM-5.

From these reported works, only one clear feature emerges about the role of the zeolite for the high activity of transition metal: to keep it in a highly dispersed state as cations or oxo-cations of low nuclearity. Moreover, most of the attempts to identify the nature of Fe species, as a function of zeolite structure or protocol of preparation, have been carried out ex situ before the reaction. It will be of paramount importance for future work to look at the identification of Fe species in running conditions.

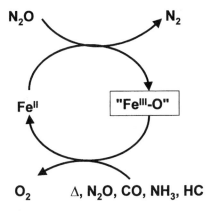

FIG. 1 Catalytic cycle of N$_2$O on Fe-zeolite catalysts.

The catalytic decomposition of N$_2$O seems regulated mainly by the cycle of Fe^{II}/Fe^{III} (Fig. 1). N$_2$O interacts with a reduced "Fe^{II}" site [Eq. (1)] to yield an extra-lattice oxygen "Fe^{III}-O" species, the so-called α-oxygen by Panov et al. [7]. Most studies have concluded that the removal of α-oxygen [Eq. (2)] exhibits the lowest rate constant. The remarkable behavior of Fe exchanged in some zeolites could be ascribed to the occurrence of "easily reduced and completely reversible" Fe oxo-cation sites [23]. TPR experiments by H$_2$ and CO of N$_2$O-treated Fe-ZSM-5 [14] and Fe-BEA [19] have shown that such "Fe^{III}-O" sites are much more reducible than those formed by O$_2$ treatment. The influence of the zeolite on the reducibility of Fe species could be understood from quantum chemical calculations (DFT method) on model clusters of FAU and BEA containing Cu^{II}, Co^{II}, and Fe^{II} cations [24, vide infra]. The calculations indicate that a charge transfer from zeolite to TMI occurs for ca. one electron and that the TMI-zeolite system behaves like a supermolecule.

However, several recent reports have provided good evidence that the interaction of N$_2$O with Fe-exchanged zeolites is not as simple as Eq. (1) indicates. It was well demonstrated that upon contacting N$_2$O with reduced Fe-ZSM-5 [14,22] or Fe-BEA [25] in the absence of O$_2$, besides the breaking of the NN–O bond, which leaves N$_2$ and adsorbed oxygen, surprisingly there is also the breaking of the stronger N–NO bond to yield absorbed nitric oxide:

$$N_2O + * \rightarrow \frac{1}{2}N_2 + NO* \tag{7}$$

Regarding the Fe-BEA sample, desorption of O_2 and NO occurred at ca. 610 and 690 K, respectively [25]. Distribution of the N_2O interaction between the two routes represented by Eqs. (1) and (7) could be estimated at ca. 70 and 30%, respectively, from the proportion of isotopomers $^{14}N_2$ (64%), $^{14}N^{15}N$ (19%), and $^{15}N_2$ (17%) found in the reduction of $^{14}N_2O$ by $^{15}NH_3$ in the absence of O_2 [25]. In the presence of oxygen, the breaking of the N–NO bond is of lower extent [25].

On the other hand, the concurrent release of O_2 and NO at the same temperature of ca. 610 K in the TPD experiments after contacting Fe-ZSM-5 with N_2O prompted El-Malki et al. [14] to suggest that O_2 and NO belong to the same ad-complex of nitrite/nitrate nature. The formation of nitrite/nitrate species has been proposed by Sang and Lund [26] for the light-off of N_2O decomposition on Fe-ZSM-5 at high N_2O pressure (>20–360 kPa). The initiation of a second nitrite/nitrate redox cycle faster than the Fe^{II}/Fe^{III} cycle would explain the light-off of N_2O conversion. These facts emphasize the great difference in behavior that can be observed for iron species hosted in different zeolite structures. On a series of Fe-exchanged zeolites the order of reactivity identified by the light-off temperature (50% N_2O conversion) was: FER > BEA > MFI,MAZ ≫ FAU [27]. However, the order did not remain the same when NH_3 was added, since all the light-off temperatures were lowered by 70–120 K, except for Fe-FER, which remained the same. The order of reactivity then became: BEA > OFF,MFI,FER > MAZ ≫ FAU.

The boosting effect of the NH_3 addition on the N_2O reduction [25] was also observed upon addition of CO [28–30] and of various hydrocarbons [12,13,30–32]. One of the possible drawbacks of hydrocarbon-assisted reduction of N_2O is the emissions of CO. Kögel et al. [13] report a selectivity to CO of 60% at 350°C for the propane-SCR of NO + N_2O on a Fe-ZSM-5 catalyst. Van den Brink et al. [33] proposed the use of a FePd-ZSM-5 to avoid CO emissions by its oxidation to CO_2 on Pd. On the other hand, the presence of highly clustered Fe^{3+} species or Fe_2O_3 particles favors the side reaction of the direct combustion of the hydrocarbon with O_2, instead of N_2O, to CO_2 [12].

Since the decomposition is regulated by the Fe^{II}/Fe^{III} redox cycle represented by Eqs. (1) and (2), the kinetics can be treated by the very classical Mars and van Krevelen model [10,29]. Moreover, it was experimentally demonstrated that inhibition by excess O_2 is of low extend for Co- and Fe-zeolites; the rate law can thus be expressed as

$$r = \frac{k_1 k_2 P_{N2O}}{k_1 + k_2} \tag{8}$$

A detailed kinetic study of N_2O decomposition over Co-, Cu-, and Fe-zeolites has been reported by Kapteijn et al. [29] with an analysis of the boosting effect

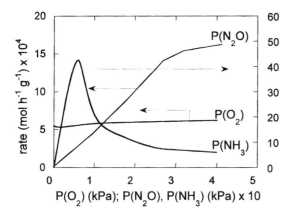

FIG. 2 Rate of N_2O reduction by NH_3 on Fe-BEA as a function of N_2O, NH_3, and O_2 pressures; T_R = 618 K, VVH = 25,000–35,000 h⁻¹. (From Ref. 25.)

of CO and NO by several rate laws. Regarding the kinetics of the N_2O + NH_3 + O_2 reaction to N_2 over Fe-BEA [25], the dependence of the rate on N_2O, NH_3, and O_2 pressures are presented in Figure 2. A rate law was proposed to account for the volcano-shaped dependence of the rate on NH_3 pressure and based on the strong adsorption of NH_3 on "Fe^{II}" sites.

An interesting point is the promoting effect of NO and N_2O conversion [29], as was pointed out earlier by Li and Armor on Co-ZSM-5 [34]. This is important because N_2O and NO are both present in comparable amount in the tail gases from nitric acid plants. That allows one to contemplate deNOx and deN₂O in one step in a single SCR reactor. The simultaneous catalytic reduction of NO and N_2O over Fe-BEA has proven to be more efficient than when processing NO and N_2O alone (Fig. 3) [35]. The synergy between NO and N_2O for their simultaneous removal is readily expressed by Eq. (1) and the following reaction:

$$NO + O^* \rightarrow NO_2 + * \qquad (9)$$

The SCR of NO_2 + NO then proceeds easily (see Section II) according to

$$NO_2 + NO + 2NH_3 \rightarrow 2N_2 + 3H_2O \qquad (10)$$

Fe-BEA and Fe-FER catalysts have been tested in pilot plant conditions in derivation of a nitric acid tail gas. At ca. 673 K and real feed conditions (NOx = 1400 ppm, N_2O = 1700 ppm, NH_3 = 2200 ppm (NH_3 slip < 5 ppm), O_2 = 2.5%, H_2O = 4%, P = 350 kPa, VVH = 18,000 h⁻¹), full NOx and 30–40% N_2O conversions were achieved [36].

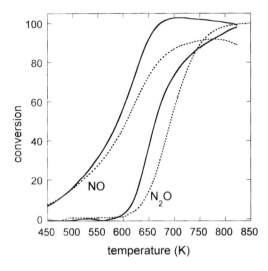

FIG. 3 Reduction of NO and N_2O (dotted lines) and NO + N_2O (solid lines) by NH_3 in the presence of O_2 on Fe-BEA. Conditions: 1500 ppm NO, 1000 ppm N_2O, 2500 ppm NH_3, 3% O_2, balance with He; space velocity: 200,000 h^{-1}; temperature: 10 K min^{-1}. (From Ref. 35.)

Recent reports described various materials based on Cu, Fe, and Ru exchanged in BEA, MCM-22, and MCM-41 active in the decomposition and/or reduction of N_2O [37–41]

III. CONTROL OF NOx EMISSIONS

A. DeNoxing from Mobile Sources

Whereas the use of three-way catalysts is an established technology for the catalytic reduction of NOx produced by gasoline engines operating at stoichiometry, there is still no appropriate catalytic technology for NOx emission abatement in vehicles with diesel and lean-burn engines. The following approaches based on catalysis and zeolites have been investigated for the NOx control in lean exhaust gases: (1) catalytic decomposition of NO; (2) selective catalytic reduction of NOx (SCR) with urea or ammonia; (3) SCR by hydrocarbon (HC), added or already present in the exhaust gas; and (4) removal through NOx storage, e.g., NOx storage and reduction system (NSR), and NOx storage and recirculation technique (SNR). More recently, new concepts have been developed where SCR by hydrocarbons remains the key step: the IAR (intermediate addition of reductant) method, where the hydrocarbon is added between an oxidation catalyst and a

SCR catalyst, and the plasma-assisted catalytic reduction of NOx, where a plasma first oxidizes NO to NO_2 in the presence of a hydrocarbon. In these two concepts the second step is the SCR by hydrocarbons of NOx, with NOx representing particularly NO_2.

Among these different possible NOx control techniques, the direct SCR of NOx by hydrocarbons requires less technology [42] and remains a promising way. The SCR of NOx by hydrocarbons has been widely investigated from the initial reports of Held et al. [43], Iwamoto et al. [44], and Ritscher and Sandner [45]. The potential of zeolite-based catalysts for practical applications, the mechanism and kinetic aspects of the SCR reaction over TMI-zeolite, and the efficiency of different zeolite-based formulations were reviewed recently by Traa et al. [46], Parvulescu et al. [47], and Fritz and Pitchon [48]. In particular, the influence of the zeolite structures, acidity, nature of the reductant, and nature of the metal were discussed.

In selective catalytic reduction, the hydrocarbon selectively reacts with NOx rather than with O_2 to form nitrogen, CO_2, and water. Traa et al. [46] classified all the mechanisms proposed in the SCR of NOx by HC over M-zeolites in three groups:

1. Catalytic decomposition of NO and subsequent regeneration of the active site from adsorbed oxygen by HC:

$$2NO \rightarrow N_2 + 2O_{ads} \tag{11}$$

$$HC + O_{ads} \rightarrow COx + H_2O \tag{12}$$

2. Oxidation of NO to NO_2, which acts as a strongly oxidizing agent of HC:

$$NO + \frac{1}{2}O_2 \text{ on "metal-oxide"} \rightarrow NO_2 \tag{13}$$

$$NO_2 + HC \rightarrow N_2 + COx + H_2O \tag{14}$$

3. Partial oxidation of HC to aldehyde, alcohol, which in turn reacts with HC:

$$HC + O_2 \text{ and/or NOx} \rightarrow \text{"HCO"} \tag{15}$$

$$\text{"HCO"} + NOx \rightarrow N_2 + COx + H_2O \tag{16}$$

In agreement with the authors, the borderline between the different types of mechanisms are not rigid, and, in practice, combinations of all three categories might operate. The main cause for the deactivation of diesel catalysts is poisoning by lubrication oil additives (phosphorus) and by SOx, and is due to hydrothermal instability. Reduction by HC is less sensitive to SOx than the NO decomposition. The Cu-based catalysts are slightly inhibited by water vapor and SOx, and suffer deactivation at elevated temperature. Noble metal catalysts such as Pt-ZSM-5

undergo less deactivation under practical conditions and are active at temperatures below 573 K, but the major and undesired reduction byproduct is N_2O [49].

There are two major requirements in these ZSM-5-based materials: An increase in hydrothermal stability and a wider temperature window with high conversion under actual diesel exhaust gas still constitute the most important challenges for a commercial application of these catalysts to reduce NOx from
mobile-source exhaust gas.

B.   DeNOxing from Stationary Sources

There are three possible technologies [50] for reducing NOx (NO, NO_2) in net
oxidizing mixtures: (1) nonselective catalytic reduction (NSCR), (2) selective
noncatalytic reduction (SNCR), and (3) selective catalytic reduction (SCR). Only
SCR uses zeolite-based catalysts. The selective catalytic reduction of NOx with
the help of a reductant (HC, NH_3, or urea) is the best control technology to achieve
near full NOx removal from off-gas of power plants, nitric acid plants, stationary
diesel engines, marine vessels, etc. The major processes use NH_3 (or urea) as
reductant. The catalyst is a combination of V_2O_5 and TiO_2 (usually promoted
with WO_3 or MoO_3) as extruded monoliths or deposited on a plate structure [50].
However, an increasing number of installations use a zeolite technology at present
[51,52]. Figure 4 presents the windows of activity of various formulations [53].
On the right-hand side of the volcano-shaped curves, the decrease of NOx conversion is due to the oxidation of NH_3 by O_2.

The basic reaction in the SCR process is

$$4NH_3 + 4NO + O_2 \rightarrow 4N_2 + 6H_2O \qquad (17)$$

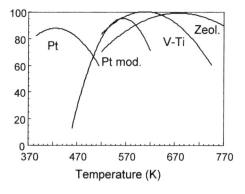

FIG. 4 Windows of activity in the SCR of NOx by NH_3 for different catalyst types.
(From Ref. 53.)

Since Pence and Thomas [54] first documented the activity of H-MOR as a catalyst for reducing NOx to N$_2$ by NH$_3$, there has been significant interest in developing zeolite-based catalysts [51,53,55,56] due to: (1) the wide SCR operating window, (2) the high NOx conversion, (3) the high selectivity toward N$_2$, (4) the very low NH$_3$ slip, (5) its sulfur tolerance above 700 K, and (6) the lessened disposal problem for the spent catalyst.

Under the same conditions, zeolites in H or Na form always exhibit much lower activity than zeolite exchanged with transition metal ions, TMI-zeolite [57,58]. Major studies in the SCR of NO by NH$_3$ over TMI-zeolite have been performed with copper, iron, and cerium. Therefore the following section will be devoted to these studies, with particular emphasis on the mechanistic aspect.

1. Cu-Containing Zeolite

Over different zeolite host structures several research groups have established a linear relationship between SCR activity and the copper ion exchange content [57–64]. It should be noticed that copper oxide aggregates are less active [58,62,65] and selective [58,65] than Cu cations or oxocations, due to their high activity for the side reaction of NH$_3$ oxidation by O$_2$ alone [58,65]. The reasons for the promotion of activity by copper are not well understood. According to Choi et al. [57], copper ions provide additional adsorption sites for NO and NH$_3$ and therefore enhance the global activity. For these authors the active sites are both Brønsted acid sites and copper ion sites. For Kieger et al. [58] the most active sites are composed of two neighboring Cu ions; their probability of occurrence is obviously more favorable in the high copper-exchanged zeolite. Nevertheless an adequate proportion of H$^+$ and copper ions also leads to very highly active catalysts in the SCR of NO by NH$_3$ [66]. Moreover, the accessibility of copper ions is an important parameter. Considering only Cu species accessible and located in the supercages, Cu-FAU exhibits a similar specific activity to Cu-MOR and Cu-MFI [58]. The "blocking" of sodalite cages of Cu-FAU by alkaline earth (Ca, Ba, etc.) or rare earth (La, Pr, etc.) elements strongly enhances the proportion of accessible copper ions and leads to a catalyst highly efficient in NOx reduction by NH$_3$ to N$_2$ [58,67]. On Cu-exchanged zeolite it is generally accepted that the rate of SCR of NO with NH$_3$ follows a Langmuir–Hinshelwood type of mechanism involving reaction on copper-containing sites of an adsorbed species derived from NO with an adsorbed species derived from NH$_3$ [58,59,62,63,68], with stronger adsorption of NH$_3$ than NO [69]. Moreover, the mechanism is regulated by a redox cycle of copper ions [58,60,62,70] in which oxidation is the rate-determining step [58]. From temperature-programmed oxidation (TPO), SCR of ^{14}NO by ^{15}NH$_3$, and previous elements of a mechanism suggested by Mizumoto et al. [71], Kieger et al. [58] proposed the mechanistic scheme shown in Figure 5.

In this mechanism, Cu^{2+} is reduced by NO + NH$_3$ to Cu$^+$, which in turn is oxidized to Cu^{2+} by NO + O$_2$. The original feature of the process demonstrated

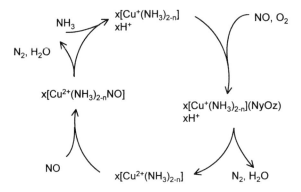

FIG. 5 Catalytic cycle of the SCR of NO by NH$_3$ on Cu-FAU.

by TPO and SCR with ^{15}NH$_3$ concerns the oxidation of Cu$^+$ complexes, which is faster with NO + O$_2$ than with O$_2$ alone [58]. In view of the correlation between copper content and SCR specific activity, the active sites below 550 K are composed of Cu ions in close vicinity, may be [CuOCu]$^{2+}$, stabilized by NH$_3$ and located in the supercages. Above 600 K, all Cu ions become active. Komatsu et al. [61] proposed [CuOCu]$^{2+}$ as the active species, with a nitrate-like species (Fig. 6a) as intermediate. Based on IR and EPR evidence, Williamson and Lunsford [62] suggested copper mononuclear species (Fig. 6b, c) as the active species under the reaction conditions performed at low temperature (283 K) and in the absence of oxygen.

Under a reaction gas mixture containing water and/or SOx, copper-zeolite catalysts exhibit a high stability if the reaction temperature is maintained above 573 K [72–74]. Below 573 K, accumulation of ammonium sulfate and/or ammonium bisulfate occurs and may cause severe blocking [73]. It should be noted that the addition of water vapor alone enhances NO conversion [72,75].

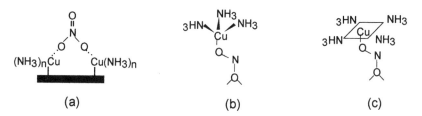

FIG. 6 Some copper-active species proposed in the SCR of NO by NH$_3$.

2. Fe-Containing Zeolite

ZSM-5 and MOR as host structures for Fe ions were found preferable to FAU in the SCR of NO by NH$_3$ in the presence of oxygen [76]. On Fe-FAU catalysts, Amiridis et al. [77] emphasize, only a fraction of Fe cations are responsible for the SCR activity, due to species located in inaccessible sites (in sodalite or hexagonal cavities). A change in the Si/Al ration from 2.4 to 4.4 has only a modest effect on the SCR catalytic activity of Fe-FAU [77]. Upon increasing the Si/Al ratio from 10 to 100 of Fe-ZSM-5 with the same Fe exchange level, the NO conversion decreased rapidly [76]. The turnover frequency, based on total Fe content, thus decreased with the increase in Fe content [76]. From this result Long and Yang [76] underline the importance of a good balance between Brønsted and iron sites. The methods of preparation of Fe-zeolite appears to be a crucial point, since overexchanged Fe-ZSM-5 prepared by sublimation technique were reported to have higher activity than overexchanged Cu-ZSM-5 prepared by conventional technique, especially at temperatures above 600 K [78].

The kinetic dependence of the SCR with respect to NO and NH$_3$, determined in the temperature range 490–545 K, was first order in NO and zero order in NH$_3$ [77]. The positive role of O$_2$ on the reaction rate was also shown. On Fe-ZSM-5, Long and Yang [79] proposed a mechanism similar to that described for H-ZSM-5 [80]. It involves the reaction between a NO$_2$-adsorbed species (NO oxidized by O$_2$ on Fe^{3+}) and two NH$_4^+$ ions to form an active intermediate that subsequently reacts with another gaseous or weakly adsorbed NO to produce N$_2$ and H$_2$O:

$$[(NH_4^+)_2NO_2]_{ads.} + NO \rightarrow 2N_2 + 3H_2O + 2H^+ \tag{18}$$

According to Schmidt et al. [81] an Fe^{2+}/Fe^{3+} redox cycle is involved in the mechanism, and binuclear Fe oxocations were suggested as active species. Binuclear Fe oxocations, $[HO-Fe-O-Fe-OH]^{2+}$, are also proposed as active species by Sun et al. [82]. From isotopic labeling, these authors conclude that the reduction of NOx with ammonia to nitrogen occurs via the formation of adsorbed ammonium nitrite. This intermediate was also suggested by Long and Yang [83] but only for nitrogen formation occurring at low temperature (<373 K). The response of activity to the presence of H$_2$O + SO$_2$ is generally similar to that described earlier for copper-zeolite catalysts.

3. Ce-Containing Zeolite

Ce-exchanged zeolite catalysts have been studied in the SCR of NO by NH$_3$ by van den Bleek and co-workers [84–87]. The SCR activity strongly increases with the cerium content up to ca. 0.75% [84]. The reaction is regulated by the redox cycle Ce^{III}/Ce^{IV} [85]. In the low-temperature region (573 K), the reaction proceeds mainly through a NO$_2$ intermediate:

$$NO + \frac{1}{2}O_2 \xrightarrow{Ce^{III}/Ce^{IV}} NO_2 \tag{19}$$

$$NO + NO_2 + 2NH_3 \rightarrow 2N_2 + 3H_2O \tag{20}$$

while at high temperature (783 K), a nitrosation reaction mechanism involving nitrosonium ion NO^+ species is assumed to contribute substantially [85]:

$$NO + Ce^{IV} \rightarrow Ce^{III}(NO)^+ \tag{21}$$

$$Ce^{III}(NO)^+ + NH_3 \rightarrow N_2 + H_2O + Ce^{III} + H^+ \tag{22}$$

$$Ce^{III} + H^+ + \frac{1}{4}O_2 \leftrightarrow Ce^{IV} + \frac{1}{4}H_2O \tag{23}$$

In simulated exhaust gas with 12% H_2O, an 11% decrease in NO conversion was found at 723 K after 116 hours of reaction on extrudates of Ce-ZSM-5/Al_2O_3 (Al_2O_3 used as binder) [86]. The deactivation was attributed to dealumination. In real diesel exhaust gas (stationary diesel engine), the decrease was 50% and due in large part to poisoning with sulfur and to a lesser extent to the deposition of elements (Ca, P, and Zn) contained in the lubricating oil [86].

The majority of the work carried out on the behavior of TMI-zeolite has emphasized the great relevance of the redox cycle of TMI in the catalytic transformation of nitrogen oxides as well as the influence of the zeolite host structure on its properties. Many important questions still remain unanswered. For instance, in the case of Cu-zeolite catalysts, the more basic ones that need to be answered before one can really understand the mechanistic aspects of the interaction of catalytic sites with NO and their subsequent reactivity are: Are there selective active sites? Is the active species Cu^+, Cu^{2+}, or $(CuOCu)^{2+}$? The same difficulty occurs with Fe-exchanged zeolites involved in N_2O decomposition. Several studies dealing with quantum chemical modeling of the TMI-exchanged zeolites have appeared recently that try to give some insight into these questions. The next section will briefly examine that point, which might become of paramount importance in the near future.

IV. QUANTUM CHEMICAL MODELING OF TMI-ZEOLITE

A. Mechanistic Studies of NO and N_2O Decomposition

A few theoretical studies have concentrated on the mechanism of NO decomposition in Cu-zeolite, assuming simple schematic models for the zeolite. Using $Si(OH)_4$ and $Al(OH)_4^-$ as zeolite models (Z), Schneider et al. [88,89] studied the binding of NO, O_2, N_2, to Cu^I-Z and the transformation of Cu-Z to CuO-Z. Adsorption of these molecules leads to a predicted partial oxidation of Cu. All reactions involving Cu-Z, CuO-Z, and the studied adsorbates are summarized in a suggested reaction scheme, which rationalizes the NO and N_2O decomposition of Eqs. (1), (11), and (12). Larger clusters were used by Trout et al. [90], including one Al (Cu^I-Z) and two Al (Cu^{II}-Z), to estimate the free energies of the reactions involving NO, NO_2, and O_2 adsorptions. These authors propose a pathway for NOx decomposition on the basis of Cu-Z + NO and CuO-Z + NO reactions, which is comparable with that proposed on the basis of much smaller models [89]. From these results, one can conclude that isolated Cu^I sites should be sufficient to explain NOx decomposition in Cu-zeolites.

However, Cu pairs were invoked first by Iwamoto et al. [91] to explain the N—N bond formation [Eq. (11)]. Actually, Cu pairs have been observed in EXAFS studies for different zeolites with high Al and Cu contents [92–94], and also suggested from IR experiments as being $[CuOCu]^{2+}$ species [95,96]. Goodman et al. have used simplified Cu-pair models [97] and specific ZSM-5 clusters [98] to study Z-CuOCu-Z and Z-CuO₂Cu-Z species. Cu—Cu distances in these models range from 2.42 to 2.83 Å, and the predicted oxocations are strongly bound. Another study using molecular dynamics (MD) and molecular orbital (MO) geometry optimizations of clusters models of ZSM-5 found $[CuOCu]^{2+}$ entities with larger Cu—Cu distances (around 3 Å) [99]. The question of the nature of the active site is thus not resolved, although one could suggest that both isolated Cu(I) and copper oxocations, for high Cu loadings, can be reasonable catalytic sites.

B. N₂O Adsorption on Fe-ZSM-5

The reaction of N_2O in Fe-ZSM-5 is known to produce a very superactive oxidizing FeO-Z center, also called α-oxygen [100]. A FeO^+ model [20] and a binuclear iron hydroxide cluster [16] were proposed to model the reactive species formed from the interaction of N_2O with Fe-ZSM-5. Both models seem incomplete, however. Indeed, according to experimental investigations of Fe-ZSM-5, the reactive species of Fe^{2+} ions, which adsorb NO or N_2O, are distributed among three types of sites [101]. In contrast, the FeO^+-Z model involves a formal Fe^{III} ion, which should be charge compensated with three Al, whereas the cluster Z contains only one Al [20]. On the other hand, the binuclear cluster proposed as active species is neutral [16] and should then be anchored in some way to the zeolite framework. As mentioned later, in section IV.D.1, the zeolite-to-metal charge transfer changes isolated Fe^{II} cations (obtained from reduction of the Fe^{III}-zeolite) into Fe^I cations [24]. Adsorption of NH_3 on these ions favors the decomposition of N_2O into FeO-Z + N_2, but also allows the N—N bond breaking of N_2O [102].

C. Siting and Coordination of Cu^I and Cu^{II}

Although simple models may account for interesting trends in the reactivity of TMI-zeolites, understanding their electronic properties depends strongly on the zeolite framework, which must be considered as a macroligand. Two main effects can influence these electronic properties: (1) The local geometry of the zeolite ligand in the vicinity of the transition metal ion (or pair of ions) may display distortions, more profound for charge-exchanged TMI^{2+} or TMI^{3+} (Cu^{2+}, Fe^{3+}); (2) the distribution of Al, defined by the synthesis in the presence of alkali or organic cations, determines the coordination with the exchanged TMIs and the positions of the nearest cocations. Quantum mechanical (QM) studies using cluster models combined or not with a molecular mechanics (MM) description of the surroundings have been devoted to the analysis of the possible sites and coordination of Cu$^+$, Cu^{2+}, and other divalent cations in ZSM-5 and of Cu^{2+} or Co^{2+} in FAU. For a review of sitings and coordinations, see Refs. 103 and 104.

Summarizing these results for Cu-zeolites, one can say that monovalent Cu^I is preferably coordinated to two framework oxygens in ZSM-5 [103], whereas Cu^{II} prefers a higher coordination, i.e., four in six-membered rings of ZSM-5 [105] and FAU [24,106,107], and four in four-membered rings of FAU [107].

D. The Zeolite as a Ligand

In organometallic chemistry, TMIs are bound to neutral or anionic molecules, and there may be a competition between outgoing and incoming ligands. A zeolite framework is a peculiar ligand, which can be mono-, bi-, . . . , hexa-dentate, according to the site where the metal ion is located. In addition, incoming molecules, like H$_2$O, NH$_3$, NO, etc., can bind to the accessible TMIs. Actually, the specificity of TMI-zeolite comes from the adaptability of the whole ensemble zeolite-TMI-molecules. Adaptability means relaxation of the framework geometry when, e.g., initial sodium cations are exchanged with TMIs and when gaseous molecules adsorb. It also means the ability to redistribute the electron density into the whole ensemble. The reasons for such a richness of applications of zeolites is probably due to this supramolecular property.

1. The Zeolite Framework to TMI Charge Transfer

Published results on cluster models of Cu-ZSM-5, Cu-FAU, Fe-BEA, and Fe-FAU show that the TMI cations receive a non-negliglible amount of charge from the zeolite framework [24,88–90,108,109], even when the zeolite is modeled with water molecules [110].

This charge transfer amounts of 0.2–0.5 electron for Cu^I models, according to the representation of the second shell of neighbors (H or Si/Al). A detailed analysis of Cu^{II} and Fe^{II} models of FAU and BEA, with various sizes and Al distributions, has shown that the charge transfer of ca. one electron is very slightly sensitive to the cluster size [24,109]. Cu^{II} with preferred tetra-coordination incorporates the transferred charge partly in its $3d$ hole and partly into the $4s$, $4p$ orbitals. The singly occupied MO (SOMO) of this doublet system is delocalized on $Cu3d$ and zeolite $O2p$ orbitals, whereas the lowest unoccupied MO (LUMO), very close in energy, is of the same nature. The next unoccupied MO corresponds, as in all zeolites (with protons and alkalis), to empty s counterion orbitals.

The oxido-reductive ability of these TMI^{II} systems can be related to the delocalization of the close SOMO and LUMO onto the full metal-framework system. Comparison between experimental and calculated EPR hyperfine coupling constants leads to the conclusion that the density of the unpaired spin is indeed delocalized on Cu and on all framework atoms, with an estimate of 0.3–0.5 unpaired electron on the Cu ions [107].

2. The Zeolite Framework as a Charge Reservoir

When incoming molecules adsorb on TMIs, a chemical binding occurs, in contrast with adsorption on alkalis [111]. The binding energies of these adsorbates are generally decreased with respect to isolated metal cations [103,109,110]. However, specific sites may favor the adsorption, due to proper positions of the zeolite oxygens with respect to the metal, as shown for NO or NO_2 on ZSM-5 models [103,108].

Even if atomic charges are not QM properties, they are useful to get a qualitative picture of the variations encountered by the electron distribution upon addition or release of ligands. It is interesting, for example, to estimate the electron repartition between the zeolite framework, the metal cation, and the adsorbates. The Mulliken net charges estimated for a Cu^{II}-FAU model (Fig. 7), including different adsorbates, are presented in Table 1.

In the bare model, the Cu^{II} ion, which compensates the charge of 2Al, has thus received 1.28 electrons from the zeolite cluster, and, due to the nature of

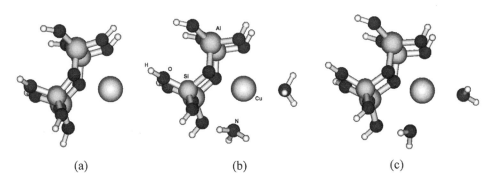

(a)   (b)   (c)

FIG. 7 Model clusters of Cu-FAU (a), $(NH_3)_2$Cu-FAU (b), and $(H_2O)_2$Cu-FAU (c).

TABLE 1 Mulliken Net Charges Calculated on the Cu,
Zeolite Framework Model and Ligands (Total Charge) for
a Cu-FAU Model

L_x	Net charges (Spin densities)		
	Cu	Zeolite	Ligands
No ligands	0.72	−0.72	
	(0.08)		
$2H_2O$	0.75	−1.01	0.26
	(0.30)		
$2NH_3$	0.74	−1.11	0.37
	(0.35)		
$3NH_3$	0.81	−1.24	0.43
	(0.43)		
NO	0.83	−1.03	0.20
NH_3			

the SOMO, the unpaired electron is distributed mostly onto the zeolite oxygens. This situation would correspond to a silent EPR ion.

When donor molecules are adsorbed, the positive net charge on the metal does not change much, whereas the zeolite framework becomes more negative. This indicates that the framework plays the role of a reservoir of electronic charge, giving it back to the adsorbates when they desorb, this process being allowed by the delocalization of the frontier MOs.

It is worth noting that, upon ligand adsorption, the nature of the SOMO changes: it becomes more localized on the metal, thus increasing its spin density. The EPR signal of such a Cu ion would then become visible. This result is in agreement with experimental observations of EPR signals appearing upon water addition in ZSM-5 [112]. Other studies available in the literature report results for Cu^I-ZSM-5 models, without and with adsorbed NO and NO_2 molecules [90,108]. The calculated net charges show that, upon adsorption of these molecules, which are electron acceptor ligands, the Cu net charge does not change much (NO) or increases (NO_2), whereas the zeolite model participates in the transfer of 0.1–0.2 electron to the adsorbates, depending on the isomers.

These examples show that there is a strong interchange of electron density between the framework and the TMIs, leading to a supermolecular picture of the full system. It is worth noting that this aspect of the exchange of electron density between the zeolite framework, the TMIs, and the adsorbates is lost if the model chosen to represent the zeolite framework is too crude, i.e., limited to Al$(OH)_4^-$ [89] or represented with water molecules [108,110].

3. Properties of the Zeolite-Containing TMIs and Other Cocations

The SCR of NO by NH_3 in Cu-FAU is favored when the cocations are protons instead of Na^+ [66]. Possible reasons for the different oxidative behaviors of HCu^I-FAU and $HNaCu^I$-FAU have been explored by quantum chemical calculations using models containing 4Al centers and four cations: $Cu^+ + 3H^+$, $Cu^+ + 2H^+ + Na^+$, $Cu^+ + H^+ + 2Na^+$, and $Cu^+ + 3Na^+$ [66]. The HCu-FAU zeolite, obtained by reduction of Cu^{II}-FAU, contains most probably Cu^+ cations at sites II (sodalite six-membered rings). The model clusters chosen thus include one six-membered ring associated with the neighboring four-membered ring, in order to accommodate enough Al/cation couples. To be realistic, the clusters have been cut into a solid structure generated by molecular mechanics (MM) energy minimization [113], followed by a partial QM optimization to allow the relaxation of the ring-oxygens in the presence of the Cu^+, H^+, and Na^+ cations [66]. Examples of the H_3Cu-FAU and Na_3Cu-FAU models are illustrated in Figure 8.

The experimental study of reactivity shows that the presence of protons would favor the oxidation of Cu^I-FAU to Cu^{II}-FAU [66]. One might thus expect the first ionization energy (IE) to be smaller for H_3Cu-FAU than for Na_3Cu-FAU, the HOMO being characteristic of $Cu3d$ and zeolite $O2p$ orbitals. In fact, the calculated IEs show the opposite trend, with 6.9 and 5.9 eV for H_3Cu-FAU and Na_3Cu-FAU, respectively, and intermediate values for Na_2HCu-FAU. In contrast, the electron affinity (EA) of H_3Cu-FAU is smaller (1.4 eV) than that of Na_3Cu-FAU (1.64 eV). Using the DFT-based concept of hardness $\chi = 1/2(\delta^2 E/\delta N^2) \approx 1/2(EI\text{-}EA)$, one finds that the hardness of the fully protonated model (2.75 eV) is much larger than that of Na_3Cu-FAU (2.13 eV). This leads one to expect a

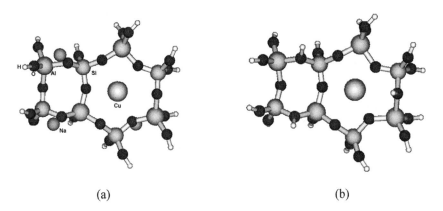

(a)   (b)

FIG. 8   Model clusters of Na_3Cu-FAU (a) and H_3Cu-FAU (b).

weaker global hardness for a system with a majority of Na cations and few protons than for a fully protonated one. This is verified when comparing the three Na$_2$HCu-FAU models with H$_3$Cu-FAU, with a global hardness of 2.2, 2.4, 2.3 for protons in various positions. Hardness is related to acidity in zeolites, and these results indicate that whatever its position, the proton is more acidic in H$_3$Cu-FAU than in Na$_2$HCu-FAU. Indeed, their calculated proton affinities are smaller by about 10 kcal mole^{-1} in the fully protonated model [66]. Combining calculated and experimental results leads one to propose that the activity of Cu-FAU is related to its acidity, i.e., the acid strength of protons surrounding Cu. This argument is in favor of a cooperative mechanism for the oxidation step.

V. CONCLUDING REMARKS

There are several issues for cleaning gaseous emissions containing nitrogen oxides by the use of zeolites. The SCR of NOx by NH$_3$ is the best control technology, but a new breakthrough would be achieved in power plants by the SCR of NOx using methane as reductant. Regarding deNOx from mobile sources, new concepts are appearing, and NOx trap and plasma-assisted catalytic reduction seem promising. The removal of N$_2$O in low concentration in flue gases will be fully efficient with the development of catalysts active below 570 K. Moreover, the simultaneous removal of NOx and N$_2$O in the same SCR reactor will also constitute a significant step forward.

   The understanding of the mechanism and of the architecture of the active site have attracted studies through quantum chemical modeling. The specificity of TMI-zeolite was demonstrated; they are well described as a supermolecules with full redistribution of electrons among all the atoms. Significant advances can be expected in the future for a better description of the active sites, and finally to achieve the design of tailored TMI-zeolite catalysts with optimal properties.

REFERENCES

1.   DS Lee, I Köhler, E Grobler, F Rohrer, R Sausen, L Gallardo-Klenner, JGJ Olivier, FJ Dentener, AF Bouwman. Atmospheric Environment 31:1735–1749, 1997.
2.   Environmental Catalysis, Handbook of Heterogeneous Catalysis. Weinheim, Germany: Wiley-VCH, 1997, p. 1559.
3.   Environmental Catalysis. London: Imperial College Press, 1999.
4.   NOxCONF 2001, La Pollution Atmosphérique d'Origine Industrielle, Paris, 21–22 March 2001.
5.   M Gubelmann, P-J Tirel. Fr. Patent 2.630.735 (1988).
6.   AS Kharitonov, TN Alexandrova, LA Vostrikova, KG Ione, GI Panov. Russ. Patent 4.445.646 (1988).

7. GI Panov, AK Uriarte, MA Rodkin, VI Sobolev. Catal. Today 41:365–385, 1998.
8. A Ribera, IWCE Arends, S de Vries, J Pérez-Ramirez, RA Sheldon. J. Catal. 195: 287–297, 2000.
9. PP Knops-Gerrits, WA Goddard III. J. Mol. Catal. A 166:135–145, 2001.
10. F Kapteijn, J Rodriguez-Mirasol, JA Moulijn. Appl. Catal. B 9:25–64, 1996.
11. G Centi, S Perathoner, F Vazzana. Chemtech. December:48–55, 1999.
12. G Centi, F Vazzana. Catal. Today 53:683–693, 1999.
13. M Kögel, R Mönnig, W Schwieger, A Tissler, T Turek. J. Catal. 182:470–478, 1999.
14. El-M El-Malki, RA van Santen, WMH Sachtler. J. Catal. 196:212–223, 2000.
15. J Pérez-Ramírez, F Kapteijn, G Mul, JA Moulijn. Chem. Commun. 693–694, 2001.
16. AV Arbuznikov, GM Zhidomirov. Catal. Lett. 40:17–23, 1996.
17. L Lázár, G Lejeune, RK Ahedi, SS Shevade, AN Kotasthane. J. Phys. Chem. B 102:4865–4870, 1998.
18. G Delahay, M Mauvezin, B Coq, S Kieger. J. Catal. 202:156–162, 2001.
19. M Mauvezin, G Delahay, B Coq, S Kieger, J-C Jumas, J Oliver-Fourcade. J. Phys Chem. B 105:928–935, 2001.
20. K Yoshizawa, Y Shiota, T Ymura, T Yamabe. J. Phys. Chem. B 104:734–740, 2000.
21. M Kasture, J Kryściak, L Matachowski, T Machej, M Derewiński. Stud. Surf. Sci. Catal. 125:579–586, 1999.
22. G Grubert, MJ Hudson, RW Joyner, M Stockenhuber. J. Catal. 196:126–133, 2000.
23. J Pérez-Ramírez, G Mul, F Kapteijn, IWCE Arends, A Ribera, JA Moulijn. Stud. Surf. Sci. Catal. 135:30-O-02, 2001.
24. D Berthomieu, A Goursot, J-M Ducéré, G Delahay, B Coq, A Martinez. Stud. Surf. Sci. Catal. 135:15-P-23, 2001.
25. B Coq, M Mauvezin, G Delahay, S Kieger. J. Catal. 195:298–303, 2000.
26. C Sang, CRF Lund. Catal. Lett. 73:73-77, 2001.
27. M Mauvezin, G Delahay, B Coq, S Kieger. Catal. Lett. 62:41–44, 1999.
28. JO Petunchi, WK Hall. J Catal. 78:327–340, 1982.
29. F Kapteijn, G Marbán, J Rodriguez-Mirasol. JA Moulijn. J. Catal. 167:256–265, 1997.
30. G Delahay, A Guzman, B Coq. Catal. Commun. In press.
31. C Pophal, T Yogo, K Tanabe, K Segawa. Catal. Lett. 44:271–274, 1997.
32. S Kameoka, K Kita, T Takeda, S Ito, K Yuzaki, T Miyadera, K Kunimori. Catal. Lett. 69:169–173, 2000.
33. RW van den Brink, S Booneveld, JR Pels, DF Bakker, MJFM Verhaak. Appl. Catal. B 32:73–81, 2001.
34. Y Li, JN Armor. Appl. Catal. B 3:55–60, 1993.
35. B Coq, M Mauvezin, G Delahay, J-B Butet, S Kieger. Appl. Catal. B 27:193–198, 2000.
36. S Kieger, L Navascués, P Gry. Proceedings NOxCONF 2001, La Pollution Atmosphérique d'Origine Industrielle, Paris, 21–22 March 2001, session 4.
37. PJ Carl, SR Larsen. J. Catal. 196:352–361, 2000.
38. S Kawi, SY Liu, S-C Chen. Catal. Today 68:237–244, 2001.
39. W-S Ju, M Matsuoka, M Anpo. Catal. Lett. 71:91–93, 2001.

40. G Fierro, G Ferrais, M Inversi, M Lo Jacono, G Moretti. Stud. Surf. Sci. Catal. 135:30-P-14, 2001.
41. AJS Mascarenhas, HMC Andrade, HO Pastore. Stud. Surf. Sci. Catal. 135:30-P-13, 2001.
42. M Shelef, RW McCabe. Catal. Today 62:35–50, 2000.
43. W Held, A König, T Richter, L Puppe. SAE Trans., Section 4, N. 900496, 209, 1990.
44. M Iwamoto, H Yahiro, Y Yu-u, S Shundo, N Mizuno. Shokubai 32:430–438, 1990.
45. JS Ritscher, MR Sander. U.S. Patent 4,297,328, 1981.
46. Y Traa, B Burger, J Weitkamp. Microporous Mesoporous Mat. 30:3–41, 1999.
47. VI Parvulescu, P Grange, B Delmon. Catal. Today 46:233–316, 1998.
48. A Fritz, V Pitchon. Appl. Catal. B 13:1–25, 1997.
49. P Gilot, M Guyon, BR Stanmore. Fuel 76:507–515, 1997.
50. H Bosch, F Janssen. Catal. Today 2:369–532, 1988.
51. P Gry. Proceedings NOxCONF 2001, La Pollution Atmosphéric d'Origine Industrielle, Paris, 21–22 March 2001, session 8.
52. JR Kiovsky, PB Koradia, CT Lim. Ind. Eng. Chem. Prod. Res. Dev. 19:218–225, 1980.
53. JW Byrne, JM Chen, BK Speronello. Catal. Today 13:33–42, 1992.
54. DT Pence, TR Thomas. Proceedings of the AEC Pollution Control Conference, Oak Ridge, TN, 1972, pp 115–121.
55. RM Heck. Catal. Today 53:519–523, 1999.
56. SA Stevenson, JC Vartuli, CF Brooks. J. Catal. 190:228–239, 2000.
57. EY Choi, IS Nam, YG Kim. J. Catal. 161:597–604, 1996.
58. S Kieger, G Delahay, B Coq, B Neveu. J. Catal. 183:267–280, 1999.
59. LAH Andersson, JGM Brandin, CUI Odenbrand. Catal. Today 4:173–185, 1989.
60. M Mitzumoto, N Yamazoe, T Seiyama. J. Catal. 55:119–128, 1978.
61. T Komatsu, M Nunokawa, IS Moon, T Takahara, S Namba, T Yashima. J. Catal. 148:427–437, 1994.
62. WB Williamson, JH Lunsford. J. Phys. Chem. 80:2664–2671, 1976.
63. T Seiyama, T Arakawa, T Matsuda, Y Takita, N Yamazoe. J. Catal. 48:1–7, 1977.
64. T Komatsu, T Ueda, T Yashima. J. Chem. Soc. Faraday Trans. 94:949–953, 1998.
65. G Delahay, B Coq, S Kieger, B Neveu. Catal. Today 54:431–438, 1999.
66. G Delahay, E Ayala Villagomez, JM Ducéré, D Berthomieu, A Goursot, B Coq. Chem. Phys. Chem. In press.
67. B Coq, G Delahay, F Fajula, S Kieger, B Neveu, J-B Peudpiece. EP0914866, 1999.
68. S Kieger. PhD dissertation, University of Montpellier (France), 1997.
69. EY Choi, IS Nam, YG Kim, JS Chung, JS Lee. J. Mol. Catal. 69:247–258, 1991.
70. SW Ham, H Choi, IS Nam, YG Kim. Catal. Lett. 42:35–40, 1996.
71. M Mizumoto, N Yamazoe, T Seiyama. J Catal. 59:319–324, 1979.
72. AV Salker, W Weisweiler. Appl. Catal. A 203:221–229, 2000.
73. SW Ham, H Choi, IS Nam, YG Kim. Catal. Today 11:611–621, 1992.
74. IS Nam, ST Choo, DJ Koh, YG Kim. Catal. Today 38:181–186, 1997.
75. JA Sullivan, J Cunningham, MA Morris, K Keneavey. Appl. Catal. B 7:137–151, 1995.

76. RQ Long, RT Yang. J. Catal. 88:332–339, 1999.
77. MD Amiridis, F Puglisi, JA Dumesic, WS Millman, NY Topsoe. J. Catal. 142: 572–584, 1993.
78. AZ Ma, W Grunert. Chem. Comm. 71–72, 1999.
79. RQ Long, RT Yang. J. Catal. 194:80–90, 2000.
80. J Eng, CH Bartholomew. J. Catal. 171:14–26, 1997.
81. R Schmidt, MD Amiridis, JA Dumesic, LM Zelewski, WS Millman. J. Phys. Chem. 96:8142–8149, 1992.
82. Q Sun, S-X Gao, H-Y Chen, WMH Sachtler. J. Catal. 201:89–99, 2001.
83. RQ Long, RT Yang. J. Catal. 198:20–28, 2001.
84. WEJ van Kooten, B Liang, HC Krijnsen, OL Oudshoorn, HPA Calis, CM van den Bleek. Appl. Catal. B 21:203–213, 1999.
85. E Ito, RJ Hultermans, PM Lugt, MHW Burgers, MS Rigutto, H van Bekkum, CM van den Bleek. Appl. Catal. B 4:95–104, 1994.
86. WEJ van Kooten, HC Krijnsen, CM van den Bleek, HPA Calis. Appl. Catal. B 25: 125–135, 2000.
87. WEJ van Kooten, J Kaptein, CM van den Bleek, HPA Calis. Catal. Lett. 63:227–231, 1999.
88. WF Schneider, KCHR Ramprasad, JB Adams. J. Phys. Chem. B 101:4353–4357, 1997.
89. WF Schneider, KCHR Ramprasad, JB Adams. J. Phys. Chem. B 102:3692–3705, 1998.
90. BL Trout, AK Chakraborty, AT Bell. J. Phys. Chem. 100:17582–17592, 1996.
91. M Iwamoto, H Yahiro, N Mizuno, WX Zhang, Y Mine, H Furukawa, S Kagawa. J. Phys. Chem. 96:9360–9366, 1992.
92. W Grünert, NW Hayes, RW Joyner, ES Shpiro, M Rafiq, H Siddiqui, GN Baeva. J. Phys. Chem. 98:10832–10846, 1994.
93. Y Kuroda, R Kumashiro, T Yoshimimoto, M Nagao. Phys. Chem. Chem. Phys. 1: 649–656, 1999.
94. C Lamberti, G Spoto, D Scarano. C Pazé, M Salvalaggio, S Bordiga, A Zecchina, G Turnes Palomino, F D'Acapito. Chem. Phys. Lett. 269:500–508, 1997.
95. J Sarkany, JL d'Itri, WMH Sachtler. Catal. Lett. 16:241–249, 1992.
96. T Beutel, J Sarkany, GD Lei, JY Jau, WMH Sachtler. J. Phys. Chem. 100:845–851, 1996.
97. BR Goodman, WF Schneider, KC Hass, JB Adams. Catal. Lett. 56:183–188, 1998.
98. BR Goodman, KC Hass, WF Schneider, JB Adams. J. Phys. Chem. B 103:10452–10460, 1999.
99. K Teraishi, M Ishida, J Irisawa, M Kume, Y Takahashi, T Nakano, H Nakamura, A Miyamoto. J. Phys. Chem. B 101:8079–8085, 1997.
100. GI Panov, VI Sobolov, AS Kharitonov. J. Mol. Catal. 61:85–97, 1990.
101. LJ Lobree, I Hwang, JA Reimer, AT Bell. J. Catal. 186:242–253, 1999.
102. A Martinez, B Coq, A Goursot. To be published.
103. J Sauer, D Nachtigallova, P Nachtigall. Catalysis by Unique Metal Ion Structures in Solid Matrices. Doordrecht (Netherlands): Kluwer, 2001, Vol. 13, pp 221–234.
104. AT Bell. Catalysis by Unique Metal Ion Structures in Solid Matrices. Doordrecht (Netherlands): Kluwer, 2001, Vol. 13, pp 55–74.

105. D Nachtigallova, P Nachtigall, J Sauer. Phys. Chem. Chem. Phys. 3:1552–1559, 2001.
106. A Delabie, K Pierlott, MH Groothaert, BM Veckhuysen, RA Schoonheydt. Microporous and Mesoporous Mat. 37:209–222, 1998.
107. D Berthomieu, J-M Ducéré, A Goursot. In press.
108. L Rodriguez-Santiago, M Sierka, V Branchadell, M Sodupe, J Sauer. J. Chem. Soc. 120:1545–1551, 1998.
109. D Berthomieu, S Krishnamurty, B Coq, G Delahay, A Goursot. J. Phys. Chem. B 105:1149–1156, 2001.
110. WF Schneider, KC Hass, R Ramprasad, JB Adams. J. Phys. Chem. 100:6032–6064, 1996.
111. A Goursot, V Vasilyev, A Arbuznikov. J. Phys. Chem. B 101:6420–6428, 1997.
112. G Turnes Palomino, P Fisicaro, S Bordiga, A Zecchina, E Giamello, C Lamberti. J. Phys. Chem. B 104:4064–4073, 2000.
113. Cerius2. Biosym Technologies: San Diego, CA, 1993.

2

Transient In Situ IR Study of Selective Catalytic Reduction of NO on Cu-ZSM-5

XIHAI KANG and STEVEN S. C. CHUANG   The University of Akron, Akron, Ohio, U.S.A.

I. INTRODUCTION

NOx is a precursor to the formation of ground-level ozone and acid rain. Concerns over the negative impact of NOx on the environment and human health led to regulation of NOx emission under the provisions of the 1990 Clean Air Act Amendments (CAAA). The CAAA require a significant decrease in the emission of NOx, hydrocarbons, and CO over the next few years. The key challenge for the automobile/truck industries and coal-fired power plants is to develop a cost-effective catalytic approach for control of NO emission [1–24]. Catalytic approaches for conversion of NO to N_2 include (1) NO decomposition [25–28], (2) the NO–CO reaction [29–31], (3) the selective reduction of NO with hydrocarbons, and (4) the selective reduction of NO with ammonia [17,32]. The direct decomposition of NO to N_2 and O_2 is the most attractive approach for NO emission control. However, no catalysts have been found to exhibit sufficient activities in the oxidizing environment where a high concentration of O_2 is present in the exhaust stream. O_2 poisons not only the NO decomposition catalysts but also the NO–CO reaction catalysts [26–28]. In contrast, the presence of O_2 results in an increase in NO conversion and N_2 yields in the selective catalytic reduction (SCR) of NO with hydrocarbons and NH_3. The interesting role of O_2 in promotion NO conversion and N_2 formation has led to a large number of postulations [1–24]: (1) the reaction of O_2 with NO to form highly reactive NO_2, (2) controlling the redox cycle of active sites and limiting coke formation, (3) the reaction of O_2 with hydrocarbons to form oxygenates that further reduce NO, and (4) enhancement of the rates of both formation and destruction nitrates—the SCR reaction intermediates.

To determine the role of O_2 in the SCR and its raction pathway, we have employed in situ IR spectroscopy and mass spectrometry to study the dynamic behaviors of adsorbed species during the step-and-pulse switch of the NO, O_2, C_3H_6 reactant flows during the SCR reaction over Cu-ZSM-5.

II. EXPERIMENTAL

A. Catalyst Preparation

The overexchanged Cu-ZSM-5-523 was prepared by repeated ion exchange of Cu-ZSM-5-83 with a 0.004 M copper acetate solution of pH = 7. Cu-ZSM-5-83 was produced by Johnson Matthey and provided through the catalyst bank of Sandia National Laboratories. The percentage copper exchange is defined as the molar ratio of the amount of Cu to that of Al multiplied by 2 [% Cu exchange = 2 (moles of Cu/moles of Al)%] [25]. The ion-exchange Cu-ZSM-5 sample was filtered, washed by distilled water, dried overnight at 373 K, and then calcined at 733 K for two hours. Inductive coupled plasma emission spectroscopy (Galbraith Laboratories, Knoxville, TN) determined Si/Al of 24.6 and Cu/Al of 2.6 in Cu-ZSM-5-523.

B. Reaction System

Figure 1 shows the reaction system, consisting of an in situ infrared (IR) reactor cell [33], a Nicolet Magna™ 550 Fourier transform infrared (FTIR) spectrometer, and a Pfeiffer PRISMA™ mass spectrometer (MS). A gas distribution system delivered the reactants and inert gases to the IR reactor cell containing the catalyst. To avoid the formation of NO_2 from the gaseous reaction of NO with O_2, the NO and O_2 flows were mixed in the vicinity of the catalyst sample in the infrared reactor cell, as shown in the inset of Figure 1. One-hundred-mg catalysts were pressed into three to four thin disks, each weighing 25 mg, by a hydraulic press at 4000–5000 psi. One of the disks was placed in the IR beam path inside the IR reactor cell, and the others were broken down into flakes and placed in the close vicinity of the IR beam path. The MS determined the changes in the concentration of the reactants and products, while the FTIR monitored the changes in the concentration of the absorbates during the transient IR experiment.

C. Transient IR Experiments

Dynamics of the formation of adsorbed species and products was studied by step and pulse experiments. The step experiment involves switching the inlet flow from He to $NO/C_3H_6/He$ and from $NO/C_3H_6/He$ to $NO/C_3H_6/O_2/He$. He was used an an inert gas to dilute the reactant stream. The switch of the flows creates a step change in the reactant concentration while maintaining the total flow rate

Flow A: 45 cm^3/min He

Flow B: 1 cm^3/min O$_2$ + 44 cm^3/min He

Flow C: 4 cm^3/min 1% NO (0.04 cm^3/min) +1 cm^3/min C$_3$H$_6$

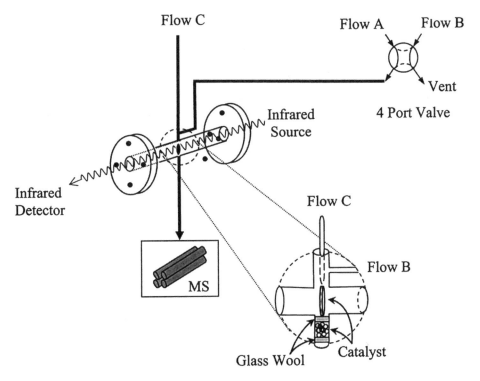

FIG. 1 Reaction system for HC-SCR.

constant at 50 cm^3/min. The pulse experiment uses a six-port gas chromatograph (GC) injection valve to introduce a C$_3$H$_6$ pulse into a steady state of NO/O$_2$/He flow over Cu-ZSM-5 at 373 and 623 K.

D.   Mass Spectrometer Analysis of the Gaseous Products

The mass-to-charge ratios (i.e., m/e or amu) for MS monitoring were $m/e = 30$ for NO, $m/e = 32$ for O$_2$, $m/e = 46$ for NO$_2$, $m/e = 41$ for C$_3$H$_6$, $m/e = 28$ for

N_2 and CO, $m/e = 12$ for CO, $m/e = 44$ for N_2O and CO_2, and $m/e = 22$ for CO_2. The responding factor of each species was determined by injecting a known amount of each species into the MS. The relative responding ratio of $m/e = 28$ to $m/e = 12$ for CO and that of $m/e = 44$ to $m/e = 22$ for CO_2 were further determined to separate the contribution of CO to $m/e = 28$ and that of CO_2 to $m/e = 44$. The responding factor obtained for each species allows for conversion of the MS profile to the corresponding molar flow rate.

III. RESULTS AND DISCUSSIONS

A. Step Transient Response

Figure 2 shows MS profiles of reactants and products during the step experiments at 623 K, providing an overall picture of the effect of O_2 on the reactant conver-

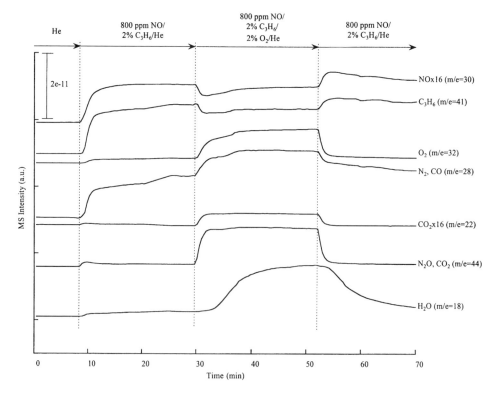

FIG. 2 MS profiles during step switch from He to 800 ppm NO/2% C_3H_6/He and from 800 ppm NO/2% C_3H_6/He to 800 ppm NO/2% C_3H_6/2% O_2/He (total flow rate 50 cm³/min) at 623 K over Cu-ZSM-5.

sion and product formation during the SCR reaction. The MS intensity of each species is proportional to its concentration in the reactor effluent. The most obvious effect of adding 2% O_2 into the 800 ppm $NO/2\%$ C_3H_6/He stream at 623 K is a significant decrease in the concentration of NO and C_3H_6 as well as an increase in N_2, CO_2, and H_2O concentrations. The former reflects an increase in the conversion of NO and C_3H_6 reactants; the latter indicates an increase in product formation.

To gain an insight into the SCR reaction, the IR spectra taken during each step switch of the inlet flow were plotted along with the variation of molar flow rate of the reactant and products. Figure 3 shows that exposure of Cu-ZSM-5 to 800 ppm $NO/2\%$ C_3H_6/He produced CH_3COO^- at 1452 cm^{-1}, $C_3H_7-NO_2$ at 1547 and 1596 cm^{-1}, Cu^+-CO at 2155 cm^{-1}, Cu^0-CN at 2198 cm^{-1}, and Cu^+-NCO at 2241 cm^{-1}. These species have also been observed during the SCR over CuO/Al_2O_3 and Pt/Al_2O_3 [34,35]. The variation of normalized infrared intensities of these IR-observable species with respect to time was plotted along with the changes in molar flow rate of a number of key species, such as NO, O_2, CO_2, and N_2, in Figure 4 to illustrate the lead/lag relationships between adsorbates and gaseous products.

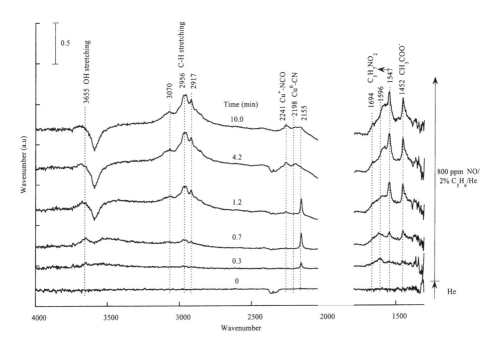

FIG. 3 IR spectra collected during step switch from He to 800 ppm $NO/2\%$ C_3H_6/He (total flow rate 50 cm^3/min) at 623 K over Cu-ZSM-5.

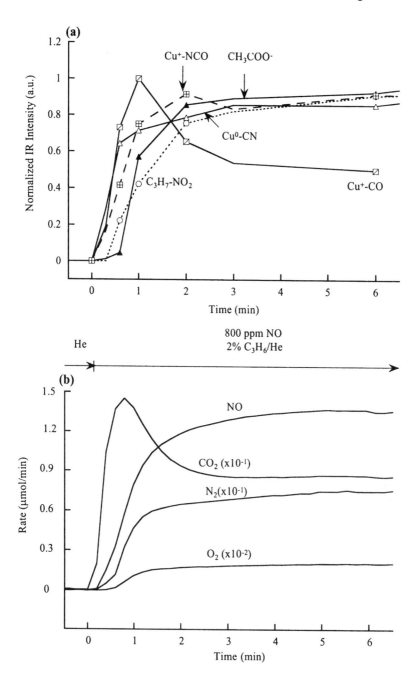

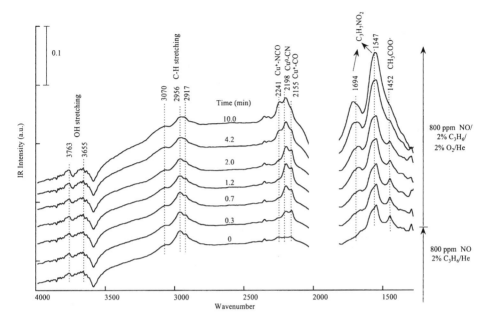

FIG. 5 IR spectra collected during step switch from 800 ppm NO/2% C₃H₆/He to 800 ppm NO/2% C₃H₆/2% O₂/He (total flow rate 50 cm³/min) at 623 K over Cu-ZSM-5.

These lead/lag relationships may allow elucidation of the sequence of their formation. Variation in the CO_2 molar flow rate followed very closely changes in the Cu^+–CO intensity, suggesting that CO_2 could be formed via Cu^+–CO. The formation of Cu^+–CO prior to that of CH_3COO^-, C_3H_7– NO_2, Cu^0–CN, and Cu^+–NCO [shown in Figs. 3 and 4(a)] indicates that the pathway for Cu^+–CO and CO_2 formation is independent of that for the adsorbates (i.e., CH_3COO^-, C_3H_7–NO_2, Cu^0–CN, and Cu^+–NCO), since these species, lagging behind CO and containing N and/or H, are not originated from CO. The same argument can also be used to conclude that the pathway for the formation of CH_3COO^- is independent of C_3H_7–NO_2.

Figure 5 shows that the addition of O_2 to the NO/C₃H₆/He flow resulted in

FIG. 4 (a) Normalized IR intensity versus time and (b) formation rate of reactants and products during step switch from He to 800 ppm NO/2% C₃H₆/He (total flow rate 50 cm³/min) at 623 K over Cu-ZSM-5 (normalized IR intensity = $(I(t)-I_0)/(I_\infty-I_0)$, where I_0 = IR intensity at $t = 0$, $I(t)$ = IR intensity at t, = I_∞ = IR intensity at $t = \infty$, i.e., final steady state).

an increase in IR intensity of C_3H_7–NO_2, Cu^+–NCO, Cu^0–CN, and Cu^+–CO. The variation of IR intensity of these species with time, plotted in Figure 6(a), shows that the formation of Cu^+–CO led that of Cu^0–CN, which further led that of C_3H_7–NO_2, and Cu^+–NCO. The O_2 addition also resulted in enhancement of CO_2 and N_2 formation as well as NO conversion, as shown in Figure 6(b). The absence of variation of CH_3COO^- suggests that its formation and destruction rates were not affected by O_2. The species may adsorb on the surface of ZSM-5 and serve as a spectator. The increase in the intensity of Cu^+–CO and Cu^0–CN revealed that O_2 did not enhance oxidation of Cu^0/Cu^+ to Cu^{2+} site. The key role of O_2 is to accelerate the rate of formation of C_3H_7–NO_2, Cu^+–NCO, Cu^0–CN, and Cu–CO adsorbates as well as gaseous N_2 and CO_2 products. The formation profile of CO_2 led that of N_2, further supporting that CO_2 and N_2 formation do not share the same reaction pathway.

B. Pulse Transient Response

Figure 7 shows (a) IR spectra and (b) MS profiles of each species during 1 cm^3 C_3H_6 pulse into the NO/O_2 flow at 373 K. Pulsing C_3H_6 caused increases in IR intensities of CH_3COO^- at 1370 cm^{-1}, $Cu^{2+}(NO_3^-)$ at 1575 and 1643 cm^{-1}, and C_3H_7–NO_2 at 1444 cm^{-1} as well as increases in MS intensities of N_2, CO_2, H_2O, and N_2O, indicating reduction of NO/O_2 by C_3H_6. Reduction of NO by C_3H_6 also caused the increase in the O_2 MS profile, suggesting that NO may have decomposed to N_2 and O_2 in the presence of C_3H_6.

Figure 8 shows (a) IR spectra and (b) MS profiles of each species during 1 cm^3 C_3H_6 pulse into the NO/O_2 flow at 623 K. Flowing NO/O_2/He over Cu-ZSM-5 produced bridged Cu^{2+} (NO_3^-). This species has also been observed during the NO decomposition reaction over Cu-ZSM-5 [26–28]. Pulsing C_3H_6 into the NO/O_2/He flow resulted in (1) the depletion of $Cu^{2+}(NO_3^-)$, (2) the emergence of Cu^+–CO, and (3) the formation of gaseous N_2 and CO_2 products. The C_3H_6 pulse also caused an initial increase in NO concentration, reflecting desorption of NO from the catalyst surface. The differences in IR-observable species during the reaction at 373 and 623 K suggest the reaction pathway for N_2 formation is strongly dependent on temperature.

FIG. 6 (a) Normalized IR intensity versus time, (b) formation rate of reactants and products, and (c) normalized formation rate during step switch from 800 ppm NO/2% C_3H_6/He to 800 ppm NO/2% C_3H_6/2% O_2/He (total flow rate 50 cm^3/min) at 623 K over Cu-ZSM-5 (normalized formation rate = $(R(t)-R_0)/(R_\infty-R_0)$, where R_0 = formation rate at $t = 0$, $R(t)$ = formation rate at t, and R_∞ = formation rate at $t = \infty$, i.e., final steady state).

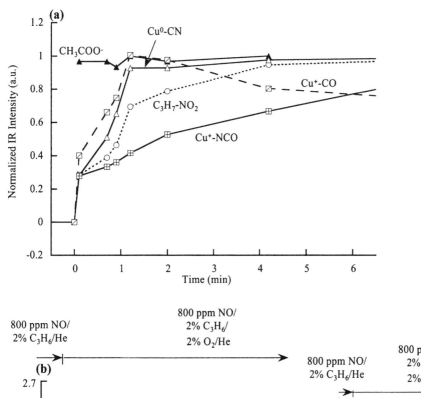

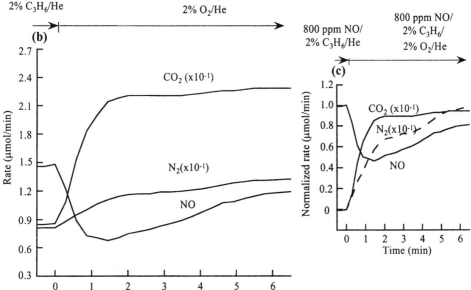

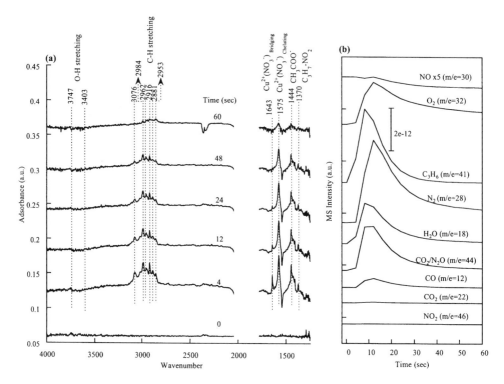

FIG. 7 (a) IR spectra collected and (b) MS profiles during 1 cm³ C₃H₆ pulse into steady flow of 800 ppm NO/2% O₂/He (total flow rate 50 cm³/min) at 373 K over Cu-ZSM-5.

C. Proposed Reaction Pathways

Figure 9 illustrates the postulated pathways of the steady-state C_3H_6-SCR reaction. CO_2 can be formed by two pathways: (1) partial oxidation of C_xH_y to Cu^+- CO followed by oxidation and (2) oxidation of $C_3H_7-NO_2$, Cu^+NCO, and Cu^0- CN. The former is much more rapid than the latter, as evidenced by the formation profiles of Cu^+-CO/CO_2 leading to that of $C_3H_7-NO_2$. $C_3H_7-NO_2$ may serve as a precursor to the formation of both CO_2 and N_2. In addition to $C_3H_7-NO_2$, the intermediates for N_2 formation may include Cu^0-CN and Cu^+-NCO, of which the IR intensity profiles parallel the N_2 molar flow rate profiles during the step switch from He to $NO/C_3H_6/He$. The nature of $C_3H_7-NO_2$, Cu^0-CN, and Cu^+- NCO intermediates can be further distinguished from the results of O_2 addition. The difference in their IR profiles during the switch from $NO/C_3H_6/He$ to $NO/$ $O_2/C_3H_6/He$ reflects their differences in reactivity toward O_2. The variation in the sequence of $C_3H_7-NO_2$ and Cu^+-NCO formation during step switch from

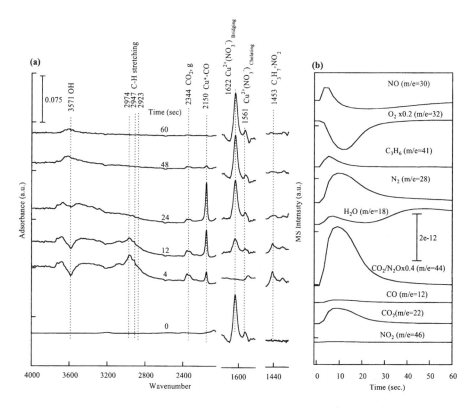

FIG. 8 (a) IR spectra collected and (b) MS profiles during 1 cm³ C₃H₆ pulse into steady flow of 800 ppm NO/2% O₂/He (total flow rate 50 cm³/min) at 623 K over Cu-ZSM-5.

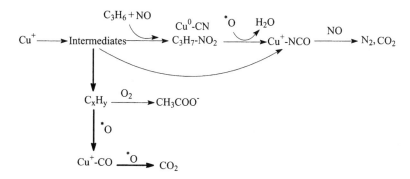

FIG. 9 Proposed pathways for the steady-state C₃H₆ SCR reaction.

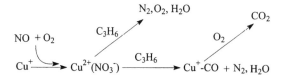

FIG. 10 Proposed pathways for the C_3/H_6 pulse SCR reaction.

He to NO/C_3H_6 and from NO/C_3H_6 to $NO/C_3H_6/O_2/He$ indicates that Cu^+-NCO does not have be formed via $C_3H_7-NO_2$. The evolution of these species is in line with that of N_2, suggesting that these species may serve as precursor for N_2 formation. In fact, it has been shown that these species react with gaseous NO to produce N_2 [36,37].

Figure 10 illustrates the postulated pathways for the C_3H_6-pulse SCR reaction where C_3H_6 is pulsed into the $NO/O_2/He$ flow. The reaction of NO with O_2 on Cu-ZSM-5 produced chelating nitrate at 373 K and bridged nitrate at 623K. Pulsing C_3H_6 led to the formation of O_2 and N_2 and CO_2 at 373 K. Formation of O_2 indicates that NO can be decomposed to N_2 and O_2 through reduction of the catalyst surface C_3H_6 at 373 K. At 623 K, O_2 is highly reactive toward C_3H_6, producing CO_2 and H_2O. No direct evidence can be observed to support the NO decomposition pathway at 623 K. The key reaction pathway for N_2 formation at this temperature is the direct reaction of nitrate with propylene, as shown in Fig. 10. The nitrate–proplyene reaction led to the formation of not only CO_2/N_2 but also CU^+-CO. The latter suggests that one of the key roles of C_3H_6 is to keep Cu in the Cu^+ state.

An effective use of hydrocarbon in HC (hydrocarbon)-SCR is to develop a catalyst that accelerates the rate of the reaction between NO species and hydrocarbon while inhibiting the direct reaction of hydrocarbons with oxygen. The different pathways for CO_2 and N_2 formation suggest that it should be possible to devise a selective poisoning approach to inhibit CO_2 formation without affecting N_2 formation. Selective inhibition of CO_2 formation should limit the direct oxidation of hydrocarbons, resulting in a significant enhancement of the selectivity toward N_2 for the HC-SCR reaction.

IV. CONCLUSIONS

Infrared spectroscopy coupled with mass spectroscopy allows determination of the dynamic behavior of adsorbates, the reaction pathways for NO reduction and C_3H_6 oxidation during SCR reaction. Transient IR studies of the $NO/C_3H_6/O_2$ reaction at 623 K showed that CO_2 and N_2 were formed from different reaction pathways. O_2 enhanced the formation of Cu^+-CO, $C_3H_7-NO_2$, and Cu^+-NCO

species and increased the overall rate of No conversion. The different N_2 and CO_2 formation pathways suggest the SCR reaction process may be further improved by a selective poisoning approach that inhibits CO_2 formation without interfering with N_2 formation.

ACKNOWLEDGMENT

This work was supported by the National Science Foundation under Grant CTS-942111996.

REFERENCES

1. J. K. Yan, G. D. Lei, W. M. H. Sachtler, H. H. Kung. J. Catal. 161:43–54, 1996.
2. I. C. Hwang, S. I. Woo. J. Phys. Chem. 101:4055–4059, 1997.
3. L. J. Lobree, A. W. Aylor, J. A. Reimer, A. T. Bell. J. Catal. 169:188–193, 1997.
4. X. Zhang, A. B. Walters, M. A. Vannice. Appl. Catal. B: Environmental, 7:321–336, 1996.
5. Q. Sun, Z. X. Gao, H. Y. Chen, W. M. H. Sachtler. J. Catal. 201:89–99, 2001.
6. V. Zuzaniuk, F. C. Meunier, J. R. H. Ross. J. Catal. 202:340–353, 2001.
7. H. Y. Chen, E. El-Malki, X. Wang, R. A. Van Santan, W. M. H. Sachtler. J. Mol. Catal. A Chemical 162:159–174, 2000.
8. L. J. Lobree, I. C. Hwang, J. A. Reimer, A. T. Bell. Catal. Lett. 63:233–240, 1999.
9. I. O. Y. Liu, N. W. Cant, B. S. Haynes, P. F. Nelson. J. Catal. 203:487–494, 2001.
10. X Wang, H. C. Chen, W. M. H. Sachtler. J. Catal. 197:281–291, 2001.
11. S. S. C. Chuang, C.-D. Tan, J. Catal. 173:95–104, 1998.
12. K. Almusaiteer, S. S. C. Chuang. J. Phys. Chem. B Environmental 104:2265–2272, 2000.
13. Z. Schay, V. S. James, G. Pal-Borbely, A. Beck, A. V. Ramaswamy, L. Guczi. J. Mol. Catal. 162:191–198, 2000.
14. J. L. d'Itri, W. M. H. Sachtler. Appl. Catal. B: Environmental 2:L7–L15, 1993.
15. K. Shimizu, J. Shibata, H. Yoshida, A. Satsuma, T. Hattori. Appl. Catal. B Environmental 30:156–162, 2001.
16. K. Nakamoto. Infrared and Raman Spectra of Inorganic and Coordination Compounds. 4th ed. New York: Wiley, 1986.
17. M. Shelef. Chem. Rev. 95:209–225, 1995.
18. A. A. Davydov. In: C. H. Rochester, ed. Infrared Spectra of Adsorbed Species on the Surface of Transition Metal Oxides. England: Wiley. 1990, pp 64–76.
19. Y. Ukisu, S. Sato, A. Abe, K. Yoshida. Appl. Catal. B Environmental 2:147–152, 1993.
20. K. K. Hansen, E. M. Skou, H. Christensen, T. Turek. J. Catal. 199:132–140, 2001.
21. D. K. Captain, M. D. Amiridis. J. Catal. 194:222–232, 2000.
22. K.-I. Shimizu, H. Kawabata, A. Satsuma, T. Hattori. J. Phys. Chem. B Environmental 103:5240–5245, 1999.
23. T. Liese, E. Loffler, W. Grunert. J. Catal. 197:123–130, 2001.

24.   H. Oka, T. Okada, K. Hori. J. Mol. Catal. A Chemical 190:51–54, 1996.
25.   M. Iwamoto, H. Yahiro, K. Tanda, N. Mizuno, Y. Mine, S. Kagawa. J. Phys. Chem. 95:3727–3730, 1991.
26.   M. V. Konduru, S. S. C. Chuang. J. Catal. 187:436–452, 1999.
27.   M. V. Konduru, S. S. C. Chung. J. Catal. 196:271–286, 2000.
28.   M. V. Konduru, S. S. C. Chung, X. Kang. J. Phys. Chem. 105:10918–10926.
29.   K. Almusaiteer, S. S. C. Chung. J. Cat. 180:161–170, 1998.
30.   K. Almusaiteer, S. S. C. Chuang. J. Cat. 184:189–201, 1999.
31.   K. Almusaiteer, S. S. C. Chung, C.-D. J. Cat. 189:247–252, 2000.
32.   M. Misano. Cattech. 2:183–195, 1998.
33.   S. S. C. Chuang, M. A. Brundage, M. Balakos, G. Srinivas. Appl. Spectrosc. 49: 1151–1163, 1995.
34.   F. Radtke, R. A. Koeppel, E. Minardi, A. Baiker. J. Catal. 167:127–141, 1997.
35.   D. K. Captain, M. D. Amiridis. J. Catal. 184:377–389, 1999.
36.   T. Beutel, B. J. Adelnman, W. M. H. Sachtler. Appl. Catal. B Environmental 9:L1–L10, 1996.

3

Comparison of Catalytic Reduction of NO by Propene on Zeolite-Based and Clay-Based Catalysts Ion-Exchanged by Cu

JOSE L. VALVERDE, FERNANDO DORADO, PAULA SÁNCHEZ, ISAAC ASENCIO, and AMAYA ROMERO   University of Castilla–La Mancha, Ciudad Real, Spain

I. INTRODUCTION

Selective catalytic reduction (SCR) of NO with hydrocarbons has been a subject of extensive study due to its potential for the effective control of NO emission in oxidant environments [1–11]. Hydrocarbons would be the preferred reducing agents over NH_3 because the practical problems associated with its use: handling and slippage through the reactor. Although many types of catalysts have been studied for this purpose, only a few copper-loaded zeolites have been demonstrated as adequate, and among them Cu-ZSM-5 gives good yields and seems to be one of the most active ones [3,5,12–15]. The majority of these catalysts are ion-exchanged zeolites, including H^+ forms. Metal oxides supported on alumina have also been studied. Shimuzu et al. [16,17] recently reported that Cu-aluminate catalysts, containing highly dispersed Cu^{2+} ions in the aluminate phase, showed high de-NOx activity comparable to Cu-ZSM-5 and higher hydrothermal stability. The activity of copper-loaded zeolites was found to depend on the Cu content. Iwamoto et al. [18] observed that the activity of Cu-ZSM-5 increased with the increment of the Cu exchange level. On the other hand, the concentration of O_2 in the NO-containing stream plays an important role in the reaction rate and product selectivity of the SCR reaction. It has been suggested that the roles of the O_2 in this reaction are: to activate NO and hydrocarbons [19], to oxidize NO to NO_2 [20], to maintain a Cu^+/Cu^{2+} site balance [21], and to react with carbonaceous deposits on Cu-ZSM-5 [20,22].

The majority of Cu ions in copper-exchanged zeolites are found as single cations located in structural cationic sites. Besides these atomically dispersed single Cu ions in cationic sites, dispersed CuO species were also detected [23–25]. In any case, the facile reduction from Cu^{2+} to Cu^+ suggested that a possible redox mechanism could be involved. Shpiro et al. [26] suggested that copper in overexchanged Cu-ZSM-5 zeolites exists as small clusters of Cu-O or as isolated ions whose oxidation state (Cu^{2+} or Cu^+) depends on the atmosphere present. Corma et al. [27/38] observed that the most active Cu-beta zeolites for SCR of NOx were those in which the conversion between Cu^{2+} and Cu^+ was quickly produced under reaction conditions. This conversion was easier in overexchanged Cu-beta samples.

Pillared clays (PILCs) are two-dimensional materials prepared by exchanging charge-compensating cations between the clay layers with large inorganic metal hydroxycations that are oligomeric and are formed by hydrolysis of metal oxides or salts. After calcination, the metal hydroxycations are decomposed into oxide pillars that keep the clay layers apart and create interlayer and interpillar spaces, thereby exposing the internal surfaces of the clay layers. The size of these oligomers appears to control the size of the pore opening in the pillared clays. It is known that the basal spacing of the natural clay is 9.6 Å and that the insertion of these hydroxycations increases it to 18–20 Å. In principle, any metal oxide or salt forming polynuclear species upon hydrolysis can be inserted as a pillar. Intercalated clays are usually natural smectites clays. Properties such as acidity, surface area, pore size distribution, and both thermal and hydrothermal stability depend on the method of synthesis as well as on the nature of the host clay. Most common ions used as pillaring agents prepared by hydrolysis of the corresponding salts in solution are polycationic species of Al, Zr, Fe, Cr, etc. Cañizares et al. [28] recently reported a comparative study in which different PILCs with single-oxide pillers of Fe, Cr, and Zr and mixed-oxide pillars of these metals and Al were prepared from two different bentonites.

One of the fields of applications of pillared clays is catalysis. More specifically, PILC-based catalysts were found to be useful for the SCR reaction by NH_3 and hydrocarbons [29–31]. Using hydrocarbons, Cu-ion exchanged PILC yielded higher than those of Cu-ZSM-5-based catalysts; and their activity was only slightly decreased by H_2O and SO_2 [32,33]. In an earlier paper, Yang et al. [34] tested different ion-exchanged pillared clays as catalysts for selective catalytic reduction of NO by ethylene. Cu-Ti-PILCs showed the highest activities at temperatures below 643 K, whereas Cu-Al-PILC was the most active at temperatures above 673 K. Ti-PILC was obtained using a procedure in which the pillaring solution of partially hydrolyzed Ti-polycations was prepared by first adding $TiCl_4$ into a HCl solution.

In spite of these relevant results, Ti-PILCs have received considerably less attention than other pillared clays. As a result, few preparation methods for these materials have been reported [34,35]. In our case, Ti metoxide was used as the source of Ti in the preparation of Ti-PILCs.

The aims of this work are:

1. To compare the catalytic reduction of NO by propene on ZSM-5 based and Ti-PILC-based catalysts ion-exchanged by Cu.
2. To characterize all the catalysts here prepared in order to relate physical and chemical properties to the catalytic behavior of the two sets of catalysts.

II. EXPERIMENTAL

A. Preparation of Catalysts

NaZSM-5 zeolite (Si/Al ratio of 20) was synthesized according to the method described elsewhere [36] using ethanol as the template. X-ray diffraction (XRD) confirmed that the product was 100% crystalline [37]. Cu was introduced by conventional ion exchange, using 25 mL of 0.1 M $Cu(CH_3COO)_2 \cdot 4H_2O$ aqueous solutions per gram of zeolite. The mixture was kept under agitation at the desired ion exchange temperature (303, 328, or 353K) for 14 h. Next, the suspension was filtered and thoroughly washed with deionized water in order to completely remove the occluded salt, and the solid was then air-dried at 393 K for 14 h. The whole procedure was repeated twice for some catalysts. Finally, the samples were calcined in air at 823 K for 4 h. Table 1 summarizes the zeolite-based catalysts here prepared. These catalysts were referred to as a function of the copper loading. For instance, CuZ-2.9 corresponds to a Cu-ZSM5 with a copper content of 2.9% by weight.

Ti-PILC was prepared as follows. The starting clay was a purified montmorillonite (purified-grade bentonite power from Fisher Company), which has a particle size of 2 μm or less, a cation exchange capacity (CEC) of 97 eq/kg dry clay and a surface area of 44.7 m^2/g. The pillaring solution of partially hydrolyzed Ti-polycations was prepared by first adding titanium metoxide to a 5 M HCl solution. The solution was aged for 3 h at room temperature. Then 1 gram of starting clay was dispersed in 1 L of deionized water for 3 h under stirring. The pillaring solution was slowly added with vigorous stirring into the clay suspension until the amount of pillaring solution reached that required to obtain a Ti/Clay ratio of 15 mM of Ti/g clay. The intercalation step took about 16 h. Subsequently, the mixture was separated by vacuum filtration or centrifugation and washed with deionized water until the liquid phase was chloride free. The sample was dried at 393 K for 12 h and calcined at 773 K for 2 h. The basal spacing of the resulting sample, measured by XRD, was 23 Å.

One gram of the Ti-pillared bentonite was added to 200 mL of 0.05 M copper acetate solution. The mixture was stirred for 6 h at room temperature. The ion

TABLE 1 Composition and Characterization of Zeolite-Based Catalysts[a]

Catalyst	Ion exchange steps and temperature (K)	Cu content (wt %)	Ion exchange level (%)	Weak acid site density (mmol NH_3/g)	Strong acid site density (mmol NH_3/g)	Surface area (m²/g)	Micropore area (m²/g)	Micropore volume (m³/g)
NaZSM-5	—	0	0	0.950 (579 K)	Not detected	369.3 (100%)	360.0 (100%)	0.158 (100%)
CuZ-2.4	1—303	2.4	94	0.881 (565 K)	Not detected	345.4 (94%)	338.8 (94%)	0.140 (89%)
CuZ-2.6	1—328	2.6	103	0.822 (573 K)	Not detected	—	—	—
CuZ-2.9	2—303	2.9	116	0.907 (555 K)	Not detected	326.4 (88%)	309.6 (86%)	0.130 (82%)
CuZ-3.7	2—328	3.7	148	0.781 (573 K)	Not detected	—	—	—
CuZ-4.4	1—353	4.4	175	0.477 (581 K)	0.440 (911 K)	321.0 (83%)	298.7 (83%)	0.126 (79%)

[a] Temperatures corresponding to the maximum of the desorption peak are included in parentheses together with the acid sites density value.

exchange product was collected by filtration or centrifugation, followed by washing five times with deionized water. The obtained solid sample was first dried at 393 K in air for 12 h and then calcined at 773 K for 2 h. After this pretreatment, the sample was ready for further experiments. Table 2 summarizes the PILC-based catalysts here prepared. Pillared clay-based catalysts were referred to as a function of the copper loading. For instance, CuTi-7.4 corresponds to a Cu-Ti-PILC with a copper content of 7.4% by weight.

B. Characterization Methods

X-ray diffraction (XRD) patterns were measured with a Philips model PW 1710 diffractometer using Ni-filtered Cu$K\alpha$ radiation. To summarize the (001) reflection intensity in PILC samples, oriented clay-aggregate specimens were prepared by drying clay suspensions on a glass slide.

Surface area and pore size distributions were determined by using nitrogen as the sorbate at 77 K in a static volumetric apparatus by using a micromeritics ASAP 2010 sorptometer. For this analysis, samples were outgassed at 453 K for 16 h under a vacuum of 5×10^{-6} torr. Specific total surface areas were calculated by using the Brunauer, Emmett, and Teller (BET) equation. The Horvath–Kawazoe method was used to determine microporous surface area and volume.

Total acid site density of the samples was measured by a temperature programmed desorption (TPD) of ammonia, by using a Micromeritics TPD-TPR analyzer. Samples were housed in a quartz tubular reactor and pretreated in flowing helium (99.999%) while heating at 15 K min^{-1} up to 773 K. After 0.5 H at 773 K, the samples were cooled to 453 K and saturated for 0.25 h in an ammonia (99.999%) stream. The sample was then allowed to equilibrate in a helium flow at 453 K for 1 h. Finally, ammonia was desorbed using a linear heating rate of 15 K min^{-1}. Temperature and detector signals were simultaneously recorded. The average relative error in the acidity determination was lower than 3%.

Temperature programmed reduction (TPR) measurements were carried out with the same apparatus previously described. After loading, the sample was outgassed by heating at 20 K min^{-1} in an argon flow to 773 K. This temperature was constant for 30 min. Next, it was cooled to 298 K and stabilized under an argon/hydrogen (99.999%, 83/17 volumetric ratio) flow. The temperature and detector signals were continuously recorded while heating at 20 K min^{-1}. A cooling trap placed between the sample and the detector retained the liquids formed during the reduction process. The TPR profiles were reproducible with an average relative error in the determination of the reduction maximum temperatures lower than 2%.

The metallic content (wt%) was determined by atomic absorption measurements by using a SpectrAA 220 FS analyzer. In all cases, calibrations from the corresponding patron solutions were performed.

TABLE 2 Composition and Characterization of PILC-Based Catalysts[a]

Catalyst	Cu content (wt%)	Ion exchange level (%)	Weak acid site density (mmol NH_3/g)	Strong acid site density (mmol NH_3/g)	Surface area (m^2/g)	Micropore area (m^2/g)	Micropore volume (m^3/g)
Ti-PILC	0	0	0.437 (579 K)	0.092 (691 K)	273.2 (100%)	224.5 (100%)	0.181 (100%)
CuTi-4.6	4.6	149	0.108 (548 K)	0.362 (628 K)	241.7 (88%)	202.2 (90%)	0.153 (85%)
CuTi-7.4	7.4	240	0.136 (543 K)	0.594 (623 K)	234.3 (86%)	189.1 (84%)	0.143 (79%)
CuTi-9.0	9.0	292	0.156 (533 K)	0.738 (613 K)	201.8 (74%)	153.7 (68%)	0.126 (69%)

[a] Temperatures corresponding to the maximum of the desorption peak are included in parentheses together with the acid sites density value.

C. Reaction Studies

The catalytic tests were carried out in a fixed-bed flow reactor. The standard reactant mixture was one constituted by NO (1000 ppm), C_3H_6 (1000 ppm), O_2 (5%), and balance He at ambient pressure. The flow rates were controlled by calibrated Brooks flowmeters. The total flow rate was 125 mL/min. The space velocity of the feed was 15,000 h^{-1} (GHSV). The effluent stream was analyzed by a chemiluminescent NO/NOx analyzer (ECO PHYSICS NO-NO_2-NOx analyzer).

III. RESULTS

A. Characterization of the Catalysts

Tables 1 and 2 list, for all the catalysts, the specific surface area, micropore area, and micropore volume, the weak and strong acid site density, and the copper content. The same tables also summarize the Cu ion exchange levels that were determined taking as a reference, in the case of zeolite-based catalysts, the number of aluminum atoms contained in the structure and, in the case of PILC-based catalysts, the cation exchange capacity (CEC) of the clay [34]. It can be observed that, except for the CuZ-2.4 sample, all the catalysts presented more Cu content than that corresponding to 100% ion exchange. In fact, all the zeolite-based catalysts contained sodium ions. As expected, the loading of these decreased with increasing copper content in zeolite (0.16 wt% of Na for the CuZ-2.4 sample and 0.07 wt% of Na for the CuZ-4.4 sample. It is also observed for these catalysts that for the same number of ion exchange steps the copper loading increased with increasing ion exchange temperatures and the acid site density progressively diminished from the value corresponding to the parent Na/ZSM5 (0.95 mmol NH_3/g) to 0.822 and 0.907 mmol NH_3/g in the ZCu-2.6 and ZCu3.7 samples, respectively. In this case, it is clear that acidity is a combination of two effects: the presence of Cu and Na on the catalyst. The first Cu ions incorporated to zeolites would occupy hidden sites (small zeolite cages) [38]. These sites may not be very accessible to NH_3 molecules. Since multiple ion exchange steps are needed to achieve the complete filling of small cages, the increase of copper content would lead Cu species to incorporate at more accessible positions. As a consequence the acidity increases. Strong acid site density was observed only for zeolite-based catalysts with copper content higher than 4.4% by weight. In contrast, all the PILC-based catalysts presented strong acid sites.

An increase of Cu content was accompanied in both sets of catalysts by a decrease of BET surface area and micropore volume, indicating that Cu introduced into the pillared matrix would preferentially occupy the interlayer area [39], whereas the Cu present in zeolite would cause the partial blocking of the zeolite channels. Temperature-programmed reduction can be used to identify and quantify the copper species in ion-exchanged samples. According to Delahay et al. [40] the reactions involved in the copper reduction process are:

$$CuO + H_2 \rightarrow Cu^0 + H_2O$$

$$Cu^{2+} + \frac{1}{2} H_2 \rightarrow Cu^+ + H^+$$

$$Cu^+ + \frac{1}{2} H_2 \rightarrow Cu^0 + H^+$$

Some authors showed that, depending on the copper content, the reduction of Cu^{2+} to Cu^+ would occur at lower temperatures while the reduction of Cu^+ to Cu^0 would occur at higher temperatures [41–43]. When the copper content in the sample is higher, the excess copper may be found as oxygenated clusters more easily reduced than the isolated copper species [24,34].

As shown in Table 3, low H_2 consumption was found in all the samples, indicating that Cu species are hard to reduce to lower valence. In the case of zeolite-based catalysts, this fact is justified considering the existence of metal ions in small zeolite cages [38]. In these small cages, Cu^{2+} ions would be favorably coordinated to the framework oxygens. This bonding is generally much stronger for multivalent than for monovalents ions. Each ion would not be readily accessible for H_2 molecules, and the activation energy for reducing such isolated metal ions would be rather high. These results are in good agreement with the measurements of acidity mentioned earlier. In PILC-based catalysts, the H_2/Cu is slightly

TABLE 3 Ratios of H_2 Consumption to Cu (H_2/Cu, mol/mol, Measured by TPR Experiments) of Zeolite-and PILC-Based Catalysts[a]

Catalyst	H_2/Cu (mol/mol) CuO to Cu^0	H_2/Cu (mol/mol) Cu^{2+} to Cu^+	H_2/Cu (mol/mol) Cu^+ to Cu^0	H_2/Cu (mol/mol) Total
NaZSM-5	—	—	—	—
CuZ-2.4	Not detected	0.169 (517 K)	0.322 (643 K)	0.491
CuZ-2.6	Not detected	0.181 (503 K)	0.271 (676 K)	0.452
CuZ-2.9	Not detected	0.249 (503 K)	0.204 (693 K)	0.453
CuZ-3.7	Not detected	0.177 (450 K)	0.260 (653 K)	0.437
CuZ-4.4	0.251 (450 K)	0.134 (461 K)	0.216 (663 K)	0.601
Ti-PILC	—	—	—	—
CuTi-4.6	0.405 (473 K)	0.172 (543 K)	0.046 (663 K)	0.623
CuTi-7.4	0.321 (431 K)	0.170 (507 K)	0.051 (685 K)	0.542
CuTi-9.0	0.395 (426 K)	0.107 (510 K)	0.036 (693 K)	0.538

[a] Temperatures corresponding to the maximum of the reduction peak are included in parentheses.

higher than that of the zeolite, also showing that in these samples an important part of Cu species is not accessible for H_2 molecules.

For the zeolite-based catalysts with a copper content lower than 4.4 wt%, the TPR profiles showed two reduction peaks (Table 3) which suggests a two-step reduction process of isolated Cu^{2+} species. The peak at a lower temperature would indicate that the process of Cu^{2+} to Cu^+ occurred. The other peak at a higher temperature suggests that the produced Cu^+ was further reduced to Cu^0. For the sample CuZ-4.4, a peak existing at 450 K would be related to the presence of CuO aggregates. Due to the absence of a diffraction line of CuO species in XRD patterns of this sample, the occurrence of CuO aggregates larger than 3 mm can be ruled out [40]. According to the measurements of acidity, the presence of CuO aggregates would be related to the occurrence of strong acid sites. On the other hand, there is a clear shift to lower temperatures of the second peak (Cu^{2+} to Cu^+) that showed that Cu^+ becomes more difficult to reduce as Cu content decreases.

In the case of PILC-based catalysts, the peaks corresponding to the three reactions involved in the copper reduction are present. Again, there is a clear shift to a lower temperature of the second peak (Cu^{2+} to Cu^+) and the first (CuO to Cu^0) that showed that Cu^{2+} and CuO get easier to reduce as Cu content increases. This fact would indicate that the higher the Cu content is, the lower CuO species dispersion that is observed. In a similar way as observed in zeolite-based catalysts, no diffraction of CuO species in XRD were detected. Figure 1 compares the TPR-profiles of the CuZ-2.9 and CuTi-7.4 samples.

Figure 2 shows, for all the catalysts, the H_2 consumption for the Cu^{2+} to Cu^+ process as a function of the Cu loading. With an increase in Cu content, H_2 consumption increased, passing through a maximum, and then decreased at higher loadings. It can be verified that the maximum in the case of zeolite-based catalysts corresponds to the CuZ-2.9 sample and in the case of PILC-based catalysts to the CuTi-7.4 sample.

B. NOx Reduction Activity

The catalytic performance of the catalysts for the SCR reaction of NOx with propene as a function of the reaction temperature is summarized in Table 4. The presence of copper in the catalysts enhanced the catalytic activity. With an increase in reaction temperature, NOx conversion increased, passing through a maximum, and then decreased at higher temperatures. According to Yang et al. [34], the decrease in NOx conversion at higher temperatures was due to the combustion of propene. In general, all the Cu-zeolites samples presented the maximum NOx conversion at the same temperature (623 K). Similar observations can be derived for all PILC-based catalysts, but in this case the corresponding maximum appeared at 523 K. It can be observed that increasing the copper loading increased NOx conversion until the copper loading reached 116% ion exchange

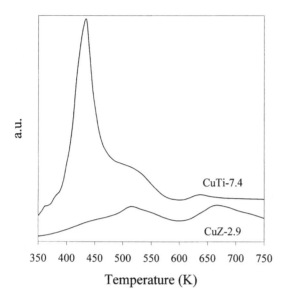

FIG. 1 TPR profiles of CuTi-7.4 and CuZ-2.9.

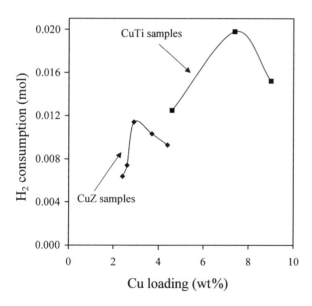

FIG. 2 Hydrogen consumption for the Cu^{2+} to Cu^+ reduction processes as a function of Cu loading.

TABLE 4 Catalytic Performance of Zeolite- and PILC-Based Catalysts Ion Exchanged by Cu

Catalyst	Maximum NOx conversion	Temperature for maximum NOx conversion (K)
CuZ-2.4	56.9	623
CuZ-2.6	62.4	623
CuZ-2.9	65.7	623
CuZ-3.7	65.6	623
CuZ-4.4	56.4	623
CuTi-4.6	26.4	523
CuTi-7.4	59.3	523
CuTi-9.0	36.7	523

Reaction conditions: $[NO] = [C_3H_6] = 1000$ ppm, $[O_2] = 5\%$, $[He] =$ balance; GHSV $= 15,000$ h^{-1}.

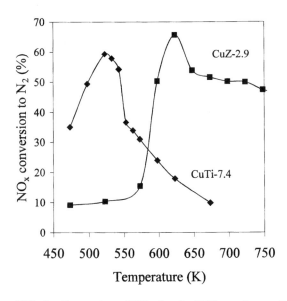

FIG. 3 Conversion of NOx for the SCR reaction on CuTi-7.4 and CuZ-2.9. Reaction conditions: $[NO] = [C_3H_6] = 1000$ ppm, $[O_2] = 5\%$, $[He] =$ balance, GHSV $= 15,000$ h^{-1}.

in zeolite-based catalysts, CuZ-2.9 sample, and 240% ion exchange in PILC-based catalysts, CuTi-7.4 sample (Fig. 3). Both CuZ-2.9 and CuTi-7.4 showed the highest H_2 consumption for the Cu^{2+} to Cu^+ process, taking as a reference the sets of zeolite-based and PILC-based catalysts, respectively. Further increase in the copper loading resulted in a decrease in NOx conversion.

IV. DISCUSSION

In this work the parent ZSM-5 was in the sodium form. According to Torre-Abreu et al. [15], Cu-MOR catalysts prepared from the sodium form exhibited much higher activity than those prepared from the acid form. Their results showed that Brönsted acidity does not promote the NO selective reduction by propene. On the other hand, Bulánek et al. [43] observed that the copper present in zeolite together with a sodium cation was more easily reduced than in CuH-zeolites. H_2 TPR results showed that the copper in zeolite-based catalysts was present mainly in the form of isolated Cu^{2+} ions. Cu^{2+} ions could be partially reduced to Cu^+ ions when the samples were treated at high temperatures. Apart from these ions, in PILC-based catalysts, CuO aggregates were also detected. As already mentioned, Cu^{2+} ions play an important role in the SCR reaction of NOx by hydrocarbons. In the presence of propene, the Cu^{2+} is reduced to Cu^+, and then Cu^+ would be oxidized back to Cu^{2+} by NOx, thus completing the catalytic cycle [40,44]. In a study reported by Attfield et al. [45] for Cu-mordenite materials, it was observed that the low coordination of the Cu^{2+} cations made them much more susceptible to redox chemistry when exposed to NO than a more highly coordinated Cu^{2+} cation. Similar reasoning was extended to other extremely active de-NOx zeolite catalysts, like ZSM-5 [45]. This pentasil zeolite has few cation sites within small inaccessible cavities in this structure, and so it will render most of the extra-framework transition metal cations poorly coordinated and highly accessible to reactant molecules. According to Bulánek et al. [43], Cu ions in zeolite matrices of MFI, FER, and MOR structures revealed two most populated Cu ions types, denoted as the Cu-II and Cu-IV types of ions (46). The Cu-II type of ions exhibited pyramidal coordination and high positive charge and are balanced by two negative framework charges, two AlO_2^-. The Cu-IV type of ions were ascribed to the Cu ions with close-up-to planar coordination and low positive charge and are adjacent to single AlO_2^- anions. At low Cu loadings and a high concentration of aluminum in the framework, the Cu-II type of ions prevails, while at high Cu loadings and a low concentration of aluminum in the framework, the Cu-IV type of ions become predominant.

The major species present in pillared montmorillonite exchanged with copper at low pH are physisorbed square planar tetra-aquo-complexes $[Cu(H_2O)_4]^{2+}$ [47]. Calcination of the samples at 673 K produced a significant change in the Cu^{2+} ion environment. According to Bahranowski et al. [47], the most characteristic feature was the increase in covalency of the Cu-O in-plane σ-bonding due to the dehydration/dehydroxilation processes accompanying high-temperature treatment, resulting in the attachment of copper centers to the lattice oxygens, forming links with the pillar rather than with the silicate sheet. It was suggested that in catalytic reactions proceeding according to the redox mechanism, the in-plane bridging oxygens played an important role in electron transfer from the electron-accepting to the electron-donating sites at the catalytic surface. According to Bahranowski et al. [48], single doping of Al-PILC with copper introduced approximately 40% more than the CEC of the Al-PILC samples, suggesting an adsorption mechanism involving the formation of a complex between copper ions and surface pillars rather than a simple ion exchange. Based on EPR studies, it was found that the majority of copper ions were associated with aluminum pillars as isolated Cu^{2+} ions, the rest forming clusters at an undefined location. Each subsequent doping increased the relative content of copper clusters [49]. A model of the calcined Cu-Al-PILC-based catalysts for oxidation of toluene and xylene with hydroperoxide was presented by Bahranowski et al. [39]. Thus, three types of copper species were proposed: isolated Cu ions anchored at the pillars rather than at the surface of the silica layer, Cu^{2+} clusters as patches of amorphous CuO, and cupric oxide. The presence of Cu^{2+} clusters increased with the copper loading on the catalysts. Because isolated Cu ions and Cu^{2+} clusters must be located in the interlayer, an increase in Cu content resulted in a reduction of pore volume due to the filling of the interlayer.

According to these studies, it seems clear that the Cu-IV type of ions, more poorly coordinated but highly accessible and reducible than the Cu-II type of ions, have to be the active sites for NO decomposition to N_2. As expected, the higher the content of these species on the catalysts is, the higher the catalytic activity that is observed (Table 4 and Fig. 2). The same maximum of temperature found in each set of catalysts (623 K for zeolite-based catalysts and 523 K for PILC-based catalysts) would indicate that there is no change in the mechanism of reaction when passing in a certain set of catalysts from one sample to another. Anyway, PILC-based catalysts appear to be more active catalytically than zeolite-based catalysts: The former required a lower reaction temperature to get the maximum of NOx conversion than did the latter. According to Yang et al. [34], Cu^{2+} on the pillared clay is more active than that on ZSM-5 due to the fact that the redox cycle occurs more easily on the pillared catalysts. On the other hand, it is also observed that the increase in Cu loading in the PILC-based catalysts resulted in an increase in CuO aggregates (Table 3). In a study of the reducibility and catalytic activity in propane oxidation [43], it was observed that the presence of disperse undefined CuO species in Cu-MFI-based catalysts resulted in a substantial increase in propane conversion as well selectivity to CO_2. This fact would explain the decided decrease in NO conversion with increasing copper content in the CuTi-7.4 sample observed at reaction temperatures higher than that corresponding to the maximum of conversion. This decided decrease was not observed in the CuZ-2.9 sample, with no occurrence of CuO species (Fig. 3).

Finally, it can be noted that the CuTi-7.4 sample, the most active catalyst found here, presented at 240% ion exchange. This value was similar to that reported by Yang et al. [34] for the most active Cu-Ti-PILC for SCR of NO by ethylene. In this case the best catalyst presented a 245% ion exchange.

IV. CONCLUSIONS

Selective catalytic reduction of NO by propene was investigated on copper ion–exchanged ZSM-5 and Ti-PILC in the presence of excess oxygen. H_2 TPR results showed that the copper in zeolite-based catalysts was present mainly in the form of isolated Cu^{2+} ions that could be partially reduced to Cu^+ ions when the samples were treated at high temperatures. Apart from these ions, in PILC-based catalysts, CuO aggregates were also detected. The presence of copper in the catalysts enhanced the catalytic activity. With the increase in reaction temperature, NOx conversion increased, passing through a maximum, and then decreased at higher temperatures. In the presence of propene, the Cu^{2+} should be reduced to Cu^+, and then Cu^+ would be oxidized back to Cu^{2+} by NOx, thus completing the catalytic cycle. In general, all the Cu-zeolite samples present the maximum NOx conversion at the same temperature (623 K). Similar observations for all PILC-based catalysts can be derived, but in this case the corresponding maximum appeared at 523 K. It was observed that increasing the copper loading increased NOx conversion until the copper loading reached 116% ion exchange in the zeolite-based catalysts, the CuZ-2.9 sample, and 240% ion exchange in PILC-based catalysts, the CuTi-7.4 sample. It seems clear that the Cu^{2+}, poorly coordinated but highly accessible and reducible, have to be the active sites for NO decomposition to N_2. As expected, the higher the content of these species on the catalysts is, the higher the catalytic activity that is observed. PILC-based catalysts appeared to be more active catalytically than zeolite-based catalysts: The former required a lower reaction temperature to get the maximum NOx conversion than did the latter. Cu^{2+} on the pillared clay would be more active than that on ZSM-5 due to the fact that the redox cycle occurs more easily on the pillared catalysts.

ACKNOWLEDGMENTS

Financial support from the European Commission (Contract ERK5-CT-1999-00001) and DGICYT (Dirección General de Investigación Científica y Técnica, Project 1FD97-1791, Ministry of Education, Spain) is gratefully acknowledged.

REFERENCES

1. M Shelef. Chem. Rev. 95:209–225, 1995.
2. M. Iwamoto. Catal. Today 29:29–35, 1996.

3. Y. Li, J.N. Armor. App. Catal. B 5:L257–L270, 1995.
4. A.W. Aylor, S.C. Larson, J.A. Reimer, A.T. Bell. J. Catal. 157:592–602, 1995.
5. B.J. Adelman, T. Bentel, G.-D. Lei, W.M.H. Sachtler. J. Catal. 158:327–335, 1996.
6. K.A. Bethke, H.H. Kung. J. Catal. 172:93–102, 1997.
7. R. Burch, A.A. Shestov, J.A. Sullivan. J. Catal. 182:497–506, 1999.
8. Y.-H. Yin, A. Pisau, L. Serventi, W.E. Alvarez, D.E. Resasco. Catal. Today 54: 419–429, 1999.
9. D.K. Captain, M.D. Amiridis. J. Catal. 184:377–389, 1999.
10. S. Xie, J.P. Rosynek, J.H. Lunsford. J. Catal. 188:24–31, 1999.
11. F.C. Meunier, J.P. Breen, V. Zuzaniuk, M. Olsson, J.R.H. Ross. J. Catal. 187:493–404, 1999.
12. J. Valyon, W.K. Hall. Stud. Surf. Sci. Cat. 75:1339–1350, 1993.
13. M. Iwamoto, N. Mizuno. J. Auto Eng. 207:23–37, 1993.
14. V.A. Matyshak, A.N. Ilichev, A.A. Ikharsky, V.N. Korchak. J. Catal. 171:245–254, 1997.
15. C. Torre-Abreu, M.F. Ribeiro, C. Henriques, F.R. Ribeiro. App. Catal. B 13:251–264, 1997.
16. K. Shimizu, H. Kawabata, H. Maeshima, A. Satsuma, T. Hattori. J. Phys. Chem. B 104:2885–2893, 2000.
17. K. Shimizu, H. Maeshima, H. Yoshida, A. Satsuma, T. Hattori. Phys. Chem. Chem. Phys. 2:2435-2439, 2000.
18. M. Iwamoto, N. Mizuno, H. Yahiro. Selective catalytic reduction of NO by hydrocarbons in oxidizing atmosphere. Proceedings of the 10th International Congress on Catalysis, Budapest, 1992, pp 213–215.
19. G.R. Bamweda, A. Obuchi, A. Ogata, J. Oi, J. Kushiyama, H. Yagita, K. Mizuno. Stud. Surf. Sci. Catal. 121:263–268, 1999.
20. C. Yokoyama, M. Misono. J. Catal. 150:9–17, 1994.
21. J.O. Petunchi, W.K. Hall. App. Catal. B 2:L17–L26, 1993.
22. J.L. d'Itri, W.M.H. Sachtler. App. Catal. B 2:L7–L15, 1993.
23. G.-D. Lei, b.J. Adelman, J. Sárkány, W.M.H. Sachtler. App. Catal. B 5:245–256, 1995.
24. Sárkány, J.L. d'Itri, W.M.H. Sachtler. Catal. Letter 16:241–257, 1992.
25. G. Centi, J. Perathoner. App. Catal. A 124:317–337, 1995.
26. E.S. Shpiro, W. Grüenert, J. Royner, G. Baeva. Catal. Letter 24:159–169, 1994.
27. A. Corma, V. Fornés, E. Palomares. Appl. Catal. B 11:233–242, 1997.
28. Cañizares, J.L. Valverde, M.R. Sun Kou, C.B. Molina. Microporous Mesoporous Materials 29:267–281, 1999.
29. R.T. Yang, J.P. Chen, E.S. Kikkinides, L.S. Cheng, J.E. Cichanowicz. Ind. Eng. Chem. Res. 31:1440–1445, 1992.
30. J.P. Chen, M.C. Hausladden, R.T. Yang. J. Catal., 151:135–146, 1995.
31. L.S. Cheng, R.T. Yang. N. Chen, J. Catal. 164:70–81, 1996.
32. R.T. Yang, W.B. Li. J. Catal. 155:414–417, 1995.
33. W.B. Li, M. Sirihungren, R.T. Yang. App. Catal. 11:347–363, 1997.
34. R.T. Yang, N. Tharappiwattananon, R.Q. Lay. App. Catal. 19:289–304, 1998.
35. H.L. del Castillo, A. Gil, P. Grange. Catal. Lett. 43:133–137, 1997.

36. M.A. Uguina, A. de Lucas, F. Ruiz, D.P. Serrano. Ind. Eng. Chem. Res. 34:451–456, 1995.
37. A. de Lucas, J.L. Valverde, L. Rodríguez, P. Sánchez, M.T. García. J. Mol. Catal. 171:195–203, 2001.
38. W.M.H. Sachtler, Z. Zhay. Adv. Catal. 39:129–219, 1993.
39. K. Bahranowski, M. Gasior, A. Kielski, J. Podobinski, E. M. Serwicka, L.A. Vartikian, R. Wodnicka. Clay Minerals 34:79–87, 1999.
40. F. Delahay, B. Coq, L. Broussons. App. Catal. B 12:49–59, 1997.
41. S. Tanabe, H. Matsumoto. Appl. Catal. 45:27–37, 1988.
42. J.A. Sullivan, J. Gunningham. App. Catal. B 15:275–289, 1998.
43. R. Bulánek, B. Wichterlová, Z. Sobalik, J. Tichý. App. Catal. B 31:13–25, 2001.
44. R.Q. Long, R.T. Yang. Ind. Eng. Chem. Res. 38:873–878, 1999.
45. M.P. Attfield, S.J. Weigel, A.K. Cheetham. J. Catal. 170:227–235, 1997.
46. Dedecek, Z. Sobálik, Z. Trarůžkova, D. Kaucký, B. Wichterlová. J. Phys. Chem. 99:16327–16337, 1995.
47. K. Bahranowski, R. Dula, M. Labanowska, E. M. Serwicka. App. Spectroscopy 50(11):1439–1445, 1996.
48. K. Bahranowski, A. Kielski, E.M. Serwicka, E. Wisla-Walsh, K. Wodnicki. Mineralogia Polonica 29(1):55–65, 1998.

4

Chemistry of Sulfur Oxides on Transition Metal Surfaces

XI LIN and BERNHARDT L. TROUT   Massachusetts Institute of Technology, Cambridge, Massachusetts, U.S.A.

I. INTRODUCTION

Transition metals are currently the most widely used heterogeneous catalysts in industry. In this article, we shall focus our attention primarily on the automobile engine-out exhaust emission catalysts for environmental concerns. In order to meet automobile emission control requirements in the United States, so-called three-way catalysts, consisting of Rh, Pt, and Pd, were selected to simultaneously convert CO, hydrocarbons, and NOx to CO_2, H_2O, and N_2 [1]. In this conversion process, both oxidation and reduction reactions take place on the same three-way catalyst surfaces. Therefore, only a narrow range of air-to-fuel (A/F) ratios around the stoichiometric point should be taken as the operating "window" of the catalytic conversion. However, it is certainly desirable to have a wider A/F range with a better rate of conversion of CO, hydrocarbons, and NOx.

Lean-burning NOx-trapping catalysts were designed for this purpose. The idea is to separate the competing oxidation and reduction reactions by time, via periodically alternating the A/F ratio between lean (high A/F ratio) and stoichiometric (A/F ratio $\sim$ 14.6) conditions in the combustion chamber. Under lean conditions, both CO and hydrocarbons can be efficiently oxidized to CO_2 and H_2O, while NOx will be oxidized to the unfavorable chemical NO_2. At the same time, by introducing NOx-trapping materials such as BaO, NO_2 is trapped on the catalyst within this lean cycle. When the stoichiometric cycle is alternatively switched on, the trapped NO_2 is reversibly released from the BaO surface and further reduced to N_2. However, this novel lean-burn NOx-trap catalysis process has considerable practical problems due to serious sulfur poisoning issues (described below).

Before getting to the sulfur poisoning problem for the lean-burn NOx-trap catalyst, let us examine how the traditional three-way catalyst could get over this sulfur problem. Ever since the early 1980s, when three-way catalysts were first introduced, a substantial amount of research has been carried out to understand the interaction of SO_2 with the catalysts, due to the concern that SO_2 might be catalytically oxidized to sulfuric acid and released to the environment together with the automobile exhaust. Note that a typical concentration of 0.03 wt% (or $\sim$200 mg L^{-1}) sulfur is present in unleaded regular gasoline, which produces about 20 ppm of SO_2 in the engine-out exhaust gas [1]. However, it is quite surprising that little sulfuric acid was actually generated by these three-way catalysts. It was thought that the use of Rh might help to lower the activity of the three-way catalysts for SO_2 oxidation, compared to a pure oxidation catalyst [2]. More importantly, the stoichiometric air-to-fuel ratio helped to suppress the SO_2 oxidation.

Under lean-burning conditions, however, full oxidation of SO_2 to SO_3 or sulfuric acid is feasible when excess oxygen exists on the three-way catalysts. Both SO_3 and sulfuric acid can severely damage NOx-trapping materials, such as BaO. This poisoning process is very difficult to reverse and therefore inhibits utilization of the lean-burning NOx-trapping catalysts. Thus, a detailed understanding of the general sulfur poisoning effect on transition metals and metal oxides is necessary for the development of next-generation automobile exhaust emission catalysts. It was proposed that sulfur poisoning will not be serious if one manages to block the oxidation channel of SO_2 to SO_3 on these catalyst surfaces under lean conditions.

Although the oxidation of SO_2 to SO_3 is not significant under stoichiometric conditions, early research did show that the presence of SO_2 in engine-out exhaust affected the reactivity of the three-way catalysts. It was demonstrated that by increasing the sulfur concentration in gasoline from zero to 0.03 wt%, one observed lowering of the conversion efficiency of CO, hydrocarbons, and NOx [3]. The effect of sulfur on the activity of three-way catalysts was found to be more pronounced under rich conditions. This was attributed to a larger coverage of catalytic sites by atomic sulfur under rich conditions than under dynamic conditions around the stoichiometric point. Laboratory durability studies also indicated a faster drop in activity with time with sulfur-containing (0.03 wt%) fuel, compared to the sulfur-free fuel [3,4].

Moreover, SO_2 has been found to influence the selectivity of three-way catalysts. The importance of sulfur chemistry on transition metal surfaces was recently highlighted in a series of extensive experimental studies geared to understanding how SO_2 poisons the oxidation of CO and propene but promotes the oxidation of alkanes, such as propane [5–7].

It is believed that the sulfur poisoning effect on transition metals is due mainly to the high reactivity of sulfur with transition metals. However, at this time, few details of the elementary reactions involving SO_2 are known. This is because these processes are complex and frequently made even more complicated by the interaction of reactants and products with coadsorbed species.

First-principles computational research has become an efficient and accurate technique, complementary to experimental work in terms of determining both the static and the dynamic properties of molecules on extended transition metal surfaces. This approach was benefited mainly by the rapid progress of density-functional theory (DFT) [8,9] and pseudo-potential [10] techniques associated with the fast increase of computational power in the past couple of decades. The static properties computed from first principles consist of a variety of electronic and geometrical structures, including adsorbate configurations, surface reconstruction, electronic spin configurations, and adiabatic potential energy surfaces. The dynamic processes comprise sticking, diffusion, desorption, and, most important and interesting, surface chemical reactions. First-principles molecular dynamics on metal surfaces are still under development, but rough estimates of entropic effects may be obtained based on static adiabatic potential energy surfaces using the harmonic approximation and transition state theory [11].

Rather generally, theoretical studies for the sulfur poisoning effect on transition metal surfaces indicated that the perturbations caused by sulfur-containing molecules on the metal electronic structure reduce the ability of these metals to adsorb CO and dissociate hydrocarbons [12–14]. Moreover, it was shown that these induced electronic perturbations could have a long-range character.

II. STATIC INTERACTIONS: EQUILIBRIUM POSITIONS AND ADIABATIC POTENTIAL ENERGY SURFACE

The major goal of research on the static properties of chemical systems is to obtain the adiabatic potential energy surface of the groundstate, which includes the following:

1. Equilibrium atomic positions, such as bond lengths, bond angles, and torsion angles, among both the adsorbates and possible substrate surface reconstruction. In surface science experiments, the atomic vibrational modes with certain underlying symmetries are the main observables and have been extensively investigated through high-resolution electron energy loss spectroscopy (HREELS), surface-extended X-ray absorption fine structure (SEXAFS), and near-edge X-ray absorption fine structure (NEXAFS) techniques. Referring to the gap-phased isolated adsorbate molecules, UV photoelectron spectroscopy (UPS), angle-resolved UPS (ARUPS), and X-ray photoelectron spectroscopy (XPS) can provide hints for identifying adsorbate species. However, the classic experimental

surface structure determination approach, low-energy electron diffraction (LEED), does not seem to be useful, because adsorbed sulfur-containing molecules, such as SO_2, do not exhibit long-range order. Even if they do, they would rapidly be destroyed upon an electron beam.

2.   The static properties also contain the electronic spin configurations, since many of the transition metals and their derivatives are ferromagnetic crystals. It is known that spin plays an essential role in the groundstate properties of small transition metal clusters [15,16]. This spin effect might not be particularly important in the case of SO_2, compared, for example, with NO [17], since the adsorbate molecule is spin-pared. To our knowledge, no experimental work, such as electron paramagnetic resonance (EPR), has been performed especially for the purpose of studying sulfur oxides on transition metal clusters.

3.   Thermodynamic properties are important by themselves, as well as serving as the basis for many dynamic approaches, such as transition state theory. Temperature-programmed desorption (TPD) is widely used, but the accuracy in many cases is only qualitative [16]. A rather wide binding energy range from 100 to 150 kJ/mol is estimated from TPD data for SO_2 on the Pt(111) surface [18]. A single crystal adsorption calorimetry study [19] on SO_2 has not been reported up to this time.

A.   Gas-Phase Sulfur Oxides

Isolated gas-phase SO is linear, possessing the $C_{\infty v}$ point group symmetry. The intramolecular S—O bond length is 1.48 Å. The spin-polarized electronic configurations around the highest occupied molecular orbital (HOMO) and the lowest unoccupied molecular orbital (LUMO) are shown in Figure 1a. From the frontier orbital point of view [20], these single-particle orbitals should be considered the most active ones in most kinds of chemical reactions, including surface reactions [21].

Gas-phase SO_2 has an intramolecular S—O bond length of 1.43 Å and an O—S—O angle of 120°. As shown in Figure 1b, both the HOMO and LUMO are localized mainly on the sulfur atom, which suggests that bonding of SO_2 via the sulfur atom to the transition metal surfaces should be expected. Note that a

---

FIG. 1   Groundstate electronic configurations. Spheres represent s-type and lobes represents p-type atomic orbitals. For the p-type atomic orbitals, white and dark regions stand for different phases of the orbitals, while bigger lobes indicate larger participation in the corresponding molecular orbitals. (a) SO (triplet, $C_{\infty v}$). Diagram on the left/right shows the majority/minority spin configurations. The HOMO is the Π bonding in the majority spin and the LUMO is the Π antibonding in the minority spin. (b) SO_2 (singlet, C_{2v}). (c) SO_3 (singlet, C_{3v}).

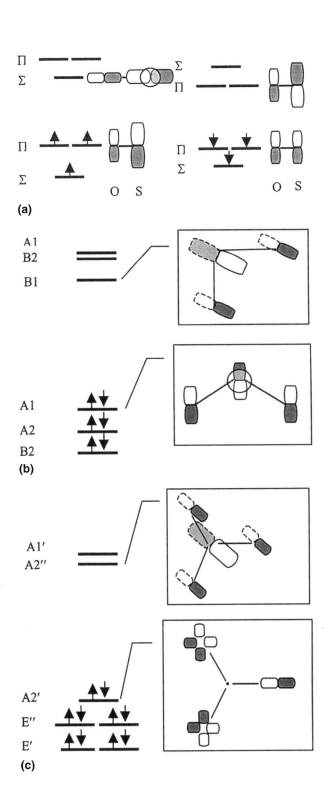

(a)

(b)

(c)

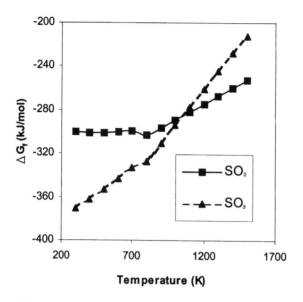

FIG. 2   Free energy/temperature plot of gas-phase SO_2 and SO_3. (From Ref. 22.)

larger overlap of the *p*-orbitals between the sulfur and oxygen atoms is expected for the HOMO, compared to the LUMO, due to the in-plane O—S—O bond angle.

Gas-phase SO_3 has the largest number of oxygen atoms among all neutral sulfur oxides. It has a planar structure and possesses a C_{3v} symmetry. The intramolecular S—O bond length is 1.43 Å, almost the same as that in the gas-phase SO_2 molecule. The O—S—O bond angles are 120°. The free energy versus temperature [22] is plotted in Figure 2 to show the thermodynamic stability of gas-phase SO_2 versus SO_3. Note that two curves cross at ~1100 K, which indicates that SO_3 is more stable than SO_2 under a typical engine-out exhaust temperature ~600 K. Therefore the experimentally observed SO_2 in engine-out exhaust gas is due to kinetic limitations under lean conditions. We notice that the LUMO level of SO_3 possesses the same phases of the *p*-orbitals from three oxygen atoms and the opposite phase of the *p*-orbital from the central sulfur atom. Therefore, one should expect a bent molecular structure (sulfur atom protruding out of the plane containing the three oxygen atoms) when the LUMO of SO_3 accepts electrons from donors of the same phase.

B. SO_2 on Pt(111)

SO_2 plays one of the most essential roles among all the sulfur oxides (SO_x, $x =$ 0, 1, 2, 3, 4), since it is readily formed by burning natural sulfur-containing materials or by roasting metal sulfides in air. The most important intermediate process in the manufacture of sulfuric acid is the oxidation of SO_2 to SO_3 in the presence of transition metal catalysts, such as platinum, because platinum is a very effective catalyst for SO_x oxidation. Thus the interaction of SO_2 on Pt(111) has received the widest attention of all of the transitions metals.

It is generally agreed that the SO_2 molecule adsorbs intact on the Pt(111) surface at low temperatures, typically 100–160 K. Through XPS, UPS, TPD, and HREELS, Sun et al. [23] found that the binding of the SO_2 molecule was through an η^2-S,O structure, while the SO_2 molecular plane was essentially perpendicular to the Pt(111) surface. Their further simple frontier molecular orbital analysis suggested a preferred configuration with the sulfur atom on a bridge site and one oxygen atom on a top site. More recently, Polcik et al. [24] claimed to have found a new flat-lying configuration of SO_2 on the Pt(111) surface at 150 K in their combined XPS and NEXAFS study and pointed out that this new flat-lying configuration was invisible in the HREELS experiments by Sun et al. [23]. But Polcik et al. did not give any detailed structural information for this new flat-lying configuration. Sellers and Shustorovich employed the empirical bond order conservation–Morse potential method [25,26] and concluded that the most stable configurations involved dicoordination binding through both η^2-S,O and η^2-O,O structures on the Pt(111) surface but no flat-lying configurations.

Our first-principles DFT calculations confirmed both of (and only) the two most stable structures found by Sun et al. [23] and Polcik et al. [24] (one perpendicular and the other parallel to the Pt(111) surface) at low temperatures. We did not find any stable η^2-O,O structure. Detailed results will be published separately.

C. SO$_2$ on Other Transition Metal Surfaces

Similar to the interaction of SO_2 on the Pt(111) surface, SO_2 follows either spontaneous or thermally activated decomposition on all of the transition metals except Ag, on which SO_2 adsorbs and desorbs only molecularly [27]. Temperature-programmed desorption, UPS, and XPS studies by Outka and Madix on Ag(110) showed that three cleanly distinct phases exist, depending on the temperature: (1) multilayer SO_2 under 120 K, (2) dual-layer SO_2 between 140 and 175 K, and (3) monolayer SO_2 between 175 and 275 K. The clean Ag(110) surface can be restored at temperatures greater than 275 K, which indicates a complete molecular desorption of SO_2 [27].

Molecular SO_2 was detected intact on Pd(100) at temperatures below 120 K in a TPD and EELS study by Burke and Madix [28]. When heated up to 135 K, multilayer SO_2 desorbed and left a single layer of SO_2 on the Pd(100) surface. The monolayer of SO_2 left consequentially decomposed at 240 K, forming chemisorbed SO, which led to atomic sulfur and oxygen on the surface at even higher temperatures. Similarly, in a combined TPD, XPS, and ARUPS study [29], Zebisch et al. showed that the SO_2 multilayer desorbed at about 130 K, and the remaining SO_2 monolayers desorbed at 360 K on a Ni(110) surface. The heating left a large number of sulfur atoms on the surface. An NEXAFS study indicated that SO_2 adsorption at 170 K is partly dissociative on Ni(110), Ni(111), and Ni(100) surfaces [30,31]. Partial dissociation of adsorbed SO_2 also occurs on Cu(100) and Cu(111) surfaces at 180 K, although more detailed measurements indicated much less dissociation on Cu(111) than on Cu(100) [32].

As for the geometric structure of the chemisorbed SO_2, it was suggested that the molecular plane of chemisorbed SO_2 aligned perpendicular to the closed-packed rows, as is the case on the Pt(111) surface. Detailed measurements also suggested that the SO_2 bonds to the surface through the S atom. The C_2 axis was shown to be perpendicular to the surface via measurements such as on Ag(110) in an NEXAFS study by Solomon et al. [33], or the C_2 axis could also be tilted within the SO_2 molecular plane on Pd(100) [28] and Cu(111) [34], similar to that on Pt(111) as described earlier. This tilted axis was attributed to the additional O–substrate bonding interaction, which might lead to the dissociation of molecular SO_2.

However, an XAFS study of low-coverage SO_2 on Ni(110), Ni(111), and Ni(100) surfaces [30,31] suggested that SO_2 species orient themselves with molecular planes approximately parallel to the surface, which is also similar to the second most stable structure on Pt(111), as discussed earlier. Therefore, one may conclude that there are in general two stable SO_2 species present on various transition metal surfaces, one perpendicular and one parallel to the surface. Depending on the symmetry restrictions in experimental techniques, one may not always be able to observe both of the species.

Unfortunately, little knowledge has been obtained directly from experiments on the surface adsorption site. Although in general it might be quite misleading, making use of the surface-cluster analogy suggested an atop site of SO_2 on the Ag(100) surface [33], and a fourfold hollow site of SO_2 with an oxygen atom close to a bridge site was suggested on the Pd(100) surface [28]. It is noted that a quite surprising location of SO_2 has been suggested in which the sulfur atom is equally distributed between the long- and short-bridge sites on Ni(100), Ni(111), and Ni(110) surfaces [30,31].

One recent NEXAFS and SEXAFS study by Polcik et al. demonstrated the presence of a SO_2-induced surface reconstruction of Cu(111) at 170 K, on which the sulfur atom of the molecular SO_2 is located at a hollow site on a locally pseudo-(100) reconstructed surface [34]. However, a later study by Jackson et al. using chemical-shift normal-incidence X-ray standing waves (CS-NIXSW) on the identical system seemed to disagree with the proposed local pseudo-(100) reconstruction [35]. A very recent scanning tunneling microscopy (STM) study by Driver and Woodruff further demonstrated that the kind of pseudo-(100) reconstructions on Cu(111) can be induced by atomic sulfur, formed by dissociated methanethiolate under an electron beam [36].

Recently Rodriguez et al. performed a DFT calculation to examine the adsorption of SO_2 on Cu(100) and showed an increasing bonding energy in the order of η^1-S $< \eta^2$-S,O $< \eta^2$-O,O $< \eta^3$-S,O,O. To make comparison with experiments, Rodriguez et al. further proposed η^2-O,O or η^2-S,O to be the most stable configuration under large coverage limit, by assuming the large surface SO_2 coverage made the η^3-S,O,O binding mode impossible [37].

D. SO on Transition Metals

Dissociation of SO_2, resulting in SO species, has been experimentally observed on Pt(111) [23,24], Pd(100) [28], Cu(100) [32], Cu(111) [32], Ni(100) [30,31], Ni(111) [30,31], and Ni(110) [29] surfaces, as discussed earlier. Our DFT studies suggested that the thermodissociation of SO_2 on Pt(111) to SO was energetically unfavorable at low temperatures. Further dissociation of molecular SO to S and O atoms would cost even more energy, therefore being even less favorable.

The dissociation of SO_2 has been observed at higher temperatures, for example, at 240–270 K on Pd(100) [28]. A similar dissociation temperature of ~300 K of SO_2 on Pt(111) and all the other transition metal surfaces is also reported. The chemisorbed SO thus formed, sequentially recombined with other surface adsorbates to form higher oxidized species, such as SO_4, at the same temperature [23].

SO_2 adsorption on Cu(100) is partly dissociative, even at about 180 K. An SEXAFS study suggested that the sulfur atom was located at a fourfold hollow site and that the oxygen atom was located at a near-bridge site [32]. The recent DFT calculations by Rodriguez et al. showed a cost or ~67–111 kJ/mol in energy for this dissociation process [37]. This is rather misleading, however, since a meaningful comparison must be done by allowing the separation of the dissociated SO and O species instead of by constraining them in one small supercell.

E. SO₃ on Transition Metals

Our DFT calculations showed that this oxidation reaction is energetically favorable at low temperatures on the Pt(111) surface. In experiments, following the dissociation of chemisorbed SO_2 on transition metal surfaces, such as Pd(100), Cu(100), and Ni(110) at ~170 K, SO_3 is formed upon adsorption as well as after heating the SO_2 layers to room temperature. On Ag(110), however, SO_2 can be oxidized to SO_3 only when preadsorbed oxygen is available.

An NEXAFS and CS-NIXSW study of SO_3 on Cu(111) shows that the C_{3v} axis of the adsorbed SO_3 is perpendicular to the surface, located at atop sites, with the sulfur atom pointing out of the plane formed by the three oxygen atoms, away from the surface [35].

The DFT calculation by Rodriguez et al. showed that the bonding of SO_3 to $Cu(100)$ was through an η^3-O,O,O configuration, with the C_{3v} axis perpendicular to the surface. They again proposed η^2-S,O as the most stable binding configuration in the high SO_3 surface coverage limit [37].

F. SO_4 on Transition Metals

The oxidation of chemisorbed SO_2 to SO_4 species has been observed on essentially all the transition metal surfaces studied. In addition to the oxidation of SO_2 to SO_3, our DFT calculations showed that this oxidation reaction, i.e., from SO_3 to SO_4, is also energetically favorable at low temperatures on the $Pt(111)$ surface.

SO_4 species have been observed via spectroscopic methods to be present on transitional metal surfaces, such as $Pt(111)$ and $Pd(100)$, at 300 K [23]. It is believed that the dissociation of SO_2 must occur first in order to provide chemisorbed atomic oxygen on the surface, if no additional gas-phase oxygen was supplied. These SO_4 species on $Pt(111)$ decompose when the temperature is above 418 K without increasing the amount of atomic sulfur on the surface [24].

Under lean conditions, when oxygen is preadsorbed on $Pt(111)$, chemisorbed SO_2 readily reacts with preadsorbed oxygen to form SO_4, which has been indicated as the key surface species responsible for SO_2-promoted catalytic oxidation of alkanes [5,6]. When CO or propene are coadsorbed, the SO_2 overlayers would be efficiently reduced to form atomic sulfur. The latter contributed to the poisoning of the oxidation of CO and propene in the presence of SO_2 under rich conditions. At ~550 K, adsorbed SO_4 is identified as the precursor to SO_3 desorption [6].

Similar to the SO_2-induced $Cu(111)$ reconstruction described earlier, it was observed in an STM study by Broekmann et al. that the topmost layer of $Cu(111)$ was reconstructed by sulfate when the $Cu(111)$ surface was exposed to a dilute sulfuric acid solution [38].

G. Modified Transition Metals Toward Designed Reactivity

Beyond the simple single-crystal transition metal surfaces, possible modifications on the activity of the transition metals toward reactions involving sulfur oxides consist of:

1. *Bimetallic or multimetallic alloys and metal oxides.* These alloys can be mixed layer by layer or mixed within layers and repeated through whole crystals. It is shown that tin, acting as a site blocker, forms a well-defined and stable alloy

with Pt(111). A much better chemical resistance of this Sn/Pt alloy is demonstrated toward SO_2, S_2, H_2S, and thiophene, compared to pure Sn and Pt [39]. Bimetallic Pd/Rh [40], Pd/Ni, and Pd/Mn systems demonstrated good catalytic activity and less sensitivity to the presence of sulfur-containing molecules than pure Pd. A general trend was obtained concerning the metal reactivity toward oxidation of SO_2, which increases in the following order: Pt $\sim$ Rh $<$ Ru $<$ Mo $<$ Cs/Mo $<$ Cs [41]. Although having served as inert supports for metal catalysts for a long time, metal oxides can actually become part of the active sites of the catalysts. Experiments on the adsorption of S_2, H_2S, CH_3SH, and thiophene on a series of oxides, Al_2O_3, ZnO, Cu_2O, MoO_2, Cr_2O_3, and CeO_2, showed that the sulfur atoms formed by the dissociation of these molecules interacted mainly with the metal centers of the surface, while SO_2 reacts preferentially with the oxygen centers to form SO_3 and SO_4 species [42]. Theoretical calculations suggest that a larger bandgap in the oxide can better prevent sulfur poisoning from the metal oxide, due to the more stable valence band and the greater difficulty in moving the valence electrons. However, this oxide support does not really help prevent the supported metals from being poisoned by sulfur. Detailed studies of various transition metals on various metal oxide supports suggested that the sulfur-containing molecules always prefer to interact with the supported transition metals.

2. *Finite-size metal particles instead of extended 2-D surfaces on supports.* The dependence of site of reactivity, such as for CO chemisorption on Cu(100) [43] and dissociation on alumina-supported Rh [44], has been known for quite a long time. Our DFT study indicated that a special tetrahedron Pt-10 cluster possessed a rather low activity toward atomic sulfur chemisorption [16].

3. *Stepped surfaces.* First-principles DFT calculations showed that the atomic oxygens are attracted to step edges on Pt(111) [45]. In ammonia synthesis catalysis, it was shown that the stepped Ru(0001) surfaces lowered the activation energy of N_2 dissociation by 1.5 eV. It is further claimed that such a low barrier at the step was due to a combination of electronic and geometric effects [46].

4. *Crystals under external fields, such as stress, electric, or magnetic fields.* It was shown by a DFT study that strained metal surfaces have significantly different chemical properties from those of unstrained surfaces. The surface reactivity can be increased by expanding the lattice, which was followed by a concurrent upward shift of the metal *d* bands. In that study, both molecular CO and atomic oxygen chemisorption energies as well as the CO dissociation barrier height varied substantially upon the strained Ru(0001) surface [47]. Recently, Xu and Mavrikakis [48] performed a DFT study on the chemisorption and dissociation of O_2 on the Cu(111) surface. They demonstrated that the parallel expansive strain to the surface led to an increase in the binding energy of both O and O_2 and a decrease in the energy barrier height of the O_2 dissociation reaction.

III. First-Principles Molecular Dynamics and Free-Energy Calculations

Experiments are performed under given external conditions, such as constant temperature, pressure, and pH. In order to mimic these experimental conditions, theoretical studies require the use of statistical ensembles. Within a particular ensemble, properties of the system are computed by properly averaging the Hamiltonian of the system over phase space. A common approach is to use computer simulations based on classical mechanics via molecular simulations (MD) or based on Metropolis' Monte Carlo (MC) method [49] to generate a probability distribution of the target statistical ensemble. Both MD and MC methods require explicit information of the potential energy surfaces.

Instead of using empirical interatomic potentials, first-principles molecular dynamics based on the Car–Parrinello approach [50] has been shown to be successful and reliable in the simulations of many complex chemical systems. In this approach both the atomic and electronic degrees of freedom are treated on an equal footing via a Lagrangian in which the dynamic variables include the coefficients of electronic wavefunction as well as the classical atomic positions and momenta in phase space. However, this Car–Parrinello molecular dynamic (CPMD) approach requires numerical integration of the CP equations of motion using a preset time step. This time step is controlled by the dynamics of the fast electronic degrees of freedom and has to be rather small, especially for zero-gap systems such as metals. A real unified and efficient first-principles molecular dynamic approach for metallic systems is still under development.

Alternatively, one could perform Born–Oppenheimer dynamics, in which the electronic degrees of freedom are relaxed to the adiabatic potential energy surface at each time step. Alternatively, one could use the harmonic approximation in approaches, such as classical transition state theory (hTST) [51–53]. This hTST approach has been shown to be a quite reasonable simplification in studies of surface events, such as diffusion and reactions, because the metal atoms in crystals and the chemisorbed molecules on the surfaces are generally tightly bound under the relevant surface temperature, compared to a typical melting temperature of the metal.

IV. CONCLUSIONS

We have performed a general review of experimental and theoretical work on the chemical properties of sulfur oxides on various transition metal surfaces, focusing on reactivity and selectivity having to do with the sulfur poisoning problem. Although it is an essential area in the field of heterogeneous catalysis, the study of sulfur oxides interacting with transition metal surfaces has started only recently, and much more work needs to be carried out in order to achieve detailed understanding and, thus, to be able to design catalysts for maximum effectiveness.

ACKNOWLEDGMENTS

This work was supported by the Ford Motor Company and NSF, under contract number CTS-9984301. Computer time was provided by the National Center for Supercomputing Applications (NCSA).

REFERENCES

1. Taylor, K. C. Automobile catalytic converters. In: Anderson, J. R., Boudart, M., eds. Catalysis Science and Technology. Vol. 5. Berlin: Springer Verlag, 1984, pp 119–170.
2. Gandhi, H. S.; Otto, K.; Piken, A. G.; Shelef, M. Environ. Sci. Technol. 11:170–174, 1977.
3. Williamson, W. B.; Gandhi, H. S.; Heyde, M. E.; Zawacki, G. A. Society of Automobile Engineering, Paper No. 790942, 1979.
4. Williamson, W. B.; Stepien, H. K.; Gandhi, H. S.; Bomback, J. L. Environ. Sci. Technol. 14:319–324, 1980.
5. Wilson, K.; Hardacre, C.; Baddeley, C. J.; Ludecke, J.; Woodfuff, D. P.; Lambert, R. M. Surface Sci. 372:279–288, 1997.
6. Wilson, K.; Hardacre, C.; Lambert, R. M. J. Phys. Chem. 99:13755–13758, 1995.
7. Lee, A. F.; Wilson, K.; Lambert, R. M.; Hubbard, C. P.; Hurley, R. G.; McCabe, R. W.; Gandhi, H. S. J. Catal. 184:491–498, 1999.
8. Hohenberg, P.; Kohn, W. Phys. Rev. 136:B864–B871, 1964.
9. Kohn, W.; Sham, L. J. Phys. Rev. 140:A1133–A1138, 1965.
10. Pickett, W. E. Computer Physics Rep. 9:115–198, 1989.
11. Truhlar, D. G.; Garrett, B. C.; Klippenstein, S. J. J. Phys. Chem. 100:12771–12800, 1996.
12. Billy, J.; Abon, M. Surface Sci. 146:L525–L532, 1984.
13. Batteas, J. D.; Dunphy, J. C.; Somorjai, G. A.; Salmeron, M. Phys. Rev. Lett. 77: 534–537, 1996.
14. Wilke, S.; Scheffler, M. Phys. Rev. Lett. 76:3380–3383, 1996.
15. Reddy, B. V.; Khanna, S. N.; Dunlap, B. I. Phys. Rev. Lett. 70:3323–3326, 1993.
16. Lin, X.; Ramer, N. J.; Rappe, A. M.; Hass, K. C.; Schneider, W. F.; Trout, B. L. J. Phys. Chem. B 105:7739–7747, 2001.
17. Hass, K. C.; Tsai, H.-H.; Kasowski, R. V. Phys. Rev. B 53:44–47, 1996.
18. Astegger, S.; Bechtold, E. Surface Sci. 122:491–504, 1982.
19. Brown, W. A.; Kose, R.; King, D. A. Chem. Rev. 98:797–831, 1998.
20. Fukui, K.; Yonezawa, T.; Shingu, H. J. Chem. Physics 20:722–725, 1952.
21. Hoffmann, R. Rev. Modern Physics 60:601–628, 1988.
22. Chase, M. W., Jr. NIST-JANAF Thermochemical Tables. 4th ed. ACS and AIP for NIST, 1999.
23. Sun, Y. M.; Sloan, D.; Alberas, D. J.; Kovar, M.; Sun, Z. J.; White, J. M. Surface Sci. 319:34–44, 1994.
24. Polcik, M.; Wilde, L.; Haase, J.; Brena, B.; Comelli, G.; Paolucci, G. Surface Sci. 381:L568–L572, 1997.
25. Sellers, H.; Shustorovich, E. Surface Sci. 346:322–336, 1996.

26. Sellers, H.; Shustorovich, E. Surface Sci. 356:209–221, 1996.

27. Outka, D. A.; Madix, R. J. Surface Sci. 137:242–260, 1984.

28. Burke, M. L.; Madix, R. J. Surface Sci. 194:223–244, 1988.

29. Zebisch, P.; Weinett, M.; Steinruck, H.-P. Surface Sci. 295:295–305, 1993.

30. Yokoyama, T.; Terada, S.; Yagi, S.; Imanishi, A.; Takenaka, S.; Kitajima, Y.; Ohta, T. Surface Sci. 324:25–34, 1995.

31. Terada, S.; Imanishi, A.; Yokoyama, T.; Takenaka, S.; Kitajima, Y.; Ohta, T. Surface Sci. 336:55–62, 1995.

32. Polcik, M.; Wilde, L.; Haase, J.; Brena, B.; Cocco, D.; Comelli, G.; Paolucci, G. Phys. Rev. B 53:13720–13724, 1996.

33. Solomon, J. L.; Madix, R. J.; Wurth, W.; Stohr, J. J. Phys. Chem. 95:3687–3691, 1991.

34. Polcik, M.; Wilde, L.; Haase, J. Phys. Rev. B 57:1868–1874, 1998.

35. Jackson, G. J.; Driver, S. M.; Woodruff, D. P.; Abrams, N.; Jones, R. G.; Butterfield, M. T.; Crapper, M. D.; Cowie, B. C. C.; Formoso, V. Surface Sci. 459:231–244, 2000.

36. Driver, S. M.; Woodruff, D. P. Surface Sci. 479:1–10, 2001.

37. Rodriguez, J. A.; Ricart, J. M.; Clotet, A.; Illas, F. J. Chemical Physics 115:454–465, 2001.

38. Broekmann, P.; Wilms, M.; Spaenig, A.; Wandelt, K. Progress Surface Sci. 67:59–77, 2001.

39. Rodriguez, J. A.; Jirsak, T.; Chaturvedi, S.; Herbek, J. J. Am. Chem. Soc. 120:11149–11157, 1998.

40. Rodriguez, J. A.; Jirsak, T.; Chaturvedi, S. J. Chem. Physics 110:3138–3147, 1999.

41. Rodriguez, J. A.; Herbek, J. Accounts Chem. Res. 32:719–728, 1999.

42. Rodriguez, J. A.; Chaturvedi, S.; Kuhn, M.; Herbek, J. J. Phys. Chemistry B 102:5511–5519, 1998.

43. te Velde, G.; Baerends, E. J. Chem. Physics 177:399–406, 1993.

44. Frank, M.; Andersson, S.; Libuda, J.; Stempel, S.; Sandell, A.; Brena, B.; Giertz, A.; Bruhwiler, P. A.; Baumer, M.; Martensson, N.; Freund, H.-J. Chem. Physics Lett. 279:92–99, 1997.

45. Feibelman, P. J.; Esch, S.; Michely, T. Phys. Rev. Lett. 77:2257–2260, 1996.

46. Dahl, S.; Logadottir, A.; Egeberg, R. C.; Larsen, J. H.; Chorkendorff, I.; Tornqvist, E.; Nørskov, J. K. Phys. Rev. Lett. 83:1814–1817, 1999.

47. Mavrikakis, M.; Hammer, B.; Nørskov, J. K. Phys. Rev. Lett. 81:2819–2822, 1998.

48. Xu, Y.; Mavrikakis, M. Surface Sci. 494:131–144, 2001.

49. Metropolis, N.; Rosenbluth, A. W.; Rosenbluth, M. N.; Teller, A. H.; Teller, E. J. Chem. Physics 85:6720–6727, 1953.

50. Car, R.; Parrinello, M. Phys. Rev. Lett. 55:2471–2474, 1985.

51. Wert, C.; Zener, C. Phys. Rev. 76:1169–1175, 1949.

52. Vineyard, G. H. J. Phys. Chem. Solids 3:121–127, 1957.

53. Voter, A. F.; Doll, D. J. Chem. Physics 82:80–92, 1985.

5

Studies on Catalysts/Additives for Gasoline Desulfurization via Catalytic Cracking

C. Y. LI, H. H. SHAN, Q. M. YUAN, C. H. YANG, J. S. ZHENG, B. Y. ZHAO, and J. F. ZHANG   University of Petroleum, Dongying, Shandong Province, People's Republic of China

I. INTRODUCTION

For a reaction catalyzed by a solid catalyst, at least one reactant must adsorb and the reaction happens on the active sites on the surface by either a Langmuir–Hinshelwood or Rideal–Eley mechanism. Obviously, the rupture of the bonds of the reactants and the formation of the bonds of the products bear close relationship to the surface properties of the catalyst. If there are no interactions between the reactants and the catalyst surface, then the catalytic reaction will not occur.

The development of new catalysts used to be for improving production efficiency, reducing production cost, or producing new products. With civilization and the advancement of humankind, however, the aims for developing catalysts have changed gradually, and more and more catalysts have been used to eliminate harmful materials. The treatment of polluted water and waste gases needs catalysts; the automotive emissions converter is a typical example. In refineries, producing low-sulfur, low-olefin, and high-octane-number environmentally benign gasoline also requires catalysts.

Sulfur in gasoline is not only a direct contributor to SO_x emissions; it is also a poison affecting the low-temperature activity of automobile catalytic converters. Therefore, it influences volatile organic compounds, NO_x, and total toxic emissions [1]. Consequently, developed countries limit the content of sulfur in gasoline stringently. In the United States, sulfur content will be lower than 30 µg/g in 2005. In China it will be reduced to 300 µg/g from the present 800 µg/g.

About 90% of sulfur in gasoline originates from FCC gasoline, so reducing the sulfur content of FCC gasoline is the main target of sulfur removal. Several

69

different routes to reduce the content of sulfur can be considered, such as hydro-treatment of FCC feed and hydrodesulfurization of FCC gasoline. The great dis-advantage of FCC feed hydrotreatment is its high operating and capital costs. Hydrodesulfurization of FCC gasoline may lead to a significant loss of octane number. If this difficulty re octane number is overcome, the process will be per-fect for sulfur removal.

The additive for sulfur reduction of FCC gasoline, invented by Wormsbecher et al. [1–3], can be added into the reaction-regeneration system of FCC expedi-ently, based on the real situation, to improve the cracking of sulfur compounds in the gasoline range. The maximum of sulfur reduction is about 40%, compared to the sulfur content of gasoline produced without adding the additive, if the additive is combined with the specially developed FCC catalyst [4].

This is a cheap sulfur-removal technique, and we have done some work on it. In this chapter, we not only introduce the results from our studies on sulfur removal additives, but also give the results on the mechanism of sulfide cracking and the catalysts of gasoline cracking desulfurization.

II. EXPERIMENTAL

A.   Materials

In evaluating the catalysts for gasoline catalytic cracking desulfurization, the feed is FCC gasoline distillate at higher than 100°C, provided by Shengli Petrochemi-cal Factory, whose sulfur content is 1650 µg/g, measured via the burning light method. The feed used to evaluate the sulfur removal additives of FCC gasoline is VGO (vacuum gas oil), supplied by Shenhua Refinery. The properties of VGO are listed in Table 1.

All the chemicals used to prepare the catalysts or additives are analytically pure.

B.   FCC Catalyst, Catalyst/Additive Preparation and Characterization

The regenerated FCC catalyst used in the experiments, also provided by Shengli Petrochemical Factory, is Vector60SL, whose BET surface area and microactivity are 112 m^2/g and 70, respectively.

Both the catalysts for gasoline catalytic cracking desulfurization and the sulfur removal additives of FCC gasoline were prepared via coprecipitation combined with impregnation. First, we used coprecipitation to make the colloid of mixed metal hydroxides; then the USY powder, bought from Zhouchun Catalyst Fac-tory, was added in with continuous stirring. After being aged for 12 h, the colloid was dried at 100°C for more than 20 h and then calcined at 700°C for 6 h.

TABLE 1 Properties of Shenghua VGO

ρ_{20}, g/cm^3		0.9197
Viscosity	50°C	67.38
(mm^2/s)	80°C	17.93
Residual coke (%)		0.25
Molecular weight		363
Distillation range	IP	235
(°C)	10%	372
	30%	410
	50%	421
	70%	448
	90%	482
	EP	512
Group composition	Saturated	65.91
(wt%)	Aromatic	26.51
	Resin and asphaltene	7.58
Metal content	Ni	0.10
(μg/g)	V	0.028
	Fe	1.30
	Na	0.26
	Cu	0.004
Element analysis	C	86.89
(wt%)	H	12.81
	N	0.30
	S	1.05

Crashing and sieving the solid to 0.078 ~ 0.18 mm, we then obtained the catalysts/additives.

X-ray diffraction and BET surface area of the catalyst/additive were measured by D/MAX-III X-ray diffractometer and ASAP2010, respectively.

C. Apparatus

1. Mechanistic Studies of Thiophene Cracking

Figure 1 presents a schematic of on-line pulse-reaction chromatography (HP4890 with PONA7531 column and FID detector). Between the sampling inlet and the column is a minireactor with a 2-mm inner diameter. The pulsed liquid sample is gasified at the sampling inlet and carried by gas to the catalyst bed to react

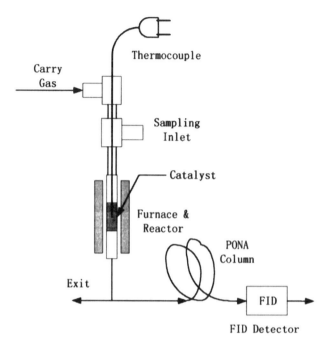

FIG. 1 Schematic of on-line pulse-reaction chromatography.

with products that go directly into the column after distributary and that are ana-lyzed with FID. To ensure that the sample was pulsed to gasify quickly and completely in the experiments of thiophene cracking, the temperature at the sam-pling inlet was controlled at 250°C. The flow rate of the carry gas (highly pure N_2), the amount of the USY zeolite used, and the quantity of thiophene pulsed were 30 mL/min, 14 mg, and 1 µL, respectively.

Furthermore, thiophene/n-heptane (sulfur content 0.33% and gasoline distil-late at over 100°C were used as the raw materials to react in a fixed-bed reactor with 25 g of catalyst to validate the results obtained from on-line pulse-reaction chromatography. Ten grams of thiophene/n-heptane or the gasoline distillate was pumped into the reactor within 1 min. Sulfides in the liquid product collected in the condenser were analyzed via Varian3800 chromatography combined with CB80 column and a PFPD detector. The sulfur content of the liquid product was also measured via the burning-light method.

The apparatus for MS transient response has been described elsewhere [5]. Thirty milligrams of USY was placed in the middle of the quartz reactor. To quicken response time, the other space of the reactor was filled with 0.3- to 0.45-

mm quartz sand. The effluents were detected with a quadrupole mass spectrometer (AMTEK QuadLink 1000) with a minimum dwell time of 3 milliseconds.

2. Evaluation of the Catalysts for Gasoline Catalytic Cracking Desulfurization

Ten grams of the FCC gasoline distillate at over 100°C was pumped into the fixed-bed reactor with 25 g of catalyst, and the liquid product was collected with a condenser immersed in ice/water bath at the outlet of the reactor. The gas from the condenser was then discharged to air after the H_2S in it was adsorbed by $Pb(Ac)_2$ solution. After reaction, N_2 was used to sweep the reactor to ensure that all the oil was out.

The octane numbers of the gasoline distillate before and after reaction were analyzed by HP5890. The sulfur content deposited on the catalysts was measured by element analyzer.

3. Evaluation of the Sulfur Removal Additives of FCC Gasoline

Mixing the additive with the regenerated FCC catalyst in a certain ratio, we then loaded the mixed catalyst into the confined fluidized-bed reactor. The catalyst was fluidized with steam. When the temperature ascended to the set value, we began to pump the VGO into the reactor. Fifty grams of the VGO was fed within 1 min. The effluent from the reactor was collected with three condensers in series immersed in ice/water bath. The uncondensed gas was collected via the draining-water method. Distilling the liquid product, we obtained the gasoline. The gas and the gasoline were all analyzed with the HP5890 to obtain the hydrocarbon composition and the octane number. The sulfur content of the gasoline is also measured via the burning-light method. The carbon content of the catalyst was determined by chromatography.

III. RESULTS AND DISCUSSION

A. Sulfur Distribution and Sulfides in FCC Gasoline

The FCC gasoline was cut to narrow distillates; the sulfur content measured via the burning-light method is listed in Table 2. The sulfur content of the distillate at 80–100°C is 507.9 µg/g, about twice that at 60–80°C, while the sulfur content at 100–120°C is almost twice that at 80–100°C. The sulfur content of the distillate at over 140°C is more than 1600 µg/g. Obviously, sulfur content increases with the boiling point of distillate and concentrates in the high-boiling-point distillates.

Because the distillate of IBP-60°C is too ''light'' and that of 160-EP is too ''heavy,'' their sulfur content is difficult to determine accurately by this method.

TABLE 2 Sulfur Distribution in FCC Gasoline

Distillation range, °C	IBP1–60	60–80	80–100	100–120	120–140	140–160	160–EP
Sulfur content, μg/g	—	252.7	507.9	961.3	1325	1604	—
Distillation range, °C		<100			>100		
Fraction, %(wt)		35%			65%		
Sulfur content, μg/g		326.2			1650		

TABLE 3 Type and Distribution of Sulfides in the Gasoline Before and After Desulfurization via Catalytic Cracking

USY/Al$_2$O$_3$/ZnO (1/2/2, wt.)	Before desulfur. (%)	After desulfur. (%)	Before desulfur. (μg/g)	After desulfur. (μg/g)	Sulfur removal (%)
Total	100	100	1650	288	82.5
Thiophene	0.69	8.02	11.5	23.1	−102
Mercaptans	0.32	0.08	5.35	0.243	95.5
2- and 3-methylthiophene	14.4	43.6	238	126	47.5
Thioethers and disulfides	11.1	3.72	184	10.7	94.2
C$_2$-substituted thiophene	34.5	30.6	569	88.1	84.5
C$_3$-substituted thiophene	26.5	10.3	438	29.6	93.3
C$_4$-substituted thiophene	12.5	3.68	201	10.6	94.8

Catalyst/oil = 2.5; temperature: 410°C. All the values in the table represent the amount of sulfur, not sulfide.

So we cut the gasoline to two distillates at 100°C and measured their sulfur content to be 326.2 μg/g for under 100°C and 1650 μg/g for over 100°C, which accounts for 83% of the total sulfur in the gasoline. As long as we reduce the sulfur content of the distillate at over 100°C to less than 800 μg/g, the overall gasoline will meet the present specification in China that limits the content of sulfur to no more than 800 μg/g.

Sulfides in the distillate at over 100°C were analyzed by chromatography with a PFPD detector; the results are shown in Table 3. In the distillate, the sulfur existing as mercaptans, thioethers, and disulfides accounts for less than 12% of the total sulfur, and the rest exists as different alkylthiophenes. The greatest amount is C$_2$-substituted thiophene (including different 2-methylthiophenes and ethylthiophenes), while the amount of thiophene is small. Therefore, to lower the sulfur content of the distillate via catalytic cracking, we must study how to make thiophene and alkylthiophenes crack effectively.

B. Cracking of Thiophene and Sulfides in FCC Gasoline

1. Cracking of Thiophene over the USY Zeolite

Fourteen milligrams of USY was placed in the reactor of on-line pulse reaction chromatography; the height of the catalyst bed was about 4 mm. When the temperature of the reactor was increased to 490°C in 30 mL/min N$_2$ gas flow and

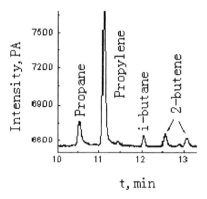

FIG. 2 Chromatograph of thiophene reacting over the USY zeolite at 490°C.

the chromatography was stable, a pure thiophene pulse was generated. The hydrocarbon products were propane, propylene, isobutane, 1-butene, and 2-butene (Fig. 2).

Thirty milligrams of the USY zeolite was used in the experiments on the MS transient response. The flow rate of the Ar carry gas was also 30 mL/min. At 490°C, 2 μL of thiophene was pulsed; the results are shown in Figure 3. The

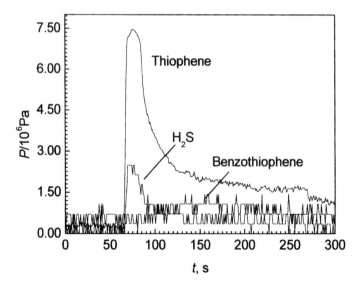

FIG. 3 Transient responses of thiophene pulsed over the USY zeolite catalyst at 490°C.

characteristic peak of H_2S ($m/e = 34$) was detected and appeared almost simultaneously with that of thiophene. This proves that thiophene can crack over the USY zeolite to produce H_2S. Furthermore, benzothiophene ($m/e = 134$) was also detected, and it appeared a little later than thiophene and H_2S.

The formations of butane, butene, and H_2S indicate that the ring of thiophene can open and that S can be removed during the cracking reaction. Furthermore, hydrogen transfer must happen simultaneously; otherwise, only high-unsaturated hydrocarbon can be produced. If a thiophene cracks to a butene and a H_2S, it must obtain six hydrogen atoms. Under the experimental conditions, no H_2 participated in the reaction. So the hydrogen can be obtained only via hydrogen transfer among thiophene molecules or thiophene and hydrocarbon fragments. In addition, after several thiophene pulses, significant coke deposited on the USY zeolite, which illustrates that dehydrogenation of hydrocarbons or hydrocarbon fragments or sulfides must take place during the reaction.

In the reaction, that propane, propylene, and benzothiophene can be formed shows that the reactions of thiophene over the USY zeolite are very complex and that maybe other sulfides can also be formed. So we performed the following experiment.

Thiophene/n-heptane (sulfur content: 0.33%) were used as the raw material to react in a fixed-bed reactor with 25 g USY zeolite at 490°C. After reaction, the sulfur content of the liquid product was reduced to 0.13%, and 61% sulfur had been removed. Obviously, the cracking desulfurization of thiophene is the dominant reaction. Sulfide analysis by chromatography with a PFPD detector shows that in the liquid product there are thiophene, 2-melthylthiophene, 3-methylthiophene, benzothiophene, and a little dimethylthiophene and trimethylthiophene, where unconverted thiophene, benzothiophene, 2-methylthiophene, and 3-methylthiophene account for 67%, 20%, 5%, and 3%, respectively (Fig. 4). That indicates that, except for cracking, thiophene can form other sulfides, and benzothiophene and 2-methylthiophene are easy to be produced.

The other conditions were the same as for Figure 2, and thiophene pulses were generated in the on-line pulse-reaction chromatography apparatus at different temperatures. The conversion of thiophene at different temperatures is depicted in Figure 5. The conversion of thiophene does not increase with temperature monotonically, but has a maximum of about 400°C. Luo et al. [6] also reported that there is a maximum conversion of thiophene at 400°C when thiophene/ethanol crack over HZSM-5. This means that hydrogen transfer may play a very important role in thiophene cracking [7]. Hydrogen transfer is an exothermic reaction, and high temperature restrains the reaction. Cracking, however, is an endothermic reaction, and high temperature promotes the reaction. That 400°C is the optimal temperature for thiophene cracking indicates that hydrogen transfer is an important elementary step of thiophene cracking. Otherwise, the conversion of thiophene should increase with temperature.

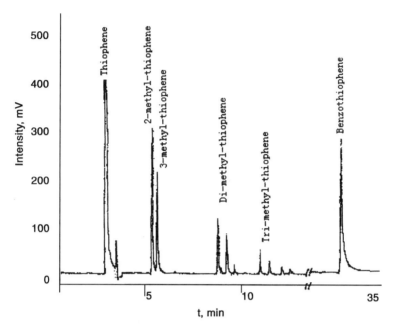

FIG. 4 Products of thiophene reacting over the USY zeolite catalyst at 490°C analyzed with a PFPD detector.

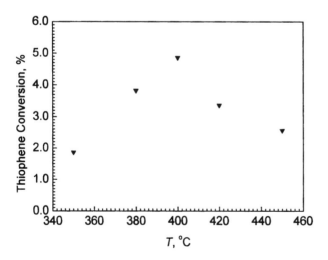

FIG. 5 Relationship between thiophene conversion and temperature.

2. Cracking of Sulfides in FCC Gasoline over a Catalyst of Gasoline Tracking Desulfurization

USY zeolite has good cracking activity for sulfides but bad selectivity (we discuss this in detail in Sec. III.C), so we chose a USY/ZnO/Al$_2$O$_3$ catalyst for gasoline cracking desulfurization to carry out the experiments that investigate the cracking of various sulfides in the FCC gasoline distillate at over 100°C.

In the gasoline distillate, more than 88% of the sulfur exists in thiophene species with different alkyl substitutions (Table 3). After the gasoline distillate reacted over the catalyst at the same conditions (400°C and catalyst/oil = 2.5), the sulfur content was reduced to 288 μg/g, and 82.5% of the sulfur was removed. In Table 3, the percentages of sulfur removed as mercaptans, thioethers, disulfides, C$_2$-substituted thiophene, C$_3$-substituted thiophene, and C$_4$-substituted thiophene are all larger than this value. This indicates that the sulfur existing in these sulfides is easier to remove via cracking. However, the sulfur existing in 2-methylthiophene and 3-methylthiophene is relatively more difficult to remove and the percent sulfur removed is only 47%. In the table we can also see that the amount of thiophene, although small, has increased more than 100%. In our opinion, this does not mean thiophene cannot desulfurize via cracking, but it may mean that alkylthiophenes can form thiophene via dealkylization.

We also investigated the effect of temperature on sulfur removal. In Figure 6, we can see that sulfur content has a minimum value between 390°C and 420°C. It seems that high temperature is not favorable for sulfide cracking. Obviously, the result is consistent with that of pure thiophene cracking.

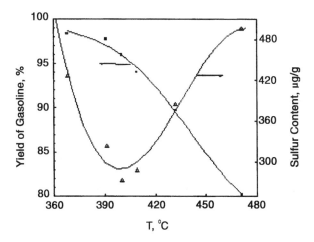

FIG. 6 Relationships between the yield and sulfur content of gasoline and temperature over the catalyst for gasoline desulfurization via catalytic cracking (catalyst/oil = 2.5).

3. Cracking Mechanisms for Thiophene and Alkyl-Thiophenes

For the thiophene desulfurization via cracking, Luo et al. [6] thought that the following species, similar to thioether in properties, is formed first:

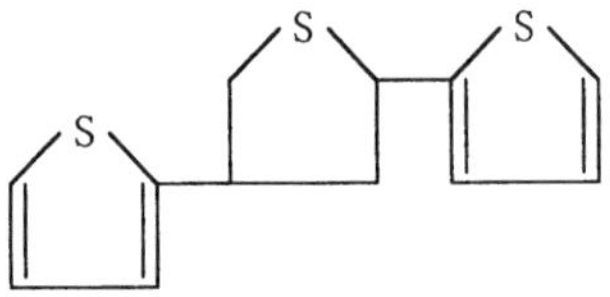

Then it decomposes to produce H_2S via pyrogenation or catalysis. The viewpoint of Wang et al. [8] is completely different. They thought the thiophene first obtains hydrogen to form tetra-hydrogen-thiophene, and then tetra-hydrogen-thiophene decomposes to produce H_2S. In our experiments, however, we did not find tetra-hydrogen-thiophene or even molecules larger than benzothiophene. Saintigny et al. [9] studied the mechanism of thiophene cracking over acid catalyst in theory and suggested that one of the $C-S$ bonds breaks first on acid sites to form surface $HC\equiv C-CH=CH-SH$, whose $C-S$ bond then breaks to $HC\equiv C-C\equiv CH$ and H_2S. In the mechanism, hydrogen transfer does not happen. If so, cracking will be the only reaction and high temperature will favor thiophene cracking. however, our experimental results show that about 400°C is the optimum temperature for thiophene (Fig. 5) desulfurization via cracking. Therefore, the cracking desulfurization of thiophene must be limited by other reactions. Based on our experimental results, we suggested that thiophene cracks by the mechanisms described in Figures 7–9.

Thiophene first obtains proton on the B acid sites of USY to form carbonium ion, and then the $C-S$ bond at the β-position breaks, for its bond energy is 268 kJ/mol, the weakest among those of $C-H$, $C-C$, and $C=C$. Thus, the ring of thiophene opens to form mercaptan species with two double bonds (Fig. 7). After carbonium ion isomerization and hydrogen transfer, the remaining $C-S$ bond of the mercaptan at the β-position breaks, and H_2S and dibutene are produced. Through hydrogen transfer, dibutene can convert to butene and even butane.

Besides butane and butene, propylene is also produced during the cracking of thiophene. In our opinion, the formation of propylene has a close relation with the formation of methylthiophene. In Figure 7, if the mercaptan with two double bonds from the ring opening of thiophene polymerizes with thiophene at the α-position, then species A is produced. After carbonium ion isomerization and hydrogen transfer (Fig. 8), the $C-S$ bond at the β-position of A breaks, and H_2S and 2-butenylthiophene are produced. 2-butenylthiophene cracks at the β-position to 2 methylthiophene and propylene after hydrogen transfer.

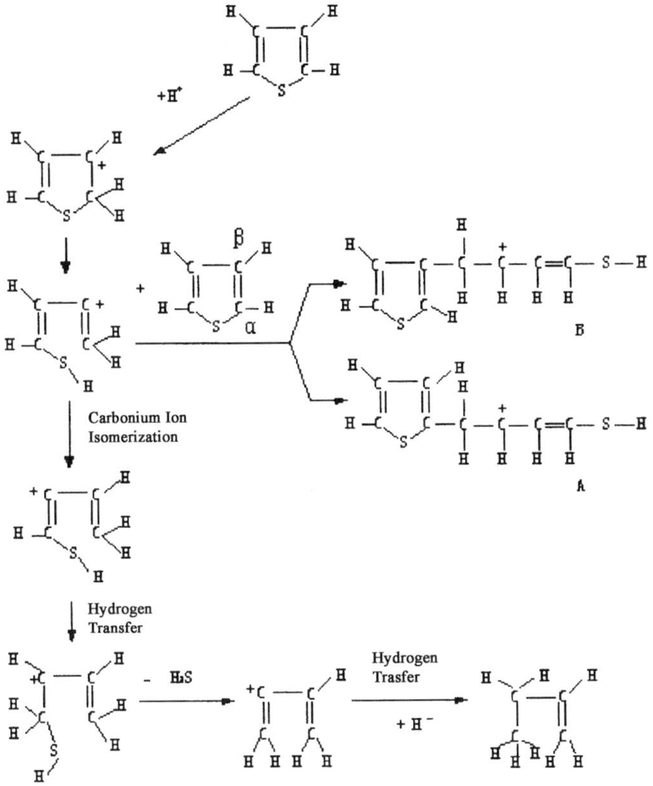

FIG. 7 Formation of butene and H_2S in the cracking of thiophene.

If polymerization between the mercaptan with two double bonds and thiophene happens at the β-position in Figure 7, then species B is formed and 3-methylthiophene is produced by the foregoing reactions. If thiophene polymerizes with two or three mercaptans with two double bonds at different positions simultaneously, then di- or trimethylthiophene is formed. Because the probability that thiophene polymerizes with two or three mercaptans simultaneously is lower than that of polymerizing with one mercaptan, even if we do not consider the effect of space obstruction, the amount of di- or trimethylthiophene is smaller than that of methylthiophene. Because the α-position of thiophene is more active than the β-position [10], the amount of 2-methylthiophene is larger than that of 3-methylthiophene in Figure 4.

Figures 7 and 8 also illustrate that cracking and hydrogen transfer are two important elementary steps. If any one of the two is blocked, then the cracking

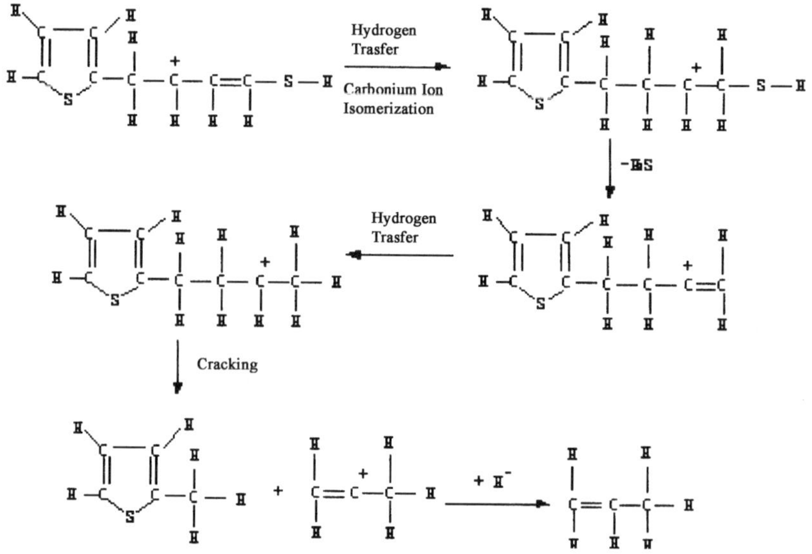

FIG. 8 Formation of propylene and 2-methylthiophene in the cracking of thiophene.

desulfurization of thiophene will be affected. Just because hydrogen transfer is also an important elementary reaction affecting thiophene and thiophene species to desulfurize via catalytic cracking, the optimum temperature appears in Figure 5, and low temperature limits the cracking reaction and high temperature does not favor hydrogen transfer.

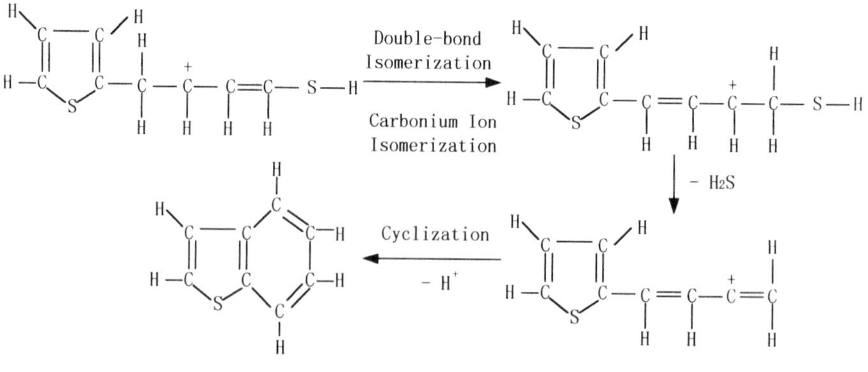

FIG. 9 Formation of benzothiophene in the cracking of thiophene.

In Figure 7, if H_2S is removed from species A via β-scission after double-bond isomerization and carbonium ion isomerization, 1,3-butenylthiophene will be formed. Benzothiophene can be produced through cyclization of 1,3-butenyl-thiophene (Fig. 9). Obviously, ring opening of thiophene is the precondition for the formation of thiophene.

In the gasoline distillate at over 100°C that we used as feedstock, the amounts of mercaptans, thioethers, and disulfides are small and easy to crack to desulfur-ize, and more than 94% are removed after the reaction (Table 3). The amount of thiophene is very small, and its sulfur content of 11.5 μg/g accounts for only 0.69% of the total sulfur. After desulfirization, it does not decrease, but increases to 23.1 μg/g. According to our experimental results with pure thiophene, it is impossible for thiophene not to crack here, and the reasonable explanation, in our opinion, is that alkyl-substituted thiophenes, which are in large amounts, can form thiophene via dealkylization.

In the gasoline distillate there are many thiophenes with different alkyl substi-tutions. Obviously the environment is different from that in which the cracking experiments of pure thiophene or high-concentration thiophene in heptane were carried out. Here the amount of thiophene is very small, and its cracking or con-version to other thiophene species may be restrained by its low concentration.

According to the results in Table 3, with increase in the carbon number of the alkyl, the conversion of alkyl-substituted thiophenes rises. So it can be con-cluded that larger substituted alkyls may favor the cracking desulfurization of thiophene species. It is well known that isoalkane is easier to crack than n-alkane. If the alkyl exists at the α-position, then the tetracarbonium shown in Figure 10 is more stable. If β-scission takes place at bond 1, then a mercaptan species is formed; if β-scission takes place at bond 2, then a thioether species is formed. No matter which one is produced, they are all easier to desulfurize via cracking. If the alkyl exists at the β-position, then through β-scission of the carbonium two kinds of mercaptan species are produced. They are also easier to desulfurize via cracking. In FCC gasoline, it is almost impossible for the thiophene with the

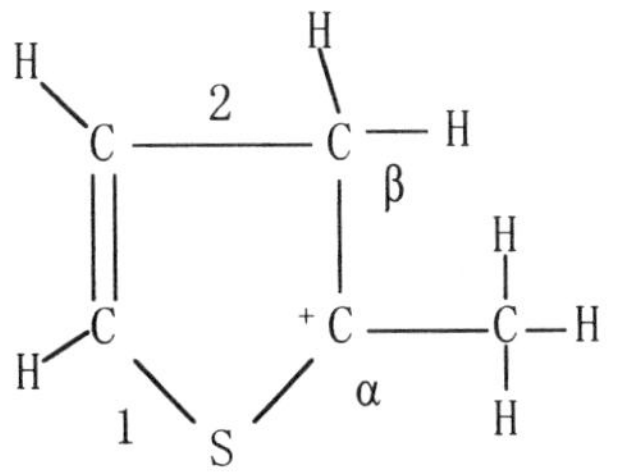

FIG. 10 Carbonium of alkyl thiophene.

longer alkyl to exist in large amounts because the long alkyl sidechain is very easy to crack under FCC conditions, so most of the C_3 or C_4 alkylthiophenes are more like alkyls thiophenes. Perhaps more alkyls can make the thiophene cycle more unstable. So the conversion of alkythiophene increases with the carbon number of the alkyl.

Furthermore, the conversion of alkyl-substituted thiophene is also the result of the synergism of cracking and hydrogen transfer. This is the same as for thiophene. Otherwise, we would not be able to explain reasonably why the optimum temperature for sulfur removal occurs.

C. Design of Catalyst/Additive and Selection of Metal Oxides for the Support

1. Design of the Catalyst/Additive

From the preceding discussion, we see that sulfides, including mercaptans, thioethers, disulfides, thiophene, and alkyl-thiophenes, can crack to desulfurize over the USY zeolite or the USY-contained catalyst. The USY zeolite has very high cracking activity, and not only sulfides, but also hydrocarbons can react over it. To remove sulfides selectively from FCC gasoline, we must keep the amount of cracked hydrocarbons as small as possible. When the FCC gasoline distillate at over 100°C reacts over the pure USY zeolite, about 96% of the sulfur can be removed. Thirty percent of the gasoline distillate, however, is cracked (Table 4). That is to say, the pure USY zeolite cannot crack sulfides selectively. Hence, to meet the requirement of cracking sulfides selectively, the catalyst/additive must

TABLE 4 Evaluation of USY Zeolite on Different Supports for Sulfur Removal from Gasoline via Catalytic Cracking

Catalyst	Gasoline yield (%)	Sulfur content (µg/g)	Sulfur removal (%)	Sulfur on catalyst (%)
USY	70.0	68.0	95.9	50.4
USY/ZnO$_2$/Al$_2$O$_3$	88.6	315.0	80.9	0
USY/ZrO$_2$/Al$_2$O$_3$	84.9	392.0	76.2	3.1
USY/MnO$_2$/Al$_2$O$_3$	91.0	273.5	83.4	Almost all
USY/CuO$_2$/Al$_2$O$_3$	86.8	145.5	91.2	Almost all
USY/La$_2$O$_3$/Al$_2$O$_3$	89.4	121.5	92.6	35.0
USY/NiO/Al$_2$O$_3$	81.8	301.0	81.8	72.0
USY/Fe$_2$O$_3$/Al$_2$O$_3$	87.2	365.0	78.0	76.5

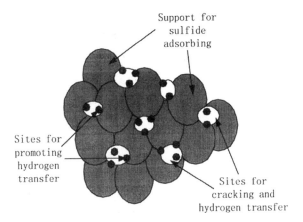

FIG. 11   Schematic of the catalyst for gasoline desulfurization via catalytic cracking.

contain, besides the cracking component, the component that can adsorb sulfides from gasoline selectively (called support, Fig. 11).

In the FCC gasoline distillate, most sulfides are thiophene and alkylthiophenes that are Lewis bases and easy to adsorb on Lewis acids [1–3]. Mixed metal oxides can form Lewis acid sites, so we choose the support materials from metal oxides. Sulfides first absorb on the metal oxide support and then crack over the USY zeolite on the support:

$$\text{Sulfides} \rightarrow \begin{array}{c} \text{Adsorbing on} \\ \text{the Support} \end{array} \rightarrow \begin{array}{c} \text{Cracking over} \\ \text{the USY zeolite} \end{array} \rightarrow \text{Hydrocarbons} + H_2S$$

Besides appropriate acidity, the support must have large surface area so as to disperse the USY zeolite better and have high ability to adsorb sulfides. The support will determine the selectivity of the catalyst to crack sulfides. So selecting appropriate metal oxides as the support is the key. Numerous metal oxides, such as ZnO, ZrO_2, MnO_2, CuO_2, La_2O_3, NiO, and Fe_2O_3, have been investigated; the results are listed in Table 4.

2.   Selection of Metal Oxides for the Support

We prepared the catalysts with a ratio of USY/an aforementioned metal oxide/ Al_2O_3 of 3/1/6 (wt). The surface areas of the catalysts depend mainly on that of the Al_2O_3 and are all about 200 m^2/g. Evaluation experiments were carried out in a fixed-bed reactor at 400°C with a catalyst/oil ratio of 2.5. The sulfur content of liquid product and coked catalyst was measured via the burning-light method and element analyzer. The results are listed in Table 4.

When the reaction proceeds over pure USY zeolite, the cracking loss of the gasoline is large, although nearly 96% of the sulfur has been removed. Furthermore, after reaction, about 50% of the total sulfur has deposited on the zeolite to form sulfur-containing coke. This is not what we had hoped. According to the mechanism we proposed, hydrogen transfer plays a very important role in the cracking of thiophene species. But if the hydrogen transfer activity is too high, then the hydrogen transfer of sulfides must be promoted greatly to form sulfur-containing coke. About half of the removed sulfur depositing on the pure USY zeolite may be due to its very high hydrogen transfer activity. Therefore, the metal oxide selected as the support component must be able to modify the acidity of the USY zeolite so that it has appropriate hydrogen transfer activity.

In Table 4 we can see that all the oxides listed have an excellent sulfur removal effect. Except for USY/ZnO/Al$_2$O$_3$ and USY/ZrO$_2$/Al$_2$O$_3$, on which no or only a small amount of sulfur is deposited, the catalysts cannot restrain effectively or even favor sulfur deposition. Over USY/La$_2$O$_3$/Al$_2$O$_3$, 92.6% of the sulfur can be removed, but 35% deposits on the catalyst. USY/MnO$_2$/Al$_2$O$_3$ and USY/CuO/Al$_2$O$_3$ can remove 83.4% and 91.2% of the sulfur, respectively; however, almost all of the removed sulfur deposits on the catalyst in the form of sulfur-containing coke or metal sulfides. Over the USY/NiO/Al$_2$O$_3$ and USY/Fe$_2$O$_3$/Al$_2$O$_3$, only a very small amount of sulfur removed gets into the cracking gas in the form of H$_2$S. Therefore, when the cracking gas passed through the Pb(Ac)$_2$ solution, a large amount of black PbS deposition was found over USY/ZnO/Al$_2$O$_3$, and no PbS was detected over USY/MnO$_2$/Al$_2$O$_3$, and USY/CuO$_2$/Al$_2$O$_3$.

During the sulfur removal reaction via catalytic cracking, we hope that the sulfur removed gets into the cracking gas in the form of H$_2$S so as to reclaim it. If the sulfur removed is deposited on the catalyst, then it will be in the flue gas on regenerating the catalyst in the form of SO$_x$, and will pollute the air. From this point of view, the best alternatives are USY/ZnO/Al$_2$O$_3$ and USY/ZrO$_2$/Al$_2$O$_3$.

Adjusting the ratio of USY/ZnO or ZrO$_2$/Al$_2$O$_3$, we prepared the catalyst for gasoline cracking desulfurization and the additive for sulfur removal of FCC gasoline.

D. Evaluation of the Catalyst for Gasoline Cracking Desulfurization

FCC contributes almost 80% of the gasoline pool in China. The sulfur content of FCC gasoline is very high. If hydrogenation is used to reduce the sulfur content, then the problem of how to abate or avoid the octane number loss due to olefin saturation is difficult to solve. So we proposed to reduce the sulfur content of FCC gasoline by letting the gasoline pass through a specially made catalyst to crack sulfides selectively. We have discussed the design of the catalyst and the selection of support materials. Based on these results, we prepared the catalyst for gasoline cracking desulfurization and investigated its performance with the FCC gasoline distillate at over 100°C, with a sulfur content of 1650 μg/g under the catalyst/oil ratio of 2.5 and different temperatures. The results are shown in Figure 6.

In the figure, the yield of gasoline drops monotonically with temperature. This shows that the cracking of hydrocarbons is quickened. However, the sulfur content of gasoline after reaction, similar to that of pure thiophene cracking, first falls and then increases. At 400°C, the sulfur content, about 280 μg/g, is the lowest, and 83% of the sulfur has been removed. At the same time, the yield of gasoline is over 96%. When the reaction is carried out at 390°C, although the sulfur content increases to 330 μg/g, the yield of gasoline is about 98%. Obviously, the catalyst has excellent sulfur removal activity and selectivity. What should be pointed out is that the cracking gas is composed mainly of C_3 and C_4 and the amount of C_1 and C_2 is very small. C_3 and C_4 are also valuable fuel and chemical raw materials.

Almost all of the sulfur removed goes into the cracking gas in the form of H_2S. First, when the cracking gas passed through the solution of $Pb(Ac)_2$, a large amount of black PbS deposition was found. Second, XRD analysis found no metal sulfides in the deactivated catalyst, and element analysis did not detect the signal for sulfur. Furthermore, TPO of the deactivated catalyst with MS as the detector did not found SO_2 (Fig. 12). It is obvious that the catalyst has very high cracking activity for sulfides.

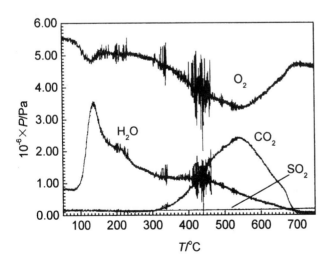

FIG. 12   TPO spectrum of the coke-deposited catalyst at 400°C with MS as the detector.

TABLE 5 Hydrocarbons and Octane Number of Gasoline Distillate Before and After Desulfurization Reaction

	Hydrocarbons					Octane number	
	n-Alkane	i-Alkane	Alkene	Naphthene	Aromatic	RON	MON
Before reaction	5.04	19.6	30.41	10.59	34.92	90.4	77.6
After reaction	5.00	26.2	21.62	9.28	37.53	91.8	78.8

Sulfur content of the feed gasoline: 1650 µg/g: temperature: 400°C; catalyst/oil: 2.5. Catalyst composition: 30 wt% USY, 60 wt% Al_2O_3, and 10 wt% other metal oxide.

Because the hydrocarbon cracking activity of the catalyst is mild, the distillation range of the liquid product does not change significantly compared to the feed. However, the hydrocarbons of the gasoline distillates before and after reaction change greatly (Table 5). After the reaction, the amounts of i-alkane and aromatic contents increase by 6.6 and 2.6 percentage points, respectively. Hence, the RON and MON increase 1.4 and 0.8, though the amount of alkene drops 8.8 percentage points. Therefore, cracking desulfurization over the catalyst not only can reduce sulfur content significantly, but also can improve the quality of the gasoline.

Now we are trying to develop the appropriate reactor and related techniques, and we hope to establish a new sulfur removal process for gasoline.

E. Evaluation of the Additive for Sulfur Removal from FCC Gasoline

The ZnO-containing additive for sulfur removal from FCC gasoline, with BET surface area of 144 m^2/g, was evaluated in a confirmed fluidized-bed reactor. The results follow.

1. Influence of the Amount of Additive in the FCC Catalyst on Sulfur Removal

Under 500°C and with a catalyst/oil ratio of 5, varying the amount of the additive in the FCC catalyst, we investigated the effect of the additive on sulfur removal. The results are shown in Figure 13. When pure regenerated FCC catalyst is used, the sulfur content of the gasoline produced is 1230 µg/g. When the additive accounts for 10% of the total catalyst (FCC catalyst + additive), the sulfur content drops to 890 µg/g and 27.6% of the sulfur is removed, compared to using pure FCC catalyst. With an increase in the amount of additive, the sulfur content of the gasoline produced falls further, due to the increased chance of contact between sulfides and additive. When 30% of the additive is added, the sulfur content

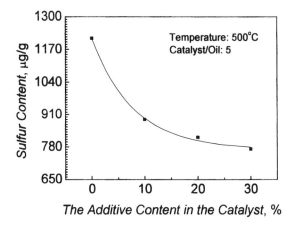

The Additive Content in the Catalyst, %

FIG. 13   Relationship between the sulfur content of gasoline and the amount of the additive in the catalyst.

is reduced to 770 μg/g and 37.40% of the sulfur is removed. Based on the relationship between the sulfur content and the amount of additive in Figure 13, however, it is unnecessary to increase the amount of additive further.

We find no paper published on how the additive works. Because of the complexity of FCC reactions, we cannot determine exactly whether it is via adsorbing and cracking sulfides in the feed or via adsorbing and cracking those in the gasoline produced. For FCC, most cracking reactions take place at the instant the feed contacts the catalyst. The cracking of sulfides, however, is relatively slow, based on the results obtained in the study of gasoline cracking desulfurization over specially made catalysts. When the residence time of gasoline in the catalyst bed is shorter than 1 s, only a little sulfur can be removed. Hence we are apt to think that the additive plays its role mainly via adsorbing and cracking sulfides selectively from the gasoline produced. Certainly, we cannot preclude the probability that the additive directly adsorbs and cracks sulfides in the feed. When the amount of the additive in the catalyst is large, it has to participate in cracking hydrocarbons, which affects the adsorbing and cracking of sulfides from the produced gasoline. Thus it is not very favorable for sulfur removal to add too much.

2.   Influence of the Catalyst/Oil Ratio on Sulfur Removal

At under 500°C and with different catalyst/oil ratios, we investigated the relationship between sulfur removal and the catalyst/oil ratio using pure FCC catalyst as well as with 30% additive. The results are shown in Figure 14. When the catalyst/oil ratio is 5, the sulfur content is 1230 μg/g using pure CC catalyst and 770 μg/g with 30% additive, respectively. On increasing the ratio to 7, the sulfur

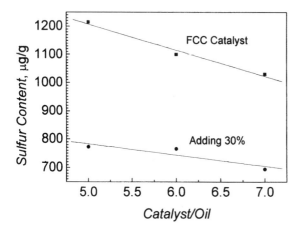

FIG. 14 Relationship between the sulfur content of gasoline and the catalyst/oil ratio.

content using pure FCC catalyst drops to 1020 µg/g, while that with 30% additive falls to 690 µg/g. Thus, whether using pure FCC catalyst or with 30% additive, the sulfur content of the produced gasoline decreases with the catalyst/oil ratio. However, the difference in sulfur content at the same ratio diminishes with the ratio of catalyst/oil.

On boosting the catalyst/oil ratio, the cracking activity of the reacting system is increased and the chances for cracking hydrocarbons and sulfides are also increased; hence, this favors sulfur removal from gasoline. However, the sulfur removal effect of the additive cannot be substituted by increasing the catalyst/oil ratio of pure FCC.

3. Influence of the Additive on the Distribution of Products

The product distributions obtained with various additive amounts at the same temperature and catalyst/oil ratio are listed in Table 6. Within the range of additive amounts from 0 to 30%, the conversions are around 64%, the yields of gasoline are between 40 and 41%, and the selectivities to gasoline are all a little more than 63%. Furthermore, the yields for dry gas, $C_3 + C_4$, and coke are all within the range of analysis error. Thus, we can conclude that the additive does not affect the distribution of products.

4. Influence of the Additive on the Hydrocarbon Composition and the Octane Number of Gasoline

The hydrocarbon composition and octane number of the gasoline produced using pure FCC catalyst and with 30% additive at 500°C and a catalyst/oil ratio of 5

TABLE 6 Product Distributions at Various Additive Amounts

Additive (%)	Conversion (%)	Yield (%)				Selectivity to gasoline (%)
		Gasoline	$C_1 + C_2$	$C_3 + C_4$	Coke	
0	64.10	40.31	2.02	13.72	8.05	63.10
10	63.83	40.72	1.78	12.88	8.45	63.22
20	64.31	40.91	2.09	13.31	8.00	63.61
30	63.99	40.67	1.86	13.26	8.20	63.55

Temperature: 500°C; catalyst/oil: 5.

are shown in Table 7. With the additive, the *i*-alkane and aromatic contents increase about 4.9 and 2.2 percentage points, respectively, while that of alkene decreases about 7 percentage points. Therefore, the MON has a slight increase, though the RON decreases 0.5 units. Thus the additive has no notably bad effect on the quality of the gasoline.

IV. CONCLUSIONS

Thiophene can react over USY zeolite to H_2S, hydrocarbons, and other methyl-thiophenes, but thiophene cracking to H_2S and hydrocarbons is the dominant reaction. Compared to thiophene, alkylthiophenes, the most abundant sulfides in FCC gasoline, are easier to desulfurize via cracking over a specially prepared sulfur removal catalyst with USY zeolite as the cracking component, and the conversion increases with the alkyl carbon number of alkylthiophene. The conversion sequence is thiophene <2- or 3-methylthiophene < C_2-substituted < C_3-substituted thiophene < C_4-substituted thiophene.

Cracking and hydrogen transfer are two important elementary reaction steps for thiophene and alkylthiophene desulfurization via cracking. Higher tempera-

TABLE 7 Hydrocarbon Composition and Octane Number of Gasoline Produced Using Pure FC Catalyst and with 30% Additive

Catalyst	Hydrocarbons					Octane number	
	n-Alkane	*i*-Alkane	Alkene	Naphthene	Aromatic	RON	MON
FCC	3.15	24.90	24.14	11.66	36.14	94.7	80.7
30% additive	2.75	29.78	17.10	12.01	38.33	94.2	80.8

Temperature: 500°C; catalyst/oil: 5.

ture favors the former, while lower temperature favors the latter. The synergism of cracking and hydrogen transfer makes about 400°C the optimum for thiophene and alkylthiophenes to desulfurize via cracking.

Based on the properties and characteristics of sulfides in FCC gasoline, we have designed a catalyst for gasoline cracking desulfurization and the additives for sulfur removal from FCC gasoline that are all composed of a sulfide-cracking component and a support that can adsorb sulfides from gasoline selectively. Both the ZrO_2 and ZnO are desired alternatives for the support because they can prevent the formation of sulfur-containing coke on the catalyst effectively, except for their excellent adsorbing performance with sulfides.

The catalyst for gasoline cracking desulfurization can remove more than 80% of the sulfur with a little cracking loss, and after desulfurization both the RON and MON rise, though the alkene content drops notably. The additive for sulfur removal from FCC gasoline also significantly affects sulfur removal. With 30% additive at 500°C and a catalyst/oil ratio of 5, the sulfur content of gasoline produced is reduced to 770 μg/g, about 37% of the sulfur is removed, compared to that using pure FCC regenerated catalyst. Furthermore, the additive has no effect on the distribution of products under experimental conditions and also no bad effect on the quality of the gasoline.

That the catalyst and additive can remove sulfur from gasoline is the result of the interaction between the special catalyst or additive surface and the sulfides in gasoline. We hope the techniques we are developing will play their roles in environmental protection.

REFERENCES

1. WC Cheng, G Kim, AW Peters, X Zhao, K Rajagopalan, MS Ziebarth, CJ Pereira. Catal Rev-Sci Eng 40(1&2): 39–79, 1998.
2. RF Wormsbecher, G Kim. US Patent: 5,376,608, 1994.
3. RF Wormsbecher, G Kim. US Patent: 5,525,210, 1996.
4. J Balko, D Podratz, J Olesen. NPRA Annual Meeting, San Antonio, Texas, 2000, AM-00-14.
5. SK Shen, CY Li, CC Yu. Stud Surf Sci Catal 119:765–770, 1998.
6. GH Luo, XQ Wang, XS Wang. Chinese J Catal 19(1):53–57, 1998.
7. Y Lu, MY He, JQ Song, XT Shu. Petroleum Refining Design 29(6):5–11, 1999.
8. P Wang, J Fu, MY He. Petroleum Processing Chemical Engineering 31(3):58–62, 2000.
9. X Saintigny, RA van Santen, S Clemendot, F Hutschka. J Catal 183:107–118, 1999.
10. SC Chen. Organic Chemistry. Beijing: Higher Education Press, 1989, p 393.

6
Removal of Heavy Metals from Aqueous Media by Ion Exchange with Y Zeolites

MARK A. KEANE   University of Kentucky, Lexington, Kentucky, U.S.A.

I. BACKGROUND: POLLUTION BY HEAVY METALS

Heavy metals (HM) is a general collective term applied to the group of metals and metalloids with an atomic density greater than 6 g cm^{-3} and includes such elements as Cu, Cd, Hg, Ni, Pb, Zn, Co, Al, and Fe [1,2]. The pollution and toxicity associated with heavy metals is now well established, with mounting evidence of adverse ecological and public health impacts [3,4]. The presence of heavy metals in water has an appreciable effect on acidity [5], and the resultant decrease in pH is known to adversely affect fish stocks and vegetation [6]. These pollutants reach the environment from a vast array of anthropogenic sources as well as natural geochemical processes. Heavy metal ions in solution are toxic to humans if the concentration is sufficiently high, and Health Agency guidelines set maximum acceptable HM concentrations in drinking water that are typically less than 3 mg dm^{-3} [7,8].

This chapter focuses on Fe, Co, Ni, Cu, Cd, and Pb as six representative HM pollutants. Iron is found naturally in large concentrations in soil and rock, normally in an insoluble form, but it can, as a result of a series of naturally occurring complex reactions, be converted to soluble forms that often result in water contamination [7]. Excess iron in groundwater can also arise from the use of iron salts as coagulants during water treatment or as a byproduct of pipe corrosion [1]. Iron is very unlikely to cause a threat to health at the concentrations typically recorded in water supplies, but excessive amounts can certainly have detrimental effects. The presence of cobalt as a water pollutant can be due to a leaching from rock/soil or the result of commercial activities relating to agriculture or the mining/metallurgical/electronics industries or as a byproduct of electroplating and pigment/paint production [6,9]. Nickel and copper are among the most toxic metals for both higher plants and many microorganisms [10,11], while copper, along with arsenic and mercury, is recognized as exhibiting the highest relative mammalian toxicity [4]. Sources of nickel and copper pollutants include mining/smelting, agricultural materials, the electronics, chemical, and metallurgical industries, as well as waste disposal in the form of leachates from landfills [6,12–14]. Cadmium is classed as a highly toxic nonessential metal that affects the action of enzymes and impedes respiration, photosynthesis, transpiration, and chlorosis [6,10]. On a comparative basis, lead is neither as toxic nor as bioavailable as cadmium but is more ubiquitous in the environment and acts as a cumulative toxin [6,15]. Sources of both cadmium and lead pollution include mining, agriculture, fossil fuel combustion, the metallurgical and electronic industries, and the manufacture and disposal of batteries, paints/pigments, polymers, and printing materials [3,6,13,16].

II. STRATEGIES FOR HEAVY METAL POLLUTION REMEDIATION

The most commonly employed treatment method for HM removal is chemical precipitation [1,17]. Although this approach is relatively simple and inexpensive, it has the decided drawback of generating a large volume of "sludge" for disposal. Iron, for instance, is soluble in the ferrous state, Fe(II), but is oxidized to the insoluble ferric form, Fe(III), in air [18], and the ferric iron hydrolyzes readily to form insoluble ferric hydroxide. Conventional water treatment for the removal of iron involves the oxidation of ferrous iron and removal of ferric hydroxide by sedimentation and filtration [18,19]. Alternative HM recovery methods include electrowinning, reverse osmosis, electrodialysis, solvent extraction, evaporation, ion exchange, and biological treatment [20,21]. The process of ion exchange, the focus of this article chapter, involves the replacement of toxic metal ions in solution by the more benign counterions that balance the surface charge of the solid exchanger. Ion exchange with aluminosilicate zeolites in batch or continuous operation, when compared with chemical precipitation (as the best established methodology), has the decided advantage of minimal associated waste generation, process simplicity, and ease of maintenance. Zeolites have been applied as ion exchangers in the removal of ammonium ions from municipal wastewater, in water softening, and, to a limited extent, in the treatment of radioactive water containing cesium and strontium [22]. However, the application of zeolites to environmental pollution control in terms of heavy metal removal from aqueous media has received scant attention.

III. APPLICATION OF ZEOLITE ION EXCHANGERS

Zeolites are crystalline aluminosilicates that are structurally unique in having cavities or pores with molecular dimensions as part of their crystalline structure.

Zeolites possess "compensating," or charge-balancing, cations (typically Na^+) that counterbalance the negative charge localized on the aluminosilicate framework, where the exchange capacity is governed by the Si/Al ratio. Because these ions are not rigidly fixed at specific locations within the hydrated unit cell, it is possible to effect exchange with external cations in solution [23]. Both synthetic [2,24–32] and naturally occurring [22,24,33–35] zeolites have been used successfully to exchange the indigenous Na^+ ions with heavy metals in aqueous solution. The application of the high-surface-area zeolite Y to the removal of Cu, Ni, Cd, and Pb from water has been reported previously [2,32]. While the process of ion exchange with zeolites has been the subject of a number of investigations, the emphasis has invariably been placed firmly on the synthesis of efficient zeolite-based catalysts [36–38]. The dearth of literature on the use of zeolites for heavy metal cleanup is possibly due to the low solution pH that is often necessary (particularly in the case of iron) to prevent metal hydroxide precipitation and ensure that ion exchange is stoichiometric; zeolites can suffer structural breakdown even under weakly acid conditions [21,28]. Hlavay et al. [39] have, however, investigated the efficiency of the zeolite clinoptilolite for the removal of iron from drinking water and found that the operation of three ion exchange columns in series reduced the initial iron content (in the range 0.7–0.9 mg dm^{-3}) to below detectable levels. The action of natural zeolite clays in lowering lead toxicity in freshwater fish [40], limiting cadmium and lead leaching in soil [41–43], and removing HM from wastewater [44–52] has received some coverage in the open literature.

IV. EXPERIMENTAL CONSIDERATIONS

The parent zeolite was Linde molecular sieve LZ-52Y, which has the nominal anhydrous unit cell composition $Na_{58}(AlO_2)_{58}(SiO_2)_{134}$: density = 1.9 g cm^{-3}; free aperture (anhydrous) = 0.74 nm; unit cell volume = 0.15 nm^3; void volume = ca. 50%. In order to obtain, as far as possible, the monoionic sodium form, the zeolite as received was contacted five times with 1 mol dm^{-3} aqueous solutions of $NaNO_3$. The zeolite was then washed briefly with deionized water, oven-dried at 363 K, and stored over saturated NH_4Cl solutions at room temperature; the water content was found by thermogravimetry (Perkin Elmer thermobalance) to be 24.8% w/w. The K-Y form was prepared by repeated exchange of the parent Na-Y with KNO_3, as described in detail elsewhere [29].

Heavy metal (Fe, Co, Ni, Cu, Cd, and Pb) removal from aqueous solution by ion exchange was conducted in the batch mode. The exchange isotherms were constructed at 293 K and 373 K (± 2 K) and at a total exchange solution concentration of 0.1 equiv. dm^{-3}, where 1 equiv. equals 1 mol of positive charge. The binary isotherm points were obtained by contacting the zeolite with aqueous (deionized water) solutions of the (divalent) heavy metal nitrate (or chloride) in the presence or absence of known concentrations of $NaNO_3$ to ensure the same initial solution-phase charge concentration. Ternary (Pb/Cd/Na and Co/Fe/Na) isotherm points were obtained in the presence of known concentrations of $NaNO_3$, where the initial individual HM solution concentration spanned the range 0.002–0.05 mol dm^{-3}. Binary HM exchange with K-Y was also performed for comparative purposes where the solution-phase charge balance was maintained with known KNO_3 concentrations. The Na-Y (or K-Y) zeolite (sieved in the mesh range 50–70 μm) was contacted with the heavy metal (HM)/Na solutions (thoroughly purged with He), and the resultant slurry was agitated at 600 rpm for three days, at which point equilibrium uptake had been achieved; the latter was ascertained from periodic sampling and analysis of the treated solution. The solution-phase pH, before and after the zeolite treatment, was measured by means of a Hanna HI 9318 Programmable Printing pH Bench-Meter. In every instance the solution phase was sufficiently acidic to ensure that HM hydroxide formation/precipitation was negligible. The zeolite was separated from solution by repeated filtration, and the metal content in the filtered liquid samples was determined after appropriate dilution. In the case of ferrous iron determination, the solution for analysis was acidified by addition of nitric acid to deliver a pH of 4.1 in order to prevent oxidation of Fe(II). The liquid-phase Na (or K) and heavy metal concentrations were measured by atomic absorption spectrophotometry (AAS, Varian SpectrAA-10), where data reproducibility was better than ±2%.

Breakthrough experiments were performed using a fixed-bed configuration, where solutions containing HM or HM/Na were passed through a packed stainless steel column (19 cm × 4.6 mm i.d.) loaded with Na-Y, employing a constant-flow pump (Hitachi Model L-7100). The breakthrough response for four selected HM (Fe, Co, Ni, and Pb) was investigated at an inlet flow rate (F_{in}) = 0.5 cm^3 min^{-1} and an HM concentration = 2 mmol dm^{-3}; the corresponding bed pressure = 17 atm. Regeneration experiments were conducted once the zeolite had been saturated with HM by contacting the Na-Y bed with 2 mol dm^{-3} $NaNO_3$ solutions delivered at a constant rate (0.5 cm^3 min^{-1}). The regenerated zeolite was washed with deionized water (flow rate = 1 cm^3 min^{-1}, volume = 60 cm^3) and the breakthrough experiments were again conducted as before.

Structural changes to the zeolite were probed by scanning electron microscopy (SEM) using a Hitachi S700 field emission SEM operated at an accelerating voltage of 25 kV. Samples (before and after ion exchange) for analysis were deposited on a standard aluminum SEM holder and double coated with gold. Treatment of Fe^{3+} solutions with Na-Y, where the initial ferric concentration was greater than 0.0033 mol dm^{-3}, proved unfeasible due to the unavoidable pH-induced precipitation of $Fe(OH)_3$; i.e., formation of the hydroxide is induced at pH = 1.7. In a Y zeolite ion exchange with an external $FeCl_3$ solution (0.0333 mol dm^{-3}) the solution pH varied from 1.8 to 2.8 and was accompanied by substantial hydroxide precipitation and zeolite structural breakdown; a loss of 69% of the initial Al component has been recorded [53]. Zeolites with higher Si/Al ratios are known to be more stable to prolonged contact with inorganic acids at pH = 2 [24], and the feasibility of stoichiometric Fe^{3+} exchange should focus on such potential candidate materials as clinoptilolite and mordenite [54]. The results presented in this chapter deal solely with divalent HM exchange. All the chemicals employed in this study were of analytical grade and were used without further purification.

V.   RESULTS AND DISCUSSION

A.   Structural Features of Zeolite Y

Zeolites as synthesized or formed in nature are crystalline, hydrated aluminosilicates of Group I and II elements. Structurally, they are made from a framework based on an infinitely extending three-dimensional network of SiO_4 and AlO_4 tetrahedra linked through common oxygen atoms. The isomorphic substitution of Si by Al gives rise to a net negative charge compensated by a cation component, i.e., the source of the ion exchange properties. The zeolites that have found the greatest application on a commercial scale belong to the family of faujasites and include zeolite X and zeolite Y. The framework structure of zeolites X and Y, shown in Figure 1, is based on a regular arrangement of truncated octahedral and sodalite cages to generate a high-surface-area microporous structure. The Y zeolite employed in this study is characterized [29] by an open framework consisting of two independent, though interconnecting, three-dimensional networks of cavities: (1) the accessible supercages of internal diameter 1.3 nm, which are

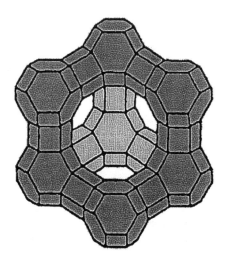

FIG. 1   Structure of faujasite.

linked by sharing rings of 12 tetrahedra (free diameter = 0.7–0.8 nm); (2) the less accessible sodalite units, which are linked through adjoining rings of six tetrahedra that form the hexagonal prisms (free diameter = 0.20–0.25 nm). Heavy metal ions are prone to precipitate from solution under alkaline or weakly acid conditions, while many HM salt solutions are sufficiently acidic to delaminate the zeolite. Representative SEM micrographs of the parent Na-Y are shown in Figure 2, where the geometrical crystalline features are evident. Routine SEM

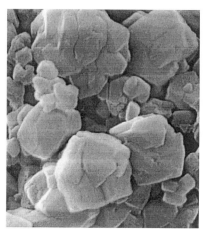

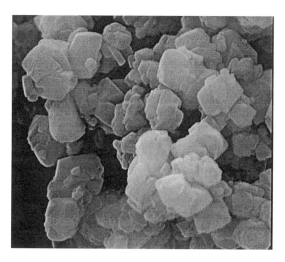

FIG. 2   SEM micrographs showing the topographical features of Na-Y.

analysis of the zeolite samples after the range of HM exchanges discussed in this chapter did not reveal any observable changes to the zeolite structure, while residual Al and Si in solution represented less than 3% of the content of the parent zeolite. Moreover, routine x-ray diffraction and IR analysis of the HM-exchanged Y zeolite showed no significant deviation from that recorded for the parent Na-Y [32]. The equilibrium solution-charge concentration varied from 0.982 to 0.118 mol equiv. dm^{-3}, and the predominant exchange process involved a direct replacement of monovalent sodium by divalent HM. There was no evidence of any appreciable overexchange or imbibition of the HM hydroxide; and while competing protonic exchange occurred to some degree, notably in the case of Na$^+$/Fe^{2+} and Na$^+$/Cu^{2+} systems, HM exchange with Na-Y was essentially stoichiometric.

B. Batch Operation

When exchanging ions of unequal charge, as in the case with exchange of the indigenous zeolitic Na$^+$ with solution-phase divalent HM cations, the exchange equilibrium can be represented by

$$HM_s^{2+} + 2Na_z^+ \rightleftharpoons HM_z^{2+} + 2Na_s^+$$

where s and z represent the solution and zeolite phases, respectively. Exchange selectivity can be quantified in terms of the separation factor, α:

$$\alpha = \frac{[HM_z]_{equil}[Na_s]_{equil}}{[HM_s]_{equil}[Na_z]_{equil}}$$

which for HM exchange with Na-Y is defined as the quotient of the equilibrium concentration ratios of HM and Na in the zeolite and in solution. If a particular entering HM cation is preferred, the value of the separation coefficient is greater than unity, and the converse holds if sodium is favored by the zeolite. The relationship between the separation factor and the equilibrium HM concentration in solution ([HM$_{soln}$]$_{equil}$) is shown in Figure 3, where the high affinity exhibited by the zeolite phase for the entering HM ions ($\alpha > 1$) is immediately evident for all the HM cations that were examined, particularly at lower concentrations. In each case, the value of α dropped with increasing starting HM solution concentration, but both Pb and Fe were favored over Na at every concentration that was considered. The other HM/Na systems are characterized by a switch in preference for the indigenous sodium at [HM$_s$]$_e$ > 0.03 mol dm^{-3}. The heavy metal removal efficiency can be conveniently quantified using the following expression:

$$\text{Removal efficiency (\%)} = \frac{[HM_s]_{initial} - [HM_s]_{equil}}{[HM_s]_{initial}} \times 100$$

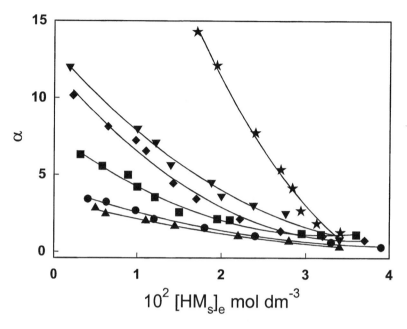

FIG. 3 Ion exchange separation factor (α) as a function of the equilibrium HM concentration ($[HM_s]_e$) for the exchange of Na-Y with external Fe ($\blacksquare$), Co ($\bullet$), Ni ($\blacktriangle$), Cu ($\blacklozenge$), Cd ($\blacktriangledown$), and Pb ($\bigstar$); $T = 293$ K.

Removal efficiencies exhibited by Na-Y for each HM are given in Table 1 at selected initial solution-phase HM concentrations ($[HM_{soln}]_{initial}$). At the lowest $[HM_{soln}]_{initial}$ values, exchange efficiency decreased in the order Pb > Cd ≥ Cu > Fe > Co > Ni. At higher concentrations, Na-Y delivered a roughly equivalent HM removal efficiency, with the exception of Pb, which exhibited a significantly higher affinity for exchange with Na-Y, as revealed in the affinity plot given in Figure 3. The foregoing affinity sequence finds support in previous reports of HM exchange with naturally occurring [34,45,55] and synthetic zeolites [25,56]. Under the stated conditions, the exchange process was operating under strong diffusion limitations, where the progress of exchange was controlled by diffusion of the HM cation within the crystal structure [23]. The effect of the aqueous environment on ion migration is pronounced, and in the aqueous exchange of zeolite Y the migrating species are cation–water complexes, where the cation in the zeolite phase is "solvated" to varying degrees by the lattice oxygens. In the hydrated zeolite, ions with a lower charge density, i.e., present in a less hydrated state, interact more strongly with the aluminosilicate framework. The observed

TABLE 1 Removal Efficiency for the Six Model HM by Exchange with Na-Y at 293 K as a Function of the Ratio of Initial HM to Zeolite in a Batch Operation

$\dfrac{[\text{HM}_s]_{\text{initial}}}{\text{zeol}}$ $(10^{-2} \text{ mol dm}^{-3} \text{ g}^{-1})$	Removal efficiency (%)					
	Fe	Co	Ni	Cu	Cd	Pb
0.5	50	40	32	58	62	100
1.0	46	32	20	50	53	82
2.0	40	21	16	42	45	59
4.0	24	16	11	26	28	38
8.0	14	11	9	16	16	24

sequence of increasing exchange efficiency can be considered to reflect an increasing effectiveness in neutralizing the negative charge on the aluminosilicate framework. Maes and Cremers [25] have viewed the neutralization of the zeolite network charge in terms of complex formation. The direct coordination of the divalent ion with the framework oxygen is equivalent to inner-sphere coordination, while the interposition of water molecules gives rise to an outer-sphere complex with respect to the zeolite lattice. The concentration of inner-sphere complexes of transition metal ions in related inorganic systems increases in the order $\text{Ni} < \text{Mn} < \text{Co} < \text{Zn} < \text{Cu} < \text{Cd}$ [25]; this increase in charge neutralization efficiency runs parallel to the exchange efficiencies recorded in this study.

Exchange with the zeolitic indigenous Na^+ ions and siting within the aluminosilicate framework must necessitate some weakening of the ion–dipole interactions between the in-going HM ions and the coordinated water molecules, where the hydration sheath is stripped and the HM ions are more effectively solvated by the zeolite framework oxygens. The enthalpy of hydration [57] of Pb^{2+} ions (-1481 kJ mol^{-1}), as the HM species that exhibited the highest affinity for exchange with Na-Y, is significantly lower than that of Cd^{2+} (-1807 kJ mol^{-1}), the second HM ion in the affinity sequence. Consequently, the Pb^{2+} ions interact more effectively with the lattice oxygens, and the efficiency of removal is the highest for the six HM toxins that have been studied. The effect of increasing the exchange temperature from 293 K to 373 K resulted in an increase in HM removal efficiency, as shown in Table 2. A similar enhancement in the degree of HM exchange has been noted elsewhere [25,30]. Such an effect can be attributed to the steric hindrance experienced by the bulky hydrated HM^{2+} ions in attempting to access the less accessible Na^+ ions [29]. At elevated exchange temperatures, the ion/dipole interaction between the HM ion and the solvent is weakened, thereby reducing the solvation coating and kinetic diameter of the in-going cation, facilitating the exchange process.

TABLE 2 Effect of Exchange
Temperature on Removal Efficiency
for the Six Model HM by Exchange
with Na-Y in a Batch Operation

HM[a]	Removal efficiency (%)	
	293 K	373 K
Fe	40	50
Co	21	40
Ni	16	38
Cu	42	58
Cd	45	63
Pb	59	69

[a] $\dfrac{[HM_s]_{initial}}{zeolite} = 0.02$ mol dm^{-3}g^{-1}

The influence of the out-going alkali metal ion (K$^+$ vs Na$^+$) on HM removal efficiency is considered in Table 3. In every instance, the Na-Y zeolite delivered (to varying degrees) higher removal efficiencies. The latter suggests that K$^+$ ions with a lower charge density interact more strongly with the aluminosilicate framework and are more resistant to exchange with HM ions in external solution. The

TABLE 3 Effect of the Nature of
the Indigenous Charge-Balancing
Alkali Metal Cation on Removal
Efficiency for the Six Model HM in a
Batch Operation at 293 K

HM[a]	Removal efficiency (%)	
	Na-Y	K-Y
Fe	33	22
Co	19	16
Ni	14	10
Cu	36	29
Cd	37	28
Pb	50	48

[a] $\dfrac{[HM_s]_{initial}}{zeolite} = 0.03$ mol dm^{-3}g^{-1}

TABLE 4   HM Removal Efficiency
Exhibited by a Fresh and a
Regenerated Sample of Na-Y
Operating in Batch Mode at 293 K

	Removal efficiency (%)	
HM[a]	Fresh Na-Y	Regenerated Na-Y
Fe	40	35
Co	21	19
Ni	16	15
Cu	42	35
Cd	45	41
Pb	59	56

[a] $\dfrac{[HM_s]_{initial}}{zeolite} = 0.02 \ mol \ dm^{-3} g^{-1}$

extent of exchange is therefore dependent on the nature of the out-going alkali metal ion and the degree of hydration of the in-going HM ion. The possibility of zeolite regeneration and reuse can be assessed from the entries in Table 4, where HM removal efficiencies (at a representative $[HM_s]_{initial}$) are recorded for the parent Na-Y and samples of HM loaded Na-Y that had been subjected to a back-exchange with Na^+. The regenerated catalyst delivered a (roughly) equivalent or lower HM removal, with a discernible loss of efficiency in the treatment of Cu- and Fe-containing solutions. The observed drop in HM removal in the second exchange cycle can be attributed to a partial zeolite decationation due to hydronium exchange and some siting of HM ions in less accessible cage sites that are not as susceptible to back-exchange with Na^+ in solution [30].

The removal of pairs of HM ions in solution by Na-Y, i.e., ternary exchange, was considered, focusing on two HM pairs as representative cases: Pb^{2+}/Cd^{2+} and Fe^{2+}/Co^{2+}. These two ternary exchange systems were chosen to check whether the observed differences in HM removal efficiency in single-component HM solutions led to some degree of selectivity in treating mixed solutions. In general, selectivity trends in ternary exchange systems can be quantified using ternary separation factors that take the form [31,58,59]

$$\underset{B,C}{\overset{A}{}}\alpha = \frac{(Na_z)^2(HM_s)(HM'_s)}{(Na_s)^2(HM_z)(HM'_z)}$$

where HM and HM′ represent the two constituent divalent ions. Appropriate combinations of the ternary separation factors generate pseudo-binary factors that describe the selectivity of the zeolite for one ion (HM) over another (HM') in the presence of a third (Na):

$$_{HM'}^{HM}\alpha = \left[\frac{_{HM',Na}^{HM}\alpha}{_{HM,Na}^{HM'}\alpha}\right]^{1/3}$$

The two pseudo-binary separations $_{Co}^{Fe}\alpha$ and $_{Cd}^{Pb}\alpha$ are plotted in Figure 4 as a function of Na_z, the fraction of zeolite charge balance contributed by the sodium component. The zeolite exhibited an overwhelming preference for exchange with Pb^{2+} over Cd^{2+} that extended over the range of Na_z values, although there was a discernible decline in selectivity at higher Na_z. The marked preference exhibited by Na-Y for exchange with Pb^{2+} observed under noncompetitive conditions, as illustrated in Figure 3 and Tables 1–4, extends to this ternary system. In treating solutions containing both lead and cadmium, Na-Y exhibits high efficiency in terms of a selective removal of the lead component. In contrast, Na-Y displayed an equivalent or marginally greater preference for iron over cobalt ($_{Co}^{Fe}\alpha > 1$) at each ternary isotherm point plotted in Figure 4. The values of $_{Na}^{Fe}\alpha$ were consistently greater than those of $_{Na}^{Co}\alpha$ at equivalent Na_z, in keeping with the higher

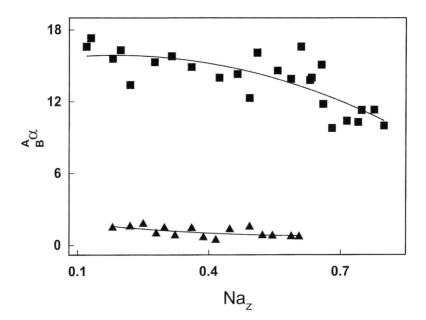

FIG. 4 Relationship between the pseudo-binary separation factor ($_B^A\alpha$) and Na_z, where A = Fe^{2+}, B = Co^{2+} (▲) and A = Pb^{2+}, B = Cd^{2+} (■).

affinity that the zeolite exhibits for exchange with iron under noncompetitive conditions (Fig. 4). Nevertheless, there is little in the way of selectivity associated with the $Na^+/Fe^{2+}/Co^{2+}$ ternary system, and the zeolite exhibits an essentially indiscriminate HM uptake.

C. Continuous Operation

The continuous removal of HM from aqueous solution using Na-Y was investigated using a fixed-zeolite-bed configuration. The breakthrough curves for four selected HM are shown in Figure 5 under identical operating conditions. Breakthrough behavior is evaluated in this study by plotting the relative HM solution concentration, i.e., the ratio of the HM concentration in the effluent ($[HM]_{out}$) to that in the inlet ($[HM]_{in}$) solution, as a function of the total number of zeolite-bed volumes that had been treated. The breakthrough response is known to be dependent on flow rate, concentration, ratio of bed length to diameter, particle size, rate of diffusion, differences in exchangeability at various sites, and the nature of the participating cations [60,61]. Taking the four HM considered in Figure 5, the removal efficiency trends observed for batch operation also extend to the continuous system. Taking, as a representative point of comparison, the processing of ca. 450 bed volumes of HM solution, $[HM]_{out}$ increased in the order

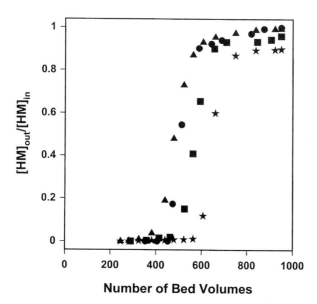

FIG. 5 Comparison of Co (●), Fe (■), Pb (★), and Ni (▲) breakthrough behavior under identical treatment conditions; $[HM]_{in} = 2$ mmol dm^{-3}, $F_{in} = 0.5$ cm^3 min^{-1}.

Pb (80 ppm) $<$ Fe (4.6 $\times$ 10^3 ppm) $<$ Co (5.9 $\times$ 10^3 ppm) $<$ Ni (1.6 $\times$ 10^4 ppm). However, the Na-Y bed was effective at keeping the HM concentration in the effluent below the target value of 3 mg dm^{-3} [7,8] in processing up to 300 bed volumes and was saturated after in excess of 700 bed volumes. As a general observation, an increase in either the inlet feed rate (decrease in contact time) or HM concentration lowered the cumulative feed volume required for saturation.

A controlled in situ elution of the HM ions from the zeolite after the break-through analysis, i.e., bed regeneration, was considered by contacting the HM-loaded zeolite bed with $NaNO_3$. Taking the Fe^{2+}/Na-Y and Co^{2+}/Na-Y exchange systems as representative, the effectiveness of this regeneration step can be as-sessed from the results presented in Table 5. The zeolitic HM component was readily released into the $NaNO_3$, and both the Fe^{2+}- and Co^{2+}-loaded zeolite exhibited an essentially identical regenerative capacity. The Na back-exchange profiles for both Fe and Co are essentially superimposable. At a regenerant con-tact in excess of five bed volumes, further HM elution was not significant. Under these treatment conditions, ca. 3 $\times$ 10^3 cm^3 (2 mmol dm^{-3}) of Co or Fe solution were treated before zeolite saturation was reached. An elution of ca. 79% was achieved in 16 cm^3 $NaNO_3$, which represents a concentration factor of ca. 160. The exchange capacities of the regenerated samples for Co and Fe were 85% and 89%, respectively, of that exhibited by the fresh Na-Y. These results indicate that the zeolite can be regenerated and reused with a reasonable degree of effi-ciency; a cyclical use of Na-Y in HM removal/zeolite regeneration shows some promise as a viable means of continuous HM uptake/concentration.

TABLE 5 Extent of HM Elution
from a Co- and Fe-loaded Zeolite as a
Function of the Number of Bed
Volumes of $NaNO_3$ Processed in a
Continuous Operation[a]

Bed volumes	% HM eluted	
	Co	Fe
2.4	75.3	74.9
4.8	79.2	80.1
9.5	80.6	82.3
21.3	82.1	82.8
60.1	83.0	83.0
85.4	83.4	83.1

[a] $[NaNO_3]$ = 2 mol dm^{-3}; F_{in} = 0.5 cm^3 min^{-1}.

VI. CONCLUSIONS

Excessive quantities of heavy metals in natural waterways arise from a range of commercial industrial activities and are a serious pollution problem with associated adverse ecological and public health impacts. Sodium-based zeolite Y with a Si/Al ratio = 2.3 is effective in removing divalent heavy metal (Fe, Co, Ni, Cu, Cd, and Pb) ions from aqueous solution in batch operations over the concentration range 0.002–0.05 mol dm^{-3}. Under conditions of stoichiometric ion exchange with no appreciable zeolite structural breakdown, the following sequence of increasing HM removal efficiencies has been established for Na-Y: Ni < Co < Fe < Cu $\leq$ Cd < Pb. The latter reflects an increasing degree of coordination to the aluminosilicate framework oxygen and a greater charge neutralization within the zeolite framework. The differences in HM affinity for exchange with Na-Y can be such as to facilitate selective HM uptake, as in the case of Pb^{2+}/Cd^{2+}, but an indiscriminate HM removal is more typical. Replacing Na^+ with K^+ as the indigenous charge-balancing species lowers the overall HM removal efficiency as the K^+ interacts more strongly with the zeolite and is more resistant to exchange with cations in external solution. An increase in exchange temperature enhanced HM removal by reducing the solvation coating and kinetic diameter of the ingoing HM cation, facilitating exchange with the univalent alkali metal. In continuous operation, the breakthrough patterns associated with HM removal reveal the same trends of exchange efficiency observed in batch operation. The HM-loaded zeolite is readily regenerated by sodium back-exchange to concentrate the initial HM solution by a factor of ca. 160, albeit that the exchange capacity of the regenerated zeolite is lower than that of the fresh sample. Regeneration/reuse of the zeolite is possible for both batch and continuous operation but is accompanied by a drop in HM removal efficiency.

REFERENCES

1. J Patterson, R Passino. Metal Speciation, Separation, and Recovery. Chelsea: Lewis, 1987.
2. S Ahmed, S Chughtai, MA Keane. Sep Purif Technol 13:57–64, 1998.
3. JE Fergusson. The Heavy Elements: Chemistry, Environmental Impact and Health Effects. Oxford: Pergamon Press, 1990.
4. HJM Bowen. The Environmental Chemistry of the Elements. London: Academic Press, 1979.
5. RW Batterbee, RJ Flower, AC Stevenson, B Rippey. Nature 314:350–352, 1985.
6. BJ Alloway, DC Ayres. Chemical Principles of Environmental Pollution. Glasgow: Blackie, 1993.
7. NF Gray. Drinking Water Quality. New York: Wiley, 1994.
8. JD Zuane. Handbook of Drinking Water Quality. New York: Van Nostrand Reinhold, 1990.

9. SL Williams, DB Aulenbach, NL Clesceri. In: AJ Rubin, ed. Aqueous-Environmental Chemistry of Metals. Ann Arbor, MI: Ann Arbor Science, 1974.
10. A Kabata-Pendias, H Pendias. Trace Elements in Soils and Plants. Boca Raton, FL: CRC Press, 1984.
11. RE Train. Quality Criteria for Water. London: Castle House, 1979.
12. WC Peters. Exploration and Mining Geology. New York: Wiley, 1978.
13. AW Rose, HE Hawkes, JS Webb. Geochemistry in Mineral Exploration. London: Academic Press, 1979.
14. P O'Neill. In: BJ Alloway, ed. Heavy Metals in Soils. Glasgow: Blackie, 1990.
15. HA Waldron. Metals in the Environment. London: Academic Press, 1980.
16. JM Pacyna. In: TC Hutchinson, KM Meema, eds. Lead, Mercury, Cadmium and Arsenic in the Environment. Chichester, UK: Wiley, 1987.
17. ER Krishnan, P Utrecht, A Patkar, J Davis, S Pour, M Foerst. Recovery of Metals from Sludges and Wastewaters. Park Ridge, IL: Noyes Data Corp., 1993.
18. G Kiely. Environmental Engineering. London: McGraw-Hill, 1997.
19. CV Tremblay, A Beaubien, P Charles, JA Nicell. Water Sci Technol 38:121–128, 1998.
20. C Liu, J Ma, J Shan, T Chai, Y Huang. Water Treat 3:303–313, 1988.
21. W Stumm, GF Lee. Ind Eng Chem 53:143–146, 1961.
22. SK Ouki, M Kavannagh. Waste Manag Res 15:383–394, 1997.
23. DW Breck. Zeolite Molecular Sieves, Structure, Chemistry and Use. New York: Wiley, 1974.
24. JD Sherman, ed. AIChE Symposium Series, Vol. 74, No. 179: Adsorption and Ion Exchange Separations, 1978.
25. A Maes, A Cremers. J Chem Soc Faraday Trans I 71:265–277, 1975.
26. PP Lai, LVC Rees. J Chem Soc Faraday Trans I 72:1809–1817, 1976.
27. RM Barrer, RP Townsend. J Chem Soc Faraday Trans I 72:661–673, 1976.
28. VA Nikashina, LI Zvereva, KM Olshanova, MA Potapova. J Chrom 120:155–158, 1976.
29. MA Keane. Microporous Mater 3:93–108, 1994.
30. MA Keane. Microporous Mater 3:385–394, 1995.
31. MA Keane. Microporous Mater 4:359–368, 1995.
32. MA Keane. Colloid Surf A 138:11–20, 1997.
33. MJ Zamzow, BR Eichbaum, KR Sandgren, DE Shanks. Sep Sci Technol 25:1555–1569, 1990.
34. E Maliou, M Malamis, PO Sakellarides. Water Sci Technol 25:133–138, 1992.
35. SK Kesraouiouki, CR Cheeseman, R Perry. J Chem Tech Biotechnol 59:121–126, 1994.
36. WN Delgass, RL Garten, M Boudart. J Phys Chem 73:2970–2979, 1969.
37. WN Delgass, RL Garten, M Boudart. J Chem Phys 50:4603–4606, 1969.
38. RL Firor, K Seff. J Phys Chem 82:1650–1655, 1978.
39. J Hlavay, G Vigh, V Olaszi, J Inczédy. Zeolites 3:188–190, 1983.
40. SK Jain, AK Raizada, S Shrivastava, K Jain. Frrenuis Environ Bull 5:466–468, 1996.
41. A Chlopecka, DC Adriano. Environ Sci Technol 30:3294–3303, 1996.
42. A Shanableh, A Kharabsheh. J Hazard Mat 45:207–217, 1996.

43. EH Rybicka, B Jedzejczyk. Appl Clay Sci 10:259–268, 1995.
44. V Albino, R Cidefi, M Pasini, C Colella. Environ Technol 16:147–156, 1995.
45. E Maliou, N Loizidou, N Spyrellis. Sci Tot Environ 149:139–144, 1994.
46. H Minamisawa, H Yamanaka, N Arai, T Okutani. Nippon Kagakj Kaishi 12:1605–1611, 1991.
47. A Assenov, C Vassilev, M Kostanova. Chem Tech 37:334–336, 1985.
48. R Cioffi, M Pansinin, D Caputo, C Colella. Environ Technol 17:1215–1224, 1996.
49. I Stefanova, Edjurova, G Gradev. J Radioanal Nucl Chem Lett 128:367–375, 1988.
50. KD Mondale, RM Carland, FF Aplan. Minerals Eng 8:535–548, 1995.
51. MJ Semmens, WP Martin. Wat Res 22:537–542, 1988.
52. P Burn, DK Ploetz, AK Saha, DC Grant, MC Skriba. AIChE Symp Ser 83:66–72, 1987.
53. JS Kim, L Zhang, MA Keane. Sep Sci Technol 36:1509–1525, 2001.
54. CV McDaniel, PK Maher. ACS Monogr 121:285–331, 1976.
55. G Blanchard, M Maunaye, G Martin. Water Res 18:1501–1507, 1984.
56. E Gallei, D Eisenbach, A Ahmed. J Catal 33:62–67, 1974.
57. J Burgess. Metal Ions in Solution. Chichester, UK: Ellis Horwood, 1978.
58. KR Franklin, RP Townsend. J Chem Soc Faraday Trans I 81:1071–1086, 1985.
59. KR Franklin, RP Townsend. J Chem Soc Faraday Trans I 84:687–702, 1988.
60. KD Mondale, RM Carland, FF Aplan. Minerals Eng 8:535–548, 1995.
61. P Burn, DK Ploetz, AK Saha, DC Grant, MC Skriba. AIChE Symp Ser 83:66–72, 1987.

7

Design and Synthesis of New Materials for Heavy Element Waste Remediation

LISA DYSLESKI, SARAH E. FRANK, STEVEN H. STRAUSS, and PETER K. DORHOUT   Colorado State University, Fort Collins, Colorado, U.S.A.

I. INTRODUCTION

Research on the separation of ionic pollutants from aqueous systems is focused on the optimization of many parameters, including cost, capacity, selectivity, and overall stability. Our labs have been interested in developing materials as ion exchange extractants for the removal and recovery of "soft" heavy metal cations such as mercury, cadmium, and lead from aqueous waste streams [1–5]. As a result of these studies, we have determined that the requirements for an ideal extractant are: selectivity for soft metal cations such as Hg^{2+}, Pb^{2+}, and Cd^{2+} over hard cations such as Na^+, Li^+, Mg^{2+}, and Cr^{3+}; the ability to recover the target soft heavy metal ion in a minimal volume of secondary waste; and the ability of the material to operate effectively through numerous extraction cycles. In order to satisfy each of these requirements, we have focused our efforts on sulfur-based materials with one or more exchangeable cations. It is our hypothesis that a material with a sulfur-rich environment will preferentially exchange hard cations held within the solid for soft cations in aqueous solution according to Lewis hard–soft acid–base theory [6]; moreover, another hypothesis is that it is important that the material has an open or flexible framework so that ion exchange is facile and so that it is able to withstand repeated ion exchange reactions.

II. LAYERED METAL CHALCOGENIDES

In the past, we have explored two types of remediation materials based on our hypotheses. The first type was based on zirconium monothiophosphate,

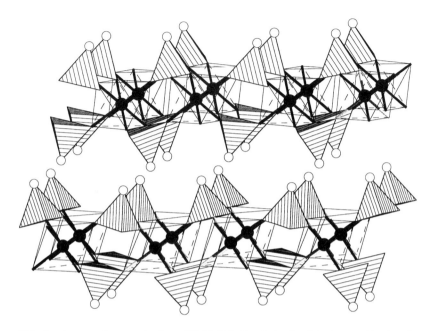

FIG. 1 Structural representation of $H_2Zr(PO_3S)_2$. Striped polyhedra are thiophosphates, filled circles are zirconium, and open circles, located in the interlayer space of the material, are sulfur atoms (with their associated protons not shown).

$H_2Zr(PO_3S)_2$, which incorporates a simple ion exchange mechanism between heavy metal cations in solution with protons in the solid. The proposed structure of $H_2Zr(PO_3S)_2$, shown in Figure 1, is related to the family of layered phosphates developed by Clearfield [7–11]. Through vibrational spectroscopy and structural analysis, it was shown that the protonated monothiophosphate groups, $(HPO_3S)^{2-}$, are oriented in the lattice so that the SH groups point into the interlayer spaces of the material. This structural characteristic provides, via ion exchange with the SH protons, an anionic, soft (i.e., sulfur-rich) Lewis basic environment for soft heavy metal cations. Consistent with our hypotheses, $H_2Zr(PO_3S)_2$ removed soft metal cations from solution according to the ion exchange equilibrium shown in Eq. (1):

$$H_2Zr(PO_3S)_2(s) + yM^{n+}(aq) \rightleftarrows H_{2-nx}M_xZr(PO_3S)_2(s)$$
$$+ nxH^+(aq) + (y - x)M^{n+}(aq) \quad (1)$$

The exchange constants, K_d, for a variety of metal ions are given in Table 1. As is consistent with Eq. (1), there was a pH dependence on the ion exchange ability; in increasingly acidic solutions, the value of x in $H_{2-nx}M_xZr(SO_3S)_2$ was

TABLE 1 Extractions of Metal Ions from Aqueous HNO_3 Using $H_2Zr(PO_3S)_2$

M^{n+}	Initial mequiv. of M^{n+}/g of extractant	K_d, mL/g^a in 0.001 M HNO_3	K_d, mL/g^a in 0.1 M HNO_3
Cd^{2+}	3.17	1800	860
Zn^{2+}	3.17	1200	350
Mn^{2+}	3.17	110	27
Co^{2+}	3.17	140	16
Ni^{2+}	3.17	76	7.0
Ca^{2+}	3.17	32	—
Na^+	6.34	16	—
Mg^{2+}	3.17	9.5	—

^{a} K_d is defined as $\{([M^{n+}_f] - [M^{n+}_i]/[M^{n+}_f])$(volume of solution, mL)/(mass of ion exchange material, g)\}.

observed to decrease. Relying on this equilibrium, the complexed metal ion could be recovered from the solid through a treatment with 3M HCl. In this reaction, the equilibrium in Eq. (1) was reversed, resulting in the reformation of $H_2Zr(PO_3S)_2$ and a solution containing dissolved M^{n+}.

Although the recyclability of this material is convenient, it is somewhat inefficient, producing an acidic secondary waste stream that has a volume that is potentially equal to or greater than that of the initial waste stream. The disposal of large volumes of acidic secondary waste can be costly; thus, in terms of recyclability, $H_2Zr(PO_3S)_2$ is an unattractive extractant. For this reason, our group has been focusing on the development of a more effective extraction approach known as redox-recyclable extraction and recovery, R^{2}ER [1,3]. This scheme combines an ion exchange mechanism with a *redox-initiated* recovery step completing the remediation/ion recovery cycle. Specifically, for soft heavy metal ion remediation, we have used MoS_2 as a solid-state extraction material. A complete R^{2}ER cycle for the MoS_2 system is illustrated in Figure 2.

MoS_2 has a layered structure that can be seen in Figure 3. It is known from the literature that *n*-butyllithium will reduce MoS_2 [12], contributing 1.3 equivalents of negative charge per formula unit into the electronic band structure of the MoS_2. This also initiates a structural phase transition from trigonal prismatic coordination to octahedrally coordinated molybdenum: a 2H-to-1T phase transition [13]. Lithium cations intercalate into the interlayer space as charge-compensating cations to retain charge neutrality in the solid.

When $Li_{1.3}MoS_2$ contacts an aqueous solution containing metal cations, the negatively charged layers reduce water to form hydrogen gas. The formation of hydrogen gas in the interlayer space causes the layers to exfoliate in solution,

(a)   **(b)**

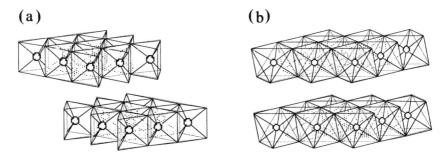

FIG. 2 Structural representation of MoS_2 used in our studies. (a) Is a representation of the 2H-MoS_2 structure, where Mo is in a trigonal prismatic site. (b) Is a representation of 1T-MoS_2 structure, where Mo is in an octahedral site. In $Li_{1.3}MoS_2$, Li^+ is located in the interlayer space of the 1T structure (not shown).

forming a colloidal solution of MoS_2 particles [14]. Through controlled experiments to measure the quantity of hydrogen gas evolved in this process, we have shown that only 0.52 electrons per formula unit are involved in the reduction of acidic solution (0.1 M H^+), leaving 0.78 equivalents of negative charge remaining on the MoS_2 layers. Thus, the colloid is nominally a suspension of $MoS_2^{0.78-}$ particles. In the presence of metal cations such as Hg^{2+}, the negatively charged $MoS_2^{0.78-}$ layers flocculated and precipitated a new solid, with the soft heavy

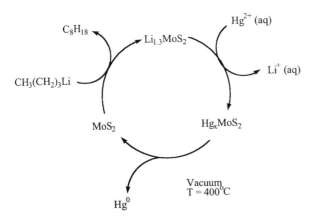

FIG. 3 Complete R^2ER cycle using $Li_{1.3}MoS_2$ showing the initial activation by butyllithium, contact with a mercury-containing solution, and deactivation by heating to isolate elemental mercury and to recover the starting material.

metal cation preferentially included within the interlayer space. The combination of this reduction and ion exchange reaction has been proposed in Eq. (2):

$$Li_{1.3}MoS_2(s) + 0.9H_3O^+ + 0.2Hg^{2+}(aq) \rightarrow (H_3O)_{0.38}Hg_{0.20}MoS_2(s)$$
$$+ 0.26H_2(g) + 1.3Li^+(aq) + 0.52H_2O \quad (2)$$

In an attempt to understand the mechanism of exchange, we discovered that $H_{0.78}MoS_2$ could be prepared independently and act as an ion exchange material. This material was synthesized and isolated using the reaction described in Eq. (3). This solid could be contacted with a solution containing contaminant metal ions that would exchange through a direct ion exchange mechanism similar to that seen with $H_2Zr(PO_3S)_2$, rather than one involving hydrogen evolution and exfoliation. This process is demonstrated in Eq. (4). Through this experimental modification, we have moved the hydrogen gas evolution into the synthetic step of the R^2ER cycle and out of the extraction step.

$$Li_{1.3}MoS_2(s) + 1.3H_3O^+ \rightarrow 0.26H_2(g) + 1.3Li^+(aq)$$
$$+ (H_3O)_{0.78}MoS_2(s) + 0.52H_2O \quad (3)$$

$$(H_3O)_{0.78}MoS_2(s) + 0.20Hg^{2+}(aq) \rightarrow (H_3O)_{0.38}Hg_{0.20}MoS_2(s)$$
$$+ 0.40H_3O^+(aq) \quad (4)$$

The extraction capabilities of $Li_{1.3}MoS_2$ and $H_{0.78}MoS_2$ for various metal ions are shown in Table 2. In each experiment, extractions were monitored for a two-hour period and the solids were then isolated from solution. Metal ion removal was determined by a comparison of the initial and final concentrations as determined by inductively coupled plasma—atomic emission spectrometry (ICP-AES). From this comparison it is obvious that $H_{0.78}MoS_2$ is not as efficient at selective ion exchange as $Li_{1.3}MoS_2$, because it does not produce a high-surface-area colloid and because ion movement through the solid (ion exchange) was slower and in some cases not observed.

TABLE 2 Comparison of the Extraction Capabilities of $Li_{1.3}MoS_2$ and $H_{0.78}MoS_2$

	K_d(mL/g)	
M^{n+}	$Li_{1.3}MoS_2$	$H_{0.78}MoS_2$
Hg^{2+}	$>10^6$	960
Pb^{2+}	3400	5
Cd^{2+}	820	—

The recovery process for the metal-containing MoS_2 solids is different from that of the metal-containing zirconium thiophosphates. The process involves a heat-induced internal redox reaction. The negative charge remaining on the layers of Hg_xMoS_2 reduced the guest ion, Hg^{2+}, to Hg^0, leaving neutral MoS_2 behind. We have demonstrated that this is accompanied by a structural-phase transition [5], wherein the octahedral MoS_2 units in the guest–host complex revert back to the trigonal prismatic units seen in native MoS_2: a 1T-to-2H phase transition for the MoS_2. The outcome of this heating process at 400°C under vacuum was the isolation of elemental mercury in the cold trap of the vacuum system. This behavior had also been seen with Ag^+ as the guest ion (although Ag^0 must be separated from the extractant in a second step). We also demonstrated the MoS_2 could be reduced again and reused through subsequent cycles. This type of cycle is beneficial over the ion exchange process demonstrated by $H_2Zr(PO_3S)_2$ due to the fact that the volume of secondary waste produced from the remediation cycle is as small as possibly attainable, elemental metals.

III. THIOSPINELS

Although activated MoS_2 and $H_2Zr(PO_3S)_2$ materials have been proven to be effective and selective extractants for soft heavy metal cations such as Hg^{2+} and Cd^{2+} from aqueous waste streams, both have problems in their extraction scheme that we sought to eliminate. In the case of the zirconium monothiophosphate, the volume of secondary waste generated is large and inherently strongly acidic. The cost and problems associated with disposing of this secondary waste become an issue. In the case of MoS_2, we have eliminated the large volume of secondary waste by producing pure metal; however, the generation of large amounts of hydrogen gas at a waste site, unless it is carefully controlled, is a concern. In addition, we have shown that $Li_{1.3}MoS_2$ is oxidized by moisture in air and will deactivate over time when exposed to atmospheric conditions. For this reason, it is less than practical to use on a large scale or under any conditions other than a strictly controlled laboratory environment. Consequently, we have been developing a new class of materials based on thiospinels with the general formula $Cu_2MSn_3S_8$ (M = Mn, Fe, Co, Ni) [15]. With this new set of compounds, we hope to incorporate the beneficial redox and ion exchange properties of the previously studied systems, without the negatives, such as recovery of a large volume of secondary waste, hydrogen evolution, and air sensitivity.

Historically, compounds adopting the spinel structure type have been considered important materials in the area of solid-state electrodes due to their open, flexible structure and ability to accept and/or exchange small ionic species [16–19]. In fact, a large body of literature is devoted to the investigation of the structure and conductivity of these materials for their potential use as lithium batteries. In 1985, Schöllhorn and Payer reported a cubic form of titanium disulfide formed by the extraction of copper from $CuTi_2S_4$. This material was then reductively intercalated with lithium cations to form $LiTi_2S_4$ [20]. Following this, attempts to further understand the structure of the solid after copper extraction and lithium insertion were made [18–22].

Copper extraction from and lithium insertion into thiospinels with the formula $Cu_2MSn_3S_8$ (M = Mn, Fe, Co, Ni) was reported in 1996 by Jumas and co-workers [15]. In order to increase the number of vacant sites in the structure, copper cations were initially removed through a reaction involving the oxidation of the transition metal, M, with I_2/CH_3CN and expulsion of Cu^+ into solution in order to maintain charge neutrality in the solid. Lithium cations were subsequently inserted, again as charge-compensating cations, as the result of reductive intercalation with n-butyllithium through which the transition metal, M, was reduced back to its former oxidation state.

An important facet of this reaction scheme is that the initial removal of copper cations increases the number of vacant interstices within the lattice and therefore facilitates lithium cation insertion. Unfortunately, copper extraction over longer periods of time, followed by lithium insertion, led to a loss in the long-range order of the solid, loss of crystallinity, and ultimately a problematic structural characterization. Since this initial work by Jumas, there have been numerous attempts at the selective exchange of cations into and out of similar structures. These have included studies on a class of thiospinels based on a Cu_2S-In_2S_3-SnS_2 mixture that suggested that selective exchange of ions into and out of thiospinel solids was possible [23–26]. As a result of these earlier studies, we believe that thiospinels would be ideal candidates for applications such as waste remediation.

The unit cell contents for a compound adopting the thiospinel structure is $A_8B_{16}S_{32}$. This unit cell contains 32 cubic-close-packed S^{2-} anions (in the $32e$ crystallographic site), 8 A cations, located in tetrahedral holes (the $8a$ sites), and 16 B cations located in octahedral holes (the $16d$ sites). Specifically, for a compound with the empirical formula $Cu_2MSn_3S_8$, Cu(I) cations are found in the $8a$ sites and M(II) and Sn(IV) cations are randomly distributed among the $16d$ sites. A depiction of the thiospinel structure is shown in Figure 4. This view illustrates the presence of both Cu(I) and M(II) cations in channels lined with Lewis-basic sulfur anions. It is this soft Lewis-basic environment that forms the basis of our hypothesis, providing a thermodynamic driving force for the preferential binding of soft heavy metals over hard ones. One final advantage of the spinel structure are the unoccupied $16c$ (octahedral), $8b$, and $48f$ (tetrahedral) sites that are available for ion movement and intercalation. These vacant sites comprise a spacious yet three-dimensionally stabilized solid that should lend itself well to repeated ion exchange reactions.

Target solids, where M = Mn, Fe, Co, and Ni, were synthesized through a solid-state reaction involving the stoichiometric combination of the elements

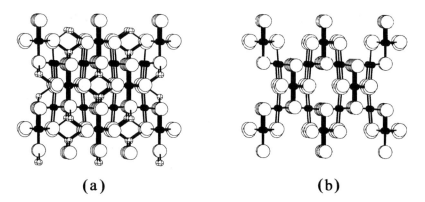

(a)   (b)

FIG. 4   Spinel structure type: (a) In $Cu_2MSn_3S_8$, Cu, crossed circles, is present in a tetrahedral coordination environment, whereas M and Sn, filled circles, are randomly distributed among the depicted octahedral sites. (b) The spinel structure in which the Cu atoms have been removed, showing channels through the structure.

sealed in a quartz ampoule, heated to 700°C for 196 hours and cooled to room temperature. A representative powder diffraction pattern of the materials with the chemical formula $Cu_2MSn_3S_8$ can be seen in Figure 5. These materials are typically highly crystalline powders displaying sharp diffraction peaks that can be indexed and compared qualitatively to the diffraction pattern of rhodostannite, $Cu_2FeSn_3S_8$ [27].

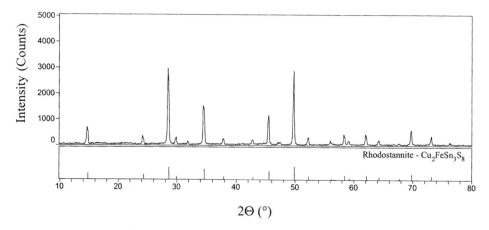

FIG. 5   PXRD analysis of our as-prepared $Cu_2FeSn_3S_8$ together with the Inorganic Crystal Diffraction Database (ICDD) standard, file # 29-0558.

Our target solids in their activated and deactivated form were studied for their ion exchange properties. In a typical exchange experiment, a given solid was contacted with 20 mL of a $M'(NO_3)_n$ solution in either 0.1 M HNO_3 or deionized water (M' = Hg, Pb, Ag, Ba, Mg). Mixtures were prepared so that the final ratio of extractant/metal ion was 5/1. Extractants were contacted for five hours, after which time they were filtered through plastic syringes fitted with 0.2-μm filters. Nonactivated materials are defined as the as-synthesized solids; activated materials, discussed later, required the removal of either Cu or M ions through a pretreatment step.

The observed K_d values of nonactivated $Cu_2MSn_3S_8$ for Hg^{2+} in water and nitric acid are presented in Table 3. From these data, it can be seen that there was a great decrease in the ion exchange ability as the pH of the solution was decreased. It is our hypothesis that this is due to the competing exchange of metal ion, M'^{n+} with H^+, similar to the behavior in $H_2Zr(PO_3S)_2$ and $H_{0.78}MoS_2$, in which we observed a decrease in the extraction abilities of the materials at lower pH governed by equilibria in Eqs. (1) and (4). From these data, it should follow that an activated material with the formula $H_xCu_yM_zSn_3S_8$ could be prepared through the reaction of $Cu_2MSn_3S_8$ with an acidic solution, similar to the MoS_2 system.

The activity of the proton-included (or activated) materials toward the removal of metal ions from solution is summarized in Table 4. These data suggest that the exchange of protons out of the solid is thermodynamically favorable, resulting in an increase in the quantity of metal ion that the material will accept in a five-hour time period. The hypothesis that hard cations will exchange more effectively than softer ones was supported by these data. However, we also observed that the absolute activity of the proton-activated material was not great.

Because of this increased affinity of the proton-included materials for soft heavy metal ions, we were interested in developing new methods of directly preparing activated materials that contained other "hard" cations, such as Na^+ and Li^+. One approach has been the synthesis of cation-deficient spinels that

TABLE 3 Extraction Capabilities of $Cu_2MSn_3S_8$ for Hg^{2+}

Extractant	K_d(mL/g)	
	H_2O	0.1 M HNO_3
$Cu_2MnSn_3S_8$	470	170
$Cu_2FeSn_3S_8$	350	0.0
$Cu_2CoSn_3S_8$	330	6.0
$Cu_2NiSn_3S_8$	62	0.0

TABLE 4 K_d Comparison of $Cu_2FeSn_3S_8$ with its
Proton-Intercalated Analog, $H_xCu_yFe_zSn_3S_8$

Metal ion	K_d(mL/g)	
	$Cu_2FeSn_3S_8$	$H_xCu_yFe_zSn_3S_8$
Hg^{2+}	350	3800
Pb^{2+}	1.0	130
Ba^{2+}	4.0	140
Mg^{2+}	1.0	0.0

could be reductively intercalated with a hard metal cation using a reducing agent containing either Na^+ or Li^+. Recently, Garg and co-workers published the synthesis of a cation-deficient thiospinel with the formula $Cu_{5.47}Fe_{2.9}Sn_{13.1}S_{32}$ [28]. In this report, single crystals were grown from a solid-state reaction containing a 1:1:3:4 molar ratio of Cu:Fe:Sn:S. In addition, a second system of cation-deficient spinels incorporating silicon as a higher-valence substituent cation has been synthesized [29]. These single crystals were found to have the defect formula $Cu_{5.52}Si_{1.04}\square_{1.44}Fe_4Sn_{12}S_{32}$ and were the product of the reaction in which the reactants Cu:Si:Fe:Sn:S were loaded in a 3.31:1:4:12:32 molar ratio.

In our laboratory, we have tried this direct synthetic approach to prepare new materials with the defect stoichiometries: $CuMnSn_3S_8$, $Cu_2Sn_3S_8$, and $CuFeSn_3S_8$. In each case, the correct stoichiometric ratio of the elements was loaded into a fused-silica ampoule and heated to 300°C for 24 hours and then to 700°C for 196 hours. The contents of the tube were then reground in an inert-atmosphere glove box and loaded into a second ampoule that was again heated to 700°C for 50 hours. The product of each reaction was a fine, black powder. Powder diffraction data revealed that these materials crystallize with the spinel structure. In the case of $CuMnSn_3S_8$, there was a significant decrease in the (220) and (422) reflections and an increase in the relative intensity of the (400) reflection. The intensities of the (220) and (422) reflections are related to the organization of copper atoms in the structure, and the decrease in intensity indicates that there are significant defects or vacancies in the copper positions. In addition, we observed that the relative intensities of the (311) and (400) reflections are reversed, compared to $Cu_2MSn_3S_8$, indicating defects or vacancies in the copper, $8a$ sites.

To activate our new solids for target ion absorption, these as-prepared defect solids were reduced using a basic solution of $Na_2S_2O_4$. A comparison of the powder diffraction data of $CuMnSn_3S_8$ and the product of the reduction reaction using $Na_2S_2O_4$ can be seen in Figure 6. As the result of the reduction and incorporation of Na^+ into the structure, the (311) and (400) reflections reversed in their

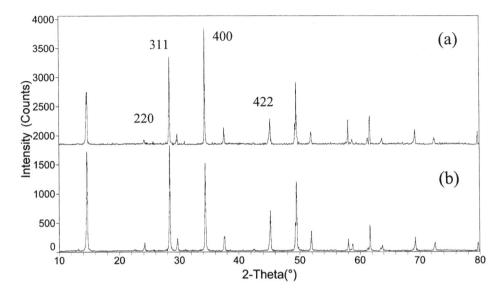

FIG. 6 Powder X-ray diffraction analysis of (a) $CuMnSn_3S_8$ and (b) $CuMnSn_3S_8$ after reduction with $Na_2S_2O_4$.

relative intensities with respect to the defect solid, and the (220) reflection increased in intensity and sharpness. Finally, there is evidence of a small (422) reflection, indicating an organization of ions in the tetrahedral interstices that were previously vacant.

A second approach we have taken has been the direct synthesis of activated materials with the stoichiometries $Na_2MSn_3S_8$ and $NaCuMSn_3S_8$. The powder diffraction pattern of a typical material resulting from the solid-state reaction of Na_2S_2, Fe, Sn, and S in a manner similar to that described earlier can be seen in Figure 7. The data suggest that the reaction to produce $NaCuMSn_3S_8$ yielded a mixture of phases, one of which has the spinel structure, matching $Cu_2FeSn_3S_8$. A second phase, which has not yet been identified, appears to be the main product of the reaction to prepare $Na_2FeSn_3S_8$. Moreover, this mixture of phases is an extremely effective extractant for mercury. The K_d values for Hg^{2+}, Pb^{2+}, and Ag^+ with "$Na_2FeSn_3S_8$" and "$NaCuFeSn_3S_8$" [30] in water are presented in Table 5. These solids show an increased affinity for the soft heavy metal ions Hg^{2+}, Pb^{2+}, and Ag^+ over that of the parent structure, $Cu_2FeSn_3S_8$. It is our belief that this is due to the presence of a hard cation, Na^+, which produces a driving force for the exchange of the soft metal cation out of solution.

In addition to the extraction abilities of these materials, we have investigated the recovery of mercury from the guest–host solids, in order to demonstrate their

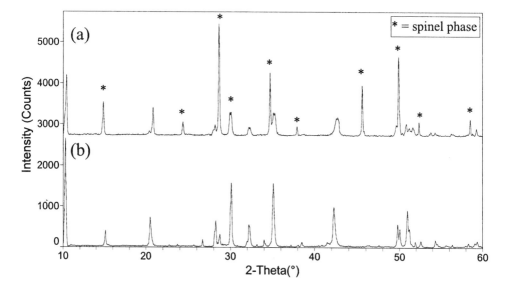

FIG. 7   PXRD analysis of (a) $NaCuFeSn_3S_8$ and (b) $Na_2FeSn_3S_8$.

benefits as ion exchange extractants for soft heavy metal ions. Over the range of thermal treatment (25–500°C), the pristine thiospinel materials display no significant thermal events. Heat treatment of the mercury-included solids, however, showed a mass loss beginning at 175°C and continuing until a 50% mass loss was observed by 360°C, Figure 8. The corresponding mass spectral analysis of the off-gases from the thermal gravimetric analysis confirmed the loss of Hg°, and powder X-ray diffraction analysis demonstrated that there was no loss in the crystallinity of the solids. X-ray photoelectron spectroscopy before and after heat treatment, shown in Figure 9, also confirmed the loss of mercury in this system.

TABLE 5   Extraction Capabilities of Thiospinels
$Na_2FeSn_3S_8$ and $CuNaFeSn_3S_8$

	$K_d(mL/g)$	
Metal ion	$Na_2FeSn_3S_8$	$CuNaFeSn_3S_8$
Hg^{2+}	8000	8700
Pb^{2+}	25000	25000
Ag^+	17000	6500

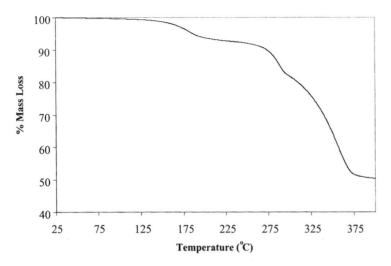

FIG. 8 Thermal gravimetric analysis of Hg-loaded $Cu_2MnSn_3S_8$.

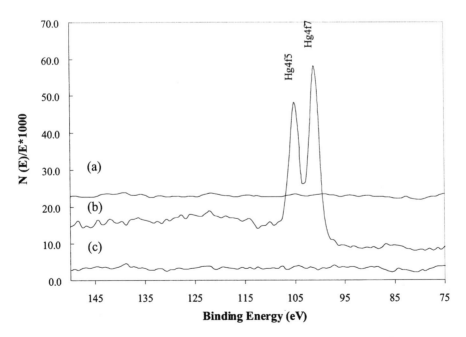

FIG. 9 XPS spectra of Hg-loaded $Cu_2FeSn_3S_8$ (a) before contact with Hg^{2+} solution, (b) after contact with Hg^{2+} solution, and (c) after heating Hg-loaded material to 500°C.

IV. CONCLUSIONS

Our laboratory has been engaged in developing materials that are now known to effectively and selectively remove soft heavy metal ions from solutions of varying pH. Earlier studies demonstrated effective ion exchange by $H_2Zr(PO_3S)_2$; however, this was accompanied by inefficient recovery of the extractant. Activated MoS_2, $Li_{1.3}MoS_2$, demonstrated effective ion removal as well as recyclability. Our new ion exchange materials, based on the thiospinel structure, were shown to be effective only when defect-laden solids could be prepared by ion removal or by direct synthesis. These materials are the basis for a new class of R^2ER solids.

ACKNOWLEDGMENT

This research was supported by the U.S. Department of Energy Environmental Management Science Program (DE-FG07-96ER14696 to P.K.D. and S.H.S). S.E.F. was an NSF-REU participant during the summer of 2001.

REFERENCES AND NOTE

1. AE Gash, AL Spain, LM Dysleski, CJ Flashenriem, A Kalaveshi, PK Dorhout, SH Strauss. Environ. Sci. Tech. 32:1007–1012, 1998.
2. PK Dorhout, SH Strauss. In: CE Winter, DM Hoffman, eds. ACS Symposium Series: Inorganic Materials Synthesis New Directions for Advanced Materials. Washington, D.C.: American Chemical Society, 1999, pp 53–68.
3. SH Strauss. In: AH Bond, ML Dietz, RD Rodgers, eds. ACS Symposium Series: Progress in Metal Ion Separation and Preconcentration. Washington, D.C.: American Chemical Society, 1999, pp 156–173.
4. AE Gash, PK Dorhout, SH Strauss. Inorg. Chem. 39:5538–5546, 2000.
5. PG Allen, AE Gash, PK Dorhout, SH Strauss. Chem. Mater. 13:2257–2265, 2001.
6. RG Pearson. In: A Scott, ed. Survey of Progress in Chemistry. New York: Academic Press, 1969, Chap 1.
7. A Clearfield, JA Stynes. J. Inorg. Nucl. Chem. 26:117–129, 1964.
8. A Clearfield, RH Blessing, JA Stynes. J. Inorg. Nuc. Chem. 30:2249–2258, 1968.
9. A Clearfield, WL Duax, AS Medina, GD Smith, JR Thomas. J. Phys. Chem. 73: 3424–3430, 1969.
10. A Clearfield, AS Medina. J. Inorg. Nucl. Chem. 32:2775–2780, 1970.
11. A Clearfield. Chem. Rev. 88:125–148, 1988.
12. MB Dines. Mater. Res. Bull. 10:287–292, 1975.
13. MA Py, RR Haering. Can. J. Phys. 61:76–84, 1983.
14. P Joensen, RF Frindt, SR Morrison. Mat. Res. Bull. 21:457–461, 1986.
15. P Lavela, JL Tirado, J Morales, J Olivier-Fourcade, J-C Jumas. J. Mater. Chem. 6: 41–47, 1996.
16. M Eisenberg. Proc. Power Sources Symp. 28:155–157, 1978.

17. M Eisenberg. J. Electrochem. Soc. 127:2382–2383, 1980.
18. S Sinha, DW Murphy. Solid State Ionics 20:81–84, 1986.
19. ACWP James, JB Goodenough, NJ Clayden. J. Solid State Chem. 77:356–365, 1988.
20. R Schöllhorn, A Payer. Angew. Chem. Int. Ed. Engl. 24:67–68, 1985.
21. ACWP James, JB Goodenough. J. Power Sources 26:277–283, 1989.
22. N Imanishi, K Inoue, Y Takeda, O Yamamoto. J. Power Sources 43–44:619–625, 1993.
23. ML Elidrissi Moubtassim, C Bousquet, J Olivier-Fourcade, JC Jumas. Chem. Mater. 10:968–973, 1998.
24. C Branci, J Sarradin, J Olivier-Fourcade, JC Jumas. Chem. Mater. 11:2846–2850, 1999.
25. R Dedryvère, J Olivier-Fourcade, JC Jumas, S Denis, C Pérez Vicente. Chem. Mater. 12:1439–1445, 2000.
26. R Dedryvère, J Olivier-Fourcade, JC Jumas, S Denis, P Lavela, P, JL Tirado. Electrochim Acta 46:127–135, 2000.
27. ICDD, file #29-0558.
28. G Garg, S Bobev, A Roy, J Ghose, D Das, AK Ganguli. Mat. Res. Bull. 36:2429–2435, 2001.
29. G Garg, S Bobev, A Roy, J Ghose, D Das, AK Ganguli. J. Solid State Chem. 161:327–331, 2001.
30. The stoichiometries of the two phases present in this material have not yet been determined. Therefore, $CuNaFeSn_3S_8$ and $Na_2FeSn_3S_8$ will be used as shorthand notation for the loaded stoichiometries representing the products of these two reactions.

8
Chemical Methods of Heavy Metal Binding

MATTHEW MATLOCK and DAVID ATWOOD   University of Kentucky, Lexington, Kentucky, U.S.A.

I. INTRODUCTION

Toxic heavy metals in air, soil, and water are problems that are constant and escalating threats to the environment in which we live. There are a multitude of sources of heavy metal pollution, most of which the general population cannot avoid [1,2]. In response to this continuing and growing threat, federal and state governments have instituted environmental regulations to protect the quality of surface water and groundwater from heavy metal pollutants [3].

To meet the federal and state guidelines for heavy metal discharge, companies commonly use methods to precipitate or chemically bind the metals before, during, or after discharge. In order to be competitive economically, many of the commercially utilized reagents are not designed to target specific metals. They are either simple to synthesize or easily obtained as byproducts of other chemical reactions.

These readily available reagents offer minimal binding sites for heavy metals. It has been found, in fact, that their mode of action is simply to act as sources of sulfide for the formation of metal sulfides. This is not an acceptable process for metal removal and deactivation. Metal sulfides are known to decompose and release heavy metals back into the environment over varying, but usually short, periods of time [4–6]. Additionally, in forming the metal sulfides, some of the reagents produce toxic byproducts, such as carbon disulfide [4,7].

In an effort to provide an environmentally sound method for chemically removing heavy metals from the environment, new, specifically designed chelates for soft heavy metals have been reported. These ligands are not only highly effective for soft metal binding, but are also economical and soon to be produced on a multiton scale.

The present overview will summarize the various chemical methods that are currently being used to chemically bind heavy metals and precipitate them from aqueous sources. Other technologies are available for heavy metal removal, such as micelle-enhanced filtration [8] and ion exchange [9] techniques, but these will not be covered here.

II. COMMERCIAL HEAVY METAL REAGENTS

A. Thiocarbonates

One class of compounds that is commonly used for the precipitation of heavy metals is the thiocarbonates. Thiocarbonates are presumed to bind heavy metals as insoluble metal thiocarbonates (that is, $CuCS_3$, $HgCS_3$, $PbCS_3$, and $ZnCS_3$). One chemical reagent able to potentially effect this precipitation is Thio-Red® ([Na, K]$_2$CS$_3$ · nH$_2$O, where $n \geq 0$) (Figs. 1a and 1b) [10]. The precipitate, however, with copper, mercury, lead, and cadmium, taken from aqueous solutions, are metal sulfides (that is, CuS, HgS, PbS, and ZnS) [4,7,10]. A byproduct of the metal sulfide precipitation with this reagent is carbon disulfide, a volatile and toxic liquid [4,7].

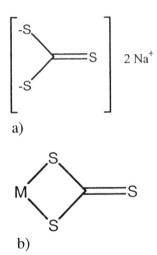

a)

b)

FIG. 1 (a) Chemical structure of Thio-Red® (a potassium or sodium thiocarbonate). (b) Predicted (and claimed) binding motif between Thio-Red® and a divalent heavy metal (M).

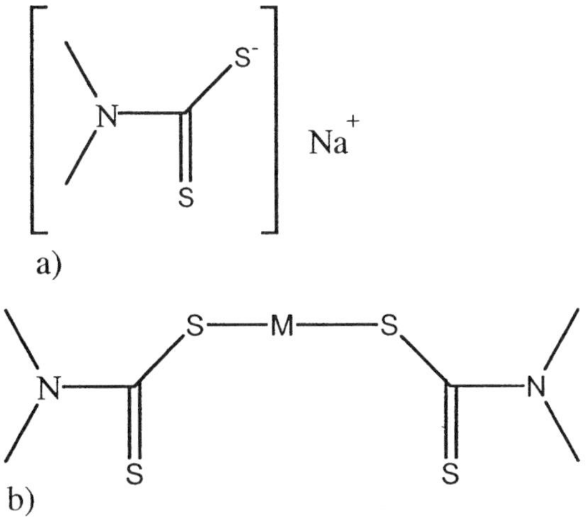

a)

b)

FIG. 2 (a) Chemical structure of HMP-2000 (a sodium dimehtyldithiocarbamate). (b) Presumed binding motif between HMP-2000 and a divalent heavy metal (M).

B. Thiocarbamates

A second class of chelating compounds is the thiocarbamates. Typically, thiocarbamates act as bidentate ligands. One example of a commercially available thiocarbamate is HMP-2000 (Figs. 2a and 2b). Several chemical distribution companies, including Ulrich Chemical Inc., distribute this sodium thiocarbamate. A serious downfall of this compound is its tendency to decompose into toxic secondary products. One major accident associated with the use of HMP-2000 occurred in December 1999, when the Guide Corporation (an auto parts manufacturing plant in Anderson, Indiana) accidentally released over 1.5 million gallons of contaminated wastewater laced with HMP-2000 into the city's wastewater system. Unable to control the chemical contaminate, the HMP-2000-laced wastewater was discharged into local state waters [11]. The compound reportedly decomposed into toxic compounds (tetramethylthiuram and thiram), which ultimately resulted in the deaths of 117 tons of fish over a 50-mile stretch from Anderson to Indianapolis, Indiana, as reported by the Indiana Department of Environmental Management [11].

C. Potentially Multidenta Ligands

A final class of ligands used for heavy metal chelation is potentially multidenta, and bridge to form polymeric compounds. One example of this type of compound

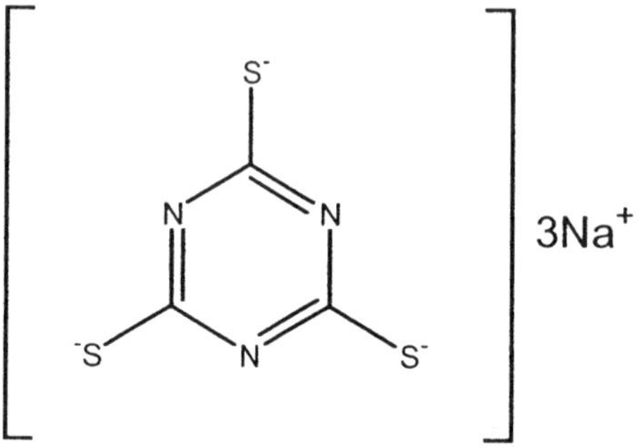

FIG. 3 (a) Chemical structure of TMT-55 (2,4,6-trimercaptotiazine, trisodium salt nonahydrate). (b) Binding motif between TMT-55 and a divalent heavy metal (M).

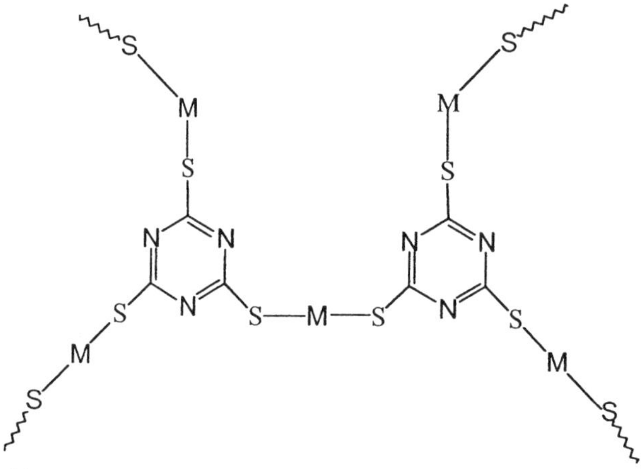

is TMT, or 2,4,6-trimercaptotiazine, trisodium salt nonahydrate, $Na_3C_3S_3 \cdot 9H_2O$ (Figs. 3a and 3b). TMT is a byproduct of chemicals manufactured and distributed by Degussa Corporation USA of Allendale and Ridgefield Park, New Jersey [12]. Despite the widespread use of TMT, only limited information has been available on how the product reacts with heavy metals in aqueous solutions and the chemistry and stability of the resulting heavy metal–TMT precipitates. Only recently has quantitative and systemic information become available on the chemistry of TMT with metals. This included dissociation constants and the controlled formation of metal complexes [5,6,13–15]. While this work demonstrates some limitations to the use of TMT as a remediation agent, one of the publications found that the reagent could be used as a sacrificial sulfide source for the formation of transition metal sulfide solid-state materials.

D. Dosage Formulas

Manufacturers of commercial compounds supply dosage formulas that allow users to quickly calculate the amount of compound needed to treat varying volumes of contaminated water. Often these dosage rates are inaccurate and generally lead to an underdosing and consequent failure to meet the permitted discharge limits. As an example, Thio-Red® is distributed as an aqueous solution and utilizes the following formula for calculating the Thio-Red® dosage for contaminated waters (Eq. 1):

$$D = pVY \tag{1}$$

where

D = dosage of Thio-Red® in mL
p = part per million (ppm) concentration of the metal contaminant
V = volume of water to be treated in gallons
Y = constant (Cd = 0.0362; Cu = 0.0641; Pb = 0.0196; Zn = 0.0622)

Previously, published reports on the Thio-Red® dosage formula have indicated serious problems that have been identified through the use of titration analyses followed by inductively coupled plasma spectroscopy (ICP-OES) or cold vapor atomic fluorescence (CVAF) spectroscopy [4,7]. In the experiments, titration experiments and computer modeling with MINTEQA2 indicate that the pH decreases as Thio-Red® is initially titrated into aqueous divalent heavy metal solutions. The pH decline associated with the precipitation of the divalent metals (M^{2+}) is explained by the following reaction involving HS^- from the Thio-Red® [4,7].

$$HS^- + M^{2+} H_2O \rightarrow MS\downarrow + H_3O^+ \tag{2}$$

Once the metals are precipitated, Reaction (2) is no longer dominant and an important source of H_3O^+ (acid) is eliminated. Without the presence of significant concentrations of dissolved metals, further addition of pH 12 Thio-Red® results in an increase in pH through the alkalinity of the Thio-Red® and the following sulfide reactions:

$$S^{2-} + H_3O^+ \rightarrow HS^- + H_2O \tag{3}$$
$$HS^- + H_3O^+ \rightarrow H_2S^0 + H_2O \tag{4}$$

The volatilization of some of the H_2S^0 from the solution would result in a further pH increase by shifting the equilibrium to the right in Reaction (4). According to Reactions (2)–(4), then, the pH minimum should coincide with the optimal Thio-Red® dosage for metal removal.

E. Effectiveness of the Commercial Compounds on Metal Stock Solutions

In an effort to compare the effectiveness of the commercial metal-binding agents, the manufacturers' dosage formulas were ignored and reactions were carried out using a stoichiometric molar ratio (and also a 10% increase in the stoichiometric molar dosage) between each commercial compound and solutions of mercury, lead, cadmium, copper, and iron(II). Tables 1–3 outline the effectiveness of the commercial compounds under pH conditions of 4.0 and 6.0. Results show that at stoichiometric doses, Thio-Red®, HMP-2000, or TMT were unable to effectively remove the cadmium, lead, copper, or iron from solutions of 50.00 ppm to meet the EPA discharge limits [3]. Even with a 10.00% molar excess, no additional significant removal was observed. HMP-2000 displayed a higher affinity for cadmium, lead, copper, and iron than the Thio-Red®, but still EPA discharge limits were not achieved. TMT displays similar results as compared with the HMP-2000, with the highest removal seen for lead and copper. Once again, it is seen that even at a 10% molar increase in dosage, TMT were unable to reduce lead or cadmium concentrations to meet EPA standards [3]. For the mercury analyses, it was found that at stoichiometric and a 10% molar dose increase (in each commercially tested compound), mercury concentrations from the 50.00-ppm stock solutions were not reduced to meet the EPA limit of 0.2 ppm [3]. Maximum results for mercury removal with Thio-Red® were seen at 20 hours at a 10% molar dose increase, with a final average value of 3.97 ppm. At one hour the results of HMP-2000, at a 10% molar dose increase, indicated a reasonably high removal of mercury, with a final concentration of 0.69 ppm. Within 20 hours at stoichiometric doses, TMT was able to reduce the 50.00-ppm mercury concentrations to an average final concentration of 9.82 ppm.

For the metal concentrations that increased over time, the formation of more soluble metal–ligand complexes or the substantial decomposition of the metal–ligand complexes may explain the increase in metal concentration. Additionally, an increase in metal concentration may also contribute to the possible high leaching rates of the metal out of the ligand complexes (or, more accurately, leached out of the metal sulfides produced by the ligand complexes) [5,6].

III. RESEARCH ADVANCES ON NOVEL LIGANDS FOR HEAVY METAL CHELATION

Immediate concerns with the commercially available ligands focus on the weak potentially multidentate binding abilities for heavy metals. Ligands with alkyl-thio

TABLE 1 ICP and CVAF Results of Thio-Red® at Stoichiometric Doses and at 10% Molar Dosage Increases

Chelating agent	Metal	Dose	Time (hours)	Solution pH	Initial metal concentration (ppm)	Final metal concentration (ppm)	EPA discharge limit (ppm) [3]
Thio-Red®							
	Pb	Stoichiometric	1	6.0	50.00	38.24	5.0
	Pb	Stoichiometric	6	6.0	50.00	44.83	5.0
	Pb	Stoichiometric	20	6.0	50.00	48.17	5.0
	Pb	10% dose increase	1	5.5	50.00	33.67	5.0
	Pb	10% dose increase	6	5.5	50.00	41.55	5.0
	Pb	10% dose increase	20	5.5	50.00	47.99	5.0
	Cu	Stoichiometric	1	5.0	50.00	27.77	
	Cu	Stoichiometric	6	5.0	50.00	28.99	
	Cu	Stoichiometric	20	5.0	50.00	28.86	
	Cu	10% dose increase	1	4.5	50.00	27.08	
	Cu	10% dose increase	6	4.5	50.00	25.77	
	Cu	10% dose increase	20	4.5	50.00	26.79	
	Cd	Stoichiometric	1	5.5	50.00	34.38	1.0
	Cd	Stoichiometric	6	5.5	50.00	39.53	1.0
	Cd	Stoichiometric	20	5.5	50.00	47.07	1.0
	Cd	10% dose increase	1	5.0	50.00	27.09	1.0
	Cd	10% dose increase	6	5.0	50.00	34.87	1.0
	Cd	10% dose increase	20	5.0	50.00	41.50	1.0
	Fe(II)	Stoichiometric	1	6.0	50.00	35.15	2.0
	Fe(II)	Stoichiometric	6	6.0	50.00	34.38	2.0
	Fe(II)	Stoichiometric	20	6.0	50.00	32.98	2.0
	Fe(II)	10% dose increase	1	5.0	50.00	34.79	2.0
	Fe(II)	10% dose increase	6	5.0	50.00	34.56	2.0
	Fe(II)	10% dose increase	20	5.0	50.00	33.56	2.0
	Hg	Stoichiometric	1	6.0	50.00	8.59	0.2
	Hg	Stoichiometric	6	6.0	50.00	8.07	0.2
	Hg	Stoichiometric	20	6.0	50.00	6.85	0.2
	Hg	10% dose increase	1	6.0	50.00	6.72	0.2
	Hg	10% dose increase	6	6.0	50.00	5.20	0.2
	Hg	10% dose increase	20	6.0	50.00	3.97	0.2

TABLE 2 ICP and CVAF Results of HMP-2000 at Stoichiometric Doses and at 10% Molar Dosage Increases

Chelating agent	Metal	Dose	Time (hours)	Solution pH	Initial metal concentration (ppm)	Final metal concentration (ppm)	EPA discharge limit (ppm) [3]
HMP-2000	Pb	Stoichiometric	1	3.5	50.00	21.90	5.0
	Pb	Stoichiometric	6	3.5	50.00	21.89	5.0
	Pb	Stoichiometric	20	3.5	50.00	23.77	5.0
	Pb	10% dose increase	1	4.0	50.00	15.46	5.0
	Pb	10% dose increase	6	4.0	50.00	16.21	5.0
	Pb	10% dose increase	20	4.0	50.00	16.31	5.0
	Cu	Stoichiometric	1	3.5	50.00	12.55	
	Cu	Stoichiometric	6	3.5	50.00	12.58	
	Cu	Stoichiometric	20	3.5	50.00	12.46	
	Cu	10% dose increase	1	4.0	50.00	7.08	
	Cu	10% dose increase	6	4.0	50.00	7.10	
	Cu	10% dose increase	20	4.0	50.00	7.19	
	Cd	Stoichiometric	1	3.0	50.00	11.52	1.0
	Cd	Stoichiometric	6	3.0	50.00	11.51	1.0
	Cd	Stoichiometric	20	3.0	50.00	12.08	1.0
	Cd	10% dose increase	1	4.0	50.00	10.47	1.0
	Cd	10% dose increase	6	4.0	50.00	10.54	1.0
	Cd	10% dose increase	20	4.0	50.00	10.96	1.0
	Fe(II)	Stoichiometric	1	4.0	50.00	25.18	2.0
	Fe(II)	Stoichiometric	6	4.0	50.00	23.07	2.0
	Fe(II)	Stoichiometric	20	4.0	50.00	23.92	2.0
	Fe(II)	10% dose increase	1	4.5	50.00	24.28	2.0
	Fe(II)	10% dose increase	6	4.5	50.00	23.21	2.0
	Fe(II)	10% dose increase	20	4.5	50.00	23.94	2.0
	Hg	Stoichiometric	1	4.0	50.00	1.01	0.2
	Hg	Stoichiometric	6	4.0	50.00	1.50	0.2
	Hg	Stoichiometric	20	4.0	50.00	2.78	0.2
	Hg	10% dose increase	1	4.0	50.00	0.69	0.2
	Hg	10% dose increase	6	4.0	50.00	1.24	0.2
	Hg	10% dose increase	20	4.0	50.00	1.63	0.2

TABLE 3 ICP and CVAF Results of TMT at Stoichiometric Doses and at 10% Molar Dosage Increases

Chelating agent	Metal	Dose	Time (hours)	Solution pH	Initial metal concentration (ppm)	Final metal concentration (ppm)	EPA discharge limit (ppm) [3]
TMT	Pb	Stoichiometric	1	5.0	50.00	18.21	5.0
	Pb	Stoichiometric	6	5.0	50.00	18.50	5.0
	Pb	Stoichiometric	20	5.0	50.00	21.05	5.0
	Pb	10% dose increase	1	5.5	50.00	16.06	5.0
	Pb	10% dose increase	6	5.5	50.00	16.58	5.0
	Pb	10% dose increase	20	5.5	50.00	17.31	5.0
	Cu	Stoichiometric	1	5.0	50.00	16.18	
	Cu	Stoichiometric	6	5.0	50.00	13.30	
	Cu	Stoichiometric	20	5.0	50.00	10.13	
	Cu	10% dose increase	1	5.5	50.00	16.19	
	Cu	10% dose increase	6	5.5	50.00	14.21	
	Cu	10% dose increase	20	5.5	50.00	12.59	
	Cd	Stoichiometric	1	5.0	50.00	37.14	1.0
	Cd	Stoichiometric	6	5.0	50.00	36.12	1.0
	Cd	Stoichiometric	20	5.0	50.00	38.22	1.0
	Cd	10% dose increase	1	5.5	50.00	21.04	1.0
	Cd	10% dose increase	6	5.5	50.00	21.04	1.0
	Cd	10% dose increase	20	5.5	50.00	21.62	1.0
	Fe(II)	Stoichiometric	1	5.0	50.00	25.04	2.0
	Fe(II)	Stoichiometric	6	5.0	50.00	25.46	2.0
	Fe(II)	Stoichiometric	20	5.0	50.00	25.27	2.0
	Fe(II)	10% dose increase	1	5.5	50.00	23.64	2.0
	Fe(II)	10% dose increase	6	5.5	50.00	22.38	2.0
	Fe(II)	10% dose increase	20	5.5	50.00	21.77	2.0
	Hg	Stoichiometric	1	5.5	50.00	18.07	0.2
	Hg	Stoichiometric	6	5.5	50.00	13.39	0.2
	Hg	Stoichiometric	20	5.5	50.00	9.82	0.2
	Hg	10% dose increase	1	5.5	50.00	15.15	0.2
	Hg	10% dose increase	6	5.5	50.00	16.90	0.2
	Hg	10% dose increase	20	5.5	50.00	10.50	0.2

chains that lack either chain length or sufficient bonding sites may produce precipitates that are unstable over time and under certain pH conditions. For this
reason our research has been focusing on the design and synthesis of ligands
with extended binding sites that not only bind heavy metals, but also produce
stable precipitates. In order to create more effective and economical ligands, we
have developed a series of new synthetic ligands which utilize biological heavy
metal binding motifs. As an example, research has led to the design and patenting
of compounds such as 2,6-pyridinediamdoethanethiol (PyDETH$_2$) [16]. This
compound utilizes two extended alkyl chains at the 2,6-position to irreversibly
bind metals. Computer modeling suggests that the designed ligands have a sufficient chain length for interactions between the metal and each terminal sulfur
group (Fig. 4).

Early results have shown that heavy metal concentrations from aqueous solutions can be reduced well below EPA discharge limits, and the resulting precipi-

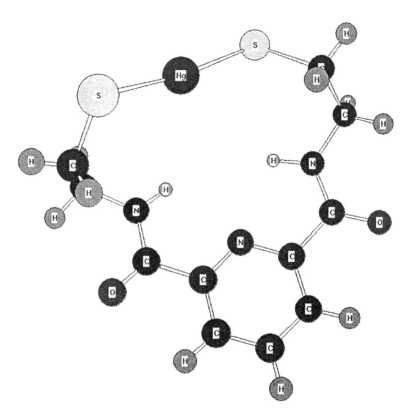

FIG. 4   Chemical structure of Hg-PyDET.

tates have shown no solubility in common organic solvents or aqueous systems over a pH range of 0.0–14.0 [17–19]. For example, the synthetic ligands have been able to reduce mercury concentrations from 50.00 ppm to 0.094 ppm and lead concentrations from 50.00 ppm to 0.050 ppm [17,18]. The ligands have also been effective in immobilizing mercury from contaminated soils. In recent tests with soils contaminated with 10,270 ppm of elemental mercury, the multidentate ligand was capable of immobilizing 99.6% of the mercury [19]. The ligands has also proven to be highly effective in the gold heap leachate solutions by selectively reducing mercury levels from an initial concentration of 34.5 ppm to 0.014 ppm within 10 minutes and to 0.008 pm within 15 minutes [20]. In addition to selectively removing mercury, the Hg-ligand compounds are highly stable even in the adverse pH conditions used during the NaCN leaching process.

IV. CONCLUSION

There is a definite need for new and more effective reagents to meet growing environmental problems. Many reagents on the market today either lack the necessary binding criteria or pose too many environmental risks to be effectively utilized. For this reason, ligands utilizing multiple binding sites for heavy metals and mimicking biological systems for metal binding look to be a possible answer to heavy metal remediation and wastewater treatment.

REFERENCES

1. BJ Alloway, ed. Heavy Metals in Soils. 2nd ed. Chapman and Hall, Glasgow, UK, 1995, Chaps 6, 8, 9, 11.
2. KR Henke, V Kühnel, DJ Stephan, RH Fraley, CM Robinson, DS Charlton, HM Gust, NS Bloom. Gas Res Inst, Chicago, IL, 1993.
3. Code of Federal Regulations (CFR). 40, 141, 261, 268.40, U.S. Government Printing Office, Superintendent of Documents, Washington, DC, 1994.
4. KR Henke. Wat Env Res 70(6):1178–1185, 1998.
5. KR Henke, D Robertson, MK Krepps, DA Atwood. Wat Res 34(11):3005–3013, 2000.
6. MM Matlock, KR Henke, DA Atwood, JD Robertson. Wat Res 35(15):3649–3655, 2001.
7. KR Henkee. Chemistry and Environmental Implications of Thio-Red® and 2,4,6-Trimercaptotriazine Compounds. PhD dissertation, University of North Dakota, Grand Forks, ND, 1984.
8. UR Dharmawardana, SD Christian, RW Taylor, JF Scamehorn. Langmuir 8:414, 1992; M Tuncay, SD Christian, EE Tucker, RW Taylor, JF Scamehorn. Langmuir 10:4688, 1994.
9. AE Gash, AL Spain, LM Dysleski, CJ Flaschenriem, A Kalaveshi, PK Dorhout,

SH Strauss. Env Sci Technol 32:1007–1012, 1998; AE Gash, PK Dorhout, SH Strauss. Chem Mater 13:2257–2265, 2001.

10. ETUS, Inc. Product Information on Thio-Red®, Sanford, FL, 1994.
11. State of Indiana's Data Facts Sheet. http://www.state.in.us/idem/macs/factsheets/whiteriver/185.
12. Degussa Corporation. Data Sheets on TMT-15 and TMT-55, Ridgefield Park, NJ, 1993.
13. MK Krepps, S Parkin, DA Atwood. Crystal Growth Design 1(4):291–297, 2001.
14. KR Henke, AR Hutchison, MK Krepps, DA Atwood. Inorg Chem 40(17):4443–4447, 2001.
15. RJ Bailey, MJ Hatfield, KR Henke, MK Krepps, J Morris, T Otieno, K Simonetti, EA Wall, DA Atwood. J Org Met Chem 623:185–190, 2001.
16. MM Matlock, BS Howerton, DA Atwood. Novel Multidentate Sulfur-Containing Ligands. University of Kentucky, 2000. U.S. (patent pending).
17. MM Matlock, BS Howerton, KR Henkee, DA Atwood. J Haz Mat 82(1):55–63, 2001.
18. MM Matlock, BS Howerton, DA Atwood. J Haz Mat B84:73–82, 2001.
19. MM Matlock, BS Howerton, DA Atwood. Adv Env Res (in press).
20. MM Matlock, BS Howerton, MA Van Aelstyn, FL Nordstrom. Env Sci Tech 36(7):1636–1639, 2002.

9
Interaction of Oil Residues in Patagonian Soil

NORMA S. NUDELMAN   University of Buenos Aires, Buenos Aires, Argentina

STELLA MARIS RÍOS   National University of Patagonia, Comodoro Rivadavia, Argentina

I. SORPTION BEHAVIOR

The sorption of hydrophobic compounds to natural solids is the dominant factor controlling their transport, biodegradation, and toxicity. The study of sorptive interactions between compounds is essential, given the prevalence of sites in the environment where multiple contaminants coexist [1]. The development of appropriate equilibrium sorption relationships for anthropogenic organic contaminants with soils and sediments is important to predict the extent of solid–water interactions in the environment [2].

In dry, low-organic-matter soils, such as Patagonian soil, sorption of nonpolar organics would likely be dominated by adsorption onto mineral surfaces, particularly clays. Since it is almost impossible to carry out sorption experiments for each field condition, the development of laboratory methodologies that gather information on this subject is essential [3–5].

The behavior of sorption of oil in environments affected by oil exploitation is complex and difficult to predict with the current state of knowledge. The quantification of this phenomenon could, in principle, be aided by applying some well-known models from physical chemistry. Although they cannot be directly extrapolated to complex systems, they do constitute an approach, however approximate, to the quantitative explanation of the problem [3].

As an example, the dual-mode (partition/hole-filling) model of soil organic matter (SOM) as a heterogeneous polymer-like sorbent of hydrophobic compounds predicts that a competing solute will accelerate diffusion of the primary solute by blocking the holes, allowing the principal solute to move faster through the SOM matrix. Thus, pyrene suppressed phenanthrene sorption and increased

139

the linearity of its isotherm [1,4]. In this context, results were reported that showed how nonlinear sorption isotherms with low-polarity organic chemicals could be modeled as a combined adsorption-partitioning process. In this case, the results confirmed the expectation that partitioning is an increasingly dominating contribution to overall sorption when cosorbates are present [2].

Petroleum, or crude oil, is a naturally occurring liquid consisting predominantly and essentially of hydrocarbon compounds, with widely varying proportions of each compound. Some of the hydrocarbons are gases and some are solids; both types are in solution in liquid hydrocarbons, which predominate. Because crude oil is a mixture, it has no definite chemical composition, nor does it have fixed physical properties; and the number of all of the individual hydrocarbon compounds that may occur in different crude oils is not yet known. It is probable that more than 600 individual compounds exist.

In this work, hydrocarbon sorption behavior in soil was determined as a contribution to the modeling, and the results were compared with artificial samples treated in the same manner. The sorption term is assumed to include both absorption and adsorption phenomena, and partitioning refers to a distribution between both phases more than to a specific absorption into the organic matter, which is indeed very low [3].

The main properties of the soils are summarized in Table 1. There are four major fractions of crude oil that are important with reference to sorption behavior:

TABLE 1   Physical and Chemical Characteristics of Soil Samples

Sand (quartz, litics, feldspars, and gypsum of eolian origin from clayed sandstones of Fm. Patagonia)	
pH[a]	7.4
Conductivity,[a] μS cm^{-1}	600
Water retention capacity, wt% (dry)[d]	43
Na$^+$,[a] meq L^{-1}	3.25
K$^+$,[a] meq L^{-1}	0.11
Ca^{+2} & Mg^{+2},[a] meq L^{-1}	<0.01
Clay (montmorillonite and illite), including silt	
Organic matter, wt% (dry)	0.02
Fe^{+3},[b] g kg^{-1}	2.5
Fe^{+2},[b] g kg^{-1}	0.4
Montmorillonite total surface,[c] cm^2 g^{-1}	600–800
Illite total surface,[c] cm^2 g^{-1}	65–100

[a] Extract 1:1 wt/wt.
[b] In clay, extract 1:200 wt/wt.
[c] *Source*: Ref. 6.
[d] *Source*: Ref. 7.

the aliphatic, aromatic, polar, and asphaltic fractions. These fractions are obtained by column chromatography of the crude oil. The aliphatic fraction contains n-alkanes, branched alkanes, cycloalkanes, isoprenoids, etc. The aromatic fraction contains monocyclic and polycyclic aromatic hydrocarbons. The polar fraction contains compounds such as thiophenes, cycloalkanecarboxylic acids, alkylpyridines, and porphyrins. And the asphaltenes are polymeric structures. The group percentages of the crude oil in this work were: aliphatic (Aliph, 41%); aromatic (Aro, 35%), polar (Pol, 17%), and asphaltenes (Asph, 7%) wt/wt.

Five samples were prepared from dry soil with different amounts of clay and moisture content, as shown in Table 2. A simulated mixture was prepared using 11 pure compounds. Table 3 shows the composition of the mixtures; the so-called "artificial sample" was designed to resemble the % fractions.

Oil uptake as a function of time was found to be bimodal: an instantaneous initial sorption, for contact times less than 1 minute, and after this time a sorption that may be represented by Eq. (1), where C_o is the initial concentration and C_t is the concentration remaining in solution at contact time t:

$$\frac{C_t}{C_o} = 1 - k_0 t \tag{1}$$

The apparent oil rate constant, k_0, for samples I–III, was obtained from Eq. (1) as the best-fit parameter by linear regression, and the results are shown in Table 4. The values of the apparent oil rate constant, k_0, gathered in Table 4 for samples I–III, show an important dependence of rate on soil moisture content.

The results in Table 4 show that the sorption rate is strongly influenced by soil water content: Dry soil favored the crude oil uptake rate, probably due to the fact that the % nonpolar components amounts, at least, to 76%, while the polar fraction is 17%. It is known that water favors sorption of polar components by H-bonding with the polar functionalities in the oil. The results in Table 4 show similar k_0 values for sample III and the artificial sample of comparable moisture and clay content, thereby giving confidence in the general treatment.

TABLE 2 Soil Sample Composition and Oil Solution Concentration Range

	Sample				
	I	II	III	IV	V
Moisture, wt%	2	2	0	0–5	0
Clay, wt%	50	50	50	0–50	0–100
Oil concentration range, mg L^{-1}	5–20	10–130	1–20	20	37

TABLE 3 Artificial Sample, Composition

Type	Name	% Subgroup	Simulated mixt., wt%	Oil wt%
Aliphatic	Cyclohexane	55		
	Heptane	20		
	Octane	12		
	Pentadecane	13		
Aliphatic fraction			44	41
Aromatic	Benzene	31		
	Naphthalene	26		
	Anthracene	13		
	Toluene	11		
	Xylenes	19		
Aromatic fraction			33	35
Polar	Iso-octanol	80		
	Phenol	20		
Polar fraction			23	17
Rest				7

An important instantaneous sorption was observed before one minute of time (the first data were taken at $t = 1$ min). Figure 1 shows the data corresponding to samples I, II, III, and IV, where the sorption percentage was plotted as a function of the initial oil concentration. It can be observed that the instantaneous sorption was in the range 10–60 wt%, and a plateau (around 60%) is reached after 20 mgL^{-1} initial oil concentration, which could be interpreted as a limiting saturation in the instantaneous sorption. The data for sample I lie slightly below the three points observed for sample III, indicating a retarding effect on sorption due to the water content. The artificial samples show an instantaneous sorption of around 30%, for a concentration similar to sample II (crude oil) (last point in

TABLE 4 Kinetic Behavior

	Crude oil sample			Artificial sample
	I	II	III	
10^4 Apparent rate constant k_0, min^{-1}	6.36	0.493	43.8	42.0[a]
Correlation coefficient, r^2	0.939	0.971	0.913	0.925

[a] For a 50% clay content and no moisture.

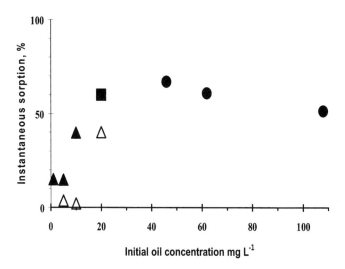

FIG. 1 Instantaneous sorption as function of initial oil concentration (mg/L). Sample I (open triangles), sample II (solid circles), sample III (solid triangles), and sample IV (solid squares).

Fig. 1). For sample II, the instantaneous sorption was around 50%; the difference could be related to the presence of the crude oil group called "the rest," which contains the most recalcitrant compounds. Due to the instantaneous sorption uptake, a single rate constant does not apply over the entire kinetic curve; this behavior has often been recognized, and most sorption kinetic models fit the data better by including an instantaneous nonkinetic fraction described by an equilibrium sorption constant.

The partition coefficient, K, for pure substances describes the distribution of chemical species between the solution and the solid; the expression for a linear sorption isotherm could be well represented by the partition coefficient. The linear and the Freundlich sorption isotherm models given by Eqs. (2) and (3), where q_e and C_e are the equilibrium solid-phase and solution-phase solute concentration, respectively, were tested.

$$q_e = K_d C_e \qquad (2)$$

$$q_e = K_F C_e^n \qquad (3)$$

In recent years several studies have reported linear sorption uptake isotherms for many compounds, which were interpreted as an indication that organic matter provides a partioning medium for organic solutes. In the present study, the linear and Freundlich sorption isotherm models were tested with regard to their fitness

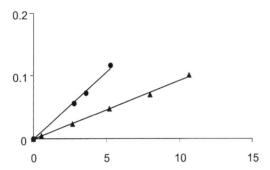

FIG. 2 Sorption isotherms, Q_e(mg/g) versus C_e(mg/L). Sample I (solid circles) and sample III (solid triangles).

to the equilibrium sorption data for samples I–V. Sorption isotherms for samples I and III are shown in Figure 2 and for sample II in Figure 3, which also includes one concentration point for each of samples IV and V. The best model was a linear distribution between the equilibrium soil-phase oil concentration, q_e, and the equilibrium organic-phase oil concentration, C_e; good correlation coefficients were obtained for long equilibrium times. The partition coefficients K_d thus obtained include properties of sorbents and of sorbates, thereby yielding more accurate partition coefficients than a single value derived from an octanol–water partition coefficient K_{ow}. Since organic matter is negligible in Patagonian soils, another model should be provided to interpret the linear isotherms.

The effects of clay and water content on the interaction of oil with soil were examined and found to be very important [Eq. (4)]. An empirical correlation of

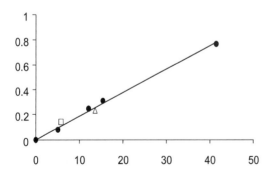

FIG. 3 Sorption isotherm, Q_e(mg/g) versus C_e(mg/L), sample II (solid circles). Single points: sample IV (square) and sample V (triangle).

K_d (in organic phase) with clay and water content was derived from the results, and it is shown by Eq. (4), which is obeyed for ranges of 0–5 wt% of water, 0–100 wt% of clays.

$$K_d(L\ kg^{-1}) = (7.41 \pm 2.19) + (4.89 \times 10^{-1} \pm 1 \times 10^{-3})\%\ clay \\ - (2.97 \pm 0.49)\%\ water \tag{4}$$

The strong inhibitory effect of water content can be interpreted as water-aided interruption of inter- and intramolecular contacts in the soil upon oil sorption. An increase in K_d when increasing the amount of clay in soil is clearly noticed in Eq. (4). These results show that when oil is loaded on dry soil with high clay and silt content, the sorption is very important and strong interactions between the oil and the soil results in loss of oil solution. It is worth mentioning that multiparametric Eq. (4) allows prediction of K_d with knowledge of the clay and water composition of the soil.

Similar studies were carried out with the artificial sample; the correlation of K_d with clay content was obeyed in the full range of 0–100 wt% of clay [Eq. (5)]. The effects of clay content on the interaction of artificial samples with soil are less important than those found for oil, probably due to the strong sorption of the asphaltenes fraction in the crude oil. The low remainder of oil in solution after soil contact cannot be attributed to biological activity. Furthermore, soil was in contact with organic solvent, such as hexane, during the experiments, which does not provide a favorable environment for microbial growth [8].

$$K_d(L\ kg^{-1}) = (2.59 \pm 0.15) + (4.83 \times 10^{-2} \pm 0.24 \times 10^{-2})\%\ clay \tag{5}$$

For soils with an important content of organic matter, the main interaction is the partition between the solution and the organic matter in the soil. A well-known correlation exists between K_p and f_{oc}, the fraction of organic matter in the soil, and the glassy/rubbery model for soil organic matter has been proposed when nonlinear sorption uptake isotherms were observed. Nevertheless, the loss of oil in the present case cannot be attributed to sorption uptake by the soil organic matter, since it is very low (0.02 wt%). The humidity of the soil has an inhibitory effect on the oil sorption when it is lower than 5%, which would be when surface coverage by water was likely less than a monolayer [8,9]. In these soils, with poor organic matter content, the main interaction is then with mineral surfaces, which may cause consequent partitioning; therefore, the reduction in soil clay contents results in an inhibitory effect on the oil sorption to mineral surfaces, as shown by the Eqs. (4) and (5).

Due to the nonpolarity of petroleum hydrocarbon molecules, only weak interactions with the clay particle surfaces are expected, such as dipole–dipole, ion–dipole, and van der Waals types of interactions. The sorption of nonionic organic compounds by clay soils is governed by the CH activity of the molecule, which arises from electrostatic activation of the methylene groups by neighboring electron-withdrawing groups, such as C=O and C=N. Molecules that have many C=O or C=N groups adjacent to methylene groups would be more polar and hence more strongly adsorbed than those compounds with fewer such groups [6].

II. AQUEOUS SOLUBILITY AND DISTRIBUTION COEFFICIENTS

Increasing evidence has made it clear that, under certain conditions, chemicals above background levels in soils may not be released easily and therefore may not have an adverse environmental effect. This has led to a broadening body of knowledge on approaches to measure or estimate the extent and rate of release of hydrocarbons from soil. It is important to have the best estimate of chemical release, because the parameters used to describe the release may also be used to make site decisions that are protective of human health and the environment. Imprecise estimates of the release parameters will result in imprecise estimates of chemical concentrations at a sensitive receptor, imprecise estimates of risk, and possibly inappropriate site remediation decisions [10].

Therefore, the behavior of the oil components in aqueous phase is of critical importance, because solute transport and transformation processes are known to occur predominantly in water. Many research efforts have been undertaken to increase understanding of the risk associated with the presence of pollutants in soil. Selection of technical options and implementation of management practices must include an understanding of the fundamental relationships between the components of the complex mixtures in the environment (soil, water, natural organic matter, contaminants, etc.) [11].

When studying oil solubility, like any other physical or chemical property, it should be presumed that being a multicomponent system, the solubility of each component should necessarily be affected by the presence of the others [1,11]. Due to its unique nature and environmental conditions, the actual composition of the oil residue in soil is strongly dependent on the specific factors affecting it since the oil spill. Therefore, the measurements in field samples are of fundamental interest, since it is impossible to reproduce similar conditions in the laboratory.

A. Organic Cosolvent Effect

The use of organic cosolvents to enhance solubilization of sparingly soluble compounds has been proposed for the environmental field for the calculation of the aqueous concentration of polynuclear aromatic hydrocarbons in complex mixtures. Some recent studies include: estimation of alcohol partition coefficients between nonaqueous-phase liquids (NAPL) and water; analyses of organic cosolvent effects on sorption equilibrium of hydrophobic organic chemicals by organoclays; and evaluation of the NAPL compositional changes in partitioning coefficients [12]. In principle, an organic cosolvent could be effectively used for estimation of the aqueous concentration of complex systems, such as the oil residual in soils.

In basic research, the enhancement of the solubilization of nonpolar solutes in water by organic cosolvents has been reported to follow a log-linear model:

$$\log S^m = \log S^w + \sigma f_c \tag{6}$$

where S^m is the solubility of the solute in the mixed solvents (cosolvent and water), S^w is the aqueous solubility, σ is the cosolvency power, and f_c is the volume fraction ($0 \leq f_c \leq 1$) of the cosolvent in the solvent mixture. Measurement of the mixed-solvent solubility (S^m) at various cosolvent fractions f_c provides a set of data that can be plotted on a log-linear scale to determine the slope (σ) and the y-intercept, S^w. The y-intercept is equal to the predicted solute concentration in pure aqueous solution (no cosolvent).

In this research, the prediction of aqueous concentrations using cosolvent mixtures has been extended to the measurement of poorly soluble compounds found in the aqueous phase of complex mixtures. In this case, the presence of one component in water phase should necessarily be affected by the presence of the others. Components will be removed according to their solubility in the specific cosolvent, which is influenced by molecular weight, functional groups, and polarity of the cosolvent.

According to Rao's solvophobic theory, the sorption coefficient K^m of a hydrophobic organic compound (HOC) decreases exponentially with increasing volume of the cosolvent (f_c) in a binary solvent mixture:

$$\ln\left(\frac{K^m}{K^w}\right) = -a\alpha\sigma f_c \tag{7}$$

where K^w is the equilibrium sorption coefficient from water (L kg^{-1}), K^m is the equilibrium sorption coefficient from mixed solvent (L kg^{-1}), a is the empirical constant accounting for water–cosolvent interactions (note that for water–methanol $a = 1$, implying ideal water–cosolvent interactions), α is the empirical constant accounting for solvent–sorbent interactions, and σ is the cosolvency power of a solvent for a solute accounting for solvent–solute interactions. At a given temperature, the parameter σ is dependent only on the sorbate and solvent properties and not on the sorbent characteristics. The value of σ for a sorbate estimated from data for different sorbents (soils, sediments) is expected to be constant if the model assumptions are valid.

Equations (6) and (7) are strictly valid for only one solute, not for a mixture of solutes of varied polarities; however, in this work the applicability of the model is tested considering the oil residual as only one solute. The aqueous concentration and the distribution coefficients in this case are global values and therefore account for the interactions among the components in the mixture and for the overall interactions of each of them with the mineral matrix. When the product $\alpha\sigma$ is small, the Eq. (7) can be expressed as

$$K^m = K^w - mf_c \tag{8}$$

where $m = K^w\alpha\sigma$. This linear approach was also tested for treating the experimental data; in all cases, the best adjustment of the experimental information with the equations was examined.

Contaminated soil samples, the product of oil spills in six different locations in the environs of Comodoro Rivadavia, were obtained. The oil spills are of different ages, crude oil sources, and environmental exposure conditions. In all cases, except for samples 1 and 6, fertilization of the affected areas was carried out to improve the general conditions of the land, to accelerate the biodegradation processes, and to favor reforestation of species adapted to the zone. Table 5 summarizes some properties of the samples.

Figure 4 shows illustrative examples of the equilibration test for samples 1 and 5. The log oil residual aqueous concentration is plotted as a function of the cosolvent fraction. The data indicate a good linear correlation, which shows good agreement with Eq. (7). Table 6 compares the measured aqueous concentrations to those calculated by Eq. (7). The values of σ_{glo} (the subscript glo is used to indicate a global behavior) correspond to the slopes of the straight line and represent the cosolvency power of the solvent for each sample. The standard deviations for the calculated log S_w values are given in Table 6 together with other statistical parameters. The relative goodness of the regression adjustment is shown by the

TABLE 5 Description of Oil-Contaminated Soil Samples

Sample	Landscape	Description	Oil spill age (years)	Conductivity (μS cm^{-1})[a,b]	pH[a]	Total oil (wt%)	Clay (wt%)
1	Meadow	Prairie	>10	9364	7.6	25.8	33
2	Coastal area	Barren soil	10	1633	7.4	16.6	22
3	Depression	Barren soil	6	646	8.0	8.7	9
4	Creek	Open shrub	3	618	7.4	8.6	8
5	Arid plateau	Shrub steppe	3	387	7.6	9.3	12
6	Meadow	Prairie	2	426	6.8	16.1	16

[a] Extract 1:5 wt/wt.
[b] 25°C.

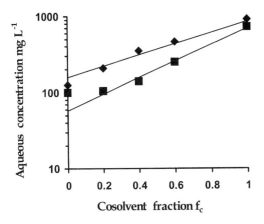

FIG. 4 Log of the oil residual aqueous concentration (mg L^{-1}), as a function of the cosolvent fractions, for samples 1 (diamonds) and 5 (squares).

r^2 coefficients and the validity of the plotting pattern by means of the critical values of F.

For the oldest samples (1, 2, and 3) the aqueous concentrations calculated according to the theory are higher than those measured, while the calculated value for the youngest samples (4, 5, and 6) is in all cases smaller than the experimental. The error in the determinations is approximately constant for the range of σ_{glo} (0.92–1.25). A good correlation exists between f_c and solubility in cosolvent mixtures (0.928 $\leq r^2 \leq$ 0.999), and the logarithmic model seems to be a good representation of the experimental data for $f_c \geq 0.2$.

TABLE 6 Equilibrium Aqueous Concentrations, Global Cosolvent Power σ_{glo}, and Statistical Regression Values

| Sample | Equilibrium aqueous concentration | | Standard deviation of log S^w | σ_{glo} | r^2 | Critical value of F (%) |
	Exptl. (mg L^{-1})	Calcd. (mg L^{-1})				
1	114.1	157.3	0.039	0.78	0.987	0.64
2	136.2	254.5	0.011	0.64	0.999	1.58
3	64.0	72.8	0.075	0.74	0.950	2.50
4	64.5	35.3	0.054	1.25	0.928	3.68
5	103.8	57.8	0.022	1.08	0.989	0.56
6	188.0	131.4	0.017	0.92	0.999	1.81

For contaminated samples 1, 2, and 3, the oil residuals contain a smaller proportion of water-soluble components when compared to the extrapolation of solubility for different cosolvent fractions. This could be interpreted by assuming that the cosolvent mobilizes the liquid-phase hydrophobic components that are not really available in the water phase. In the case of contaminated samples 4, 5, and 6, the oil residuals contain a bigger proportion of water-soluble components as compared to the extrapolated solubility to noncosolvent fractions. Although the solubilization cosolvent power is good ($0.92 \leq \sigma_{glo} \leq 1.25$), it is not possible to reproduce the aqueous concentration value by extrapolation, probably because the oil residuals should possess important hydrophilic global properties.

The reported values of cosolvent power for PAHs in soils vary between 1.63 and 9.09 when methanol is used as cosolvent; in our case the values were in the range 0.64–1.25, indicating a smaller solvent effect. This agrees with the high PAH hydrophobicity, compared to the lower hydrophobicity of hydrocarbon mixtures in oil residuals. The value of σ for naphthalene in methanol–water mixtures was estimated from Nzengung to be 8.95, and it is independent of the sorbent. But Lane shows that the σ-values were not consistent for individual compounds in different soil samples.

Table 7 shows the results obtained by applying solvophobic theory to the calculation of the distribution coefficients K_d. Although, Rao's solvophobic theory is based on the equilibrium sorption coefficient, desorption experiences have been carried out in this work. And a low hysteresis effect has been considered, due to the probable linearity of the isotherms, in case adsorption on the mineral surface was the dominant process. Table 7 shows the measured coefficients and the distribution coefficients calculated by application of both the logarithmic model, Eq. (7), and the linear approach, Eq. (8), according to the best adjustment and the interaction parameter α_{app} (the subscript is used to indicate that total interactions are taken into account). Smaller differences between the measured

TABLE 7 Distribution Coefficient K_d^w, Cosolvent Power σ_{app}, Coefficient α_{app}, and Statistical Regression Values

Sample	Model	K_d^w Exptl.	K_d^w Calcd.	r^2	σ_{app}	α_{app}	Critical value of F (%)
1	Logarithmic	2913	2272	0.993	2.31	0.97	0.33
2	Logarithmic	1337	1017	0.996	2.32	1.22	3.81
3	Linear	1387	896	0.978	2.27	0.37	1.03
4	Linear	1348	1346	0.942	2.14	0.46	2.95
5	Linear	878	1035	0.961	1.95	0.52	1.95
6	Linear	901	971	0.960	1.73	0.57	1.28

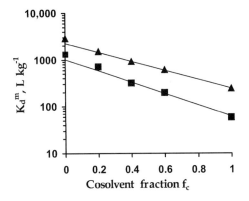

FIG. 5 Log K_d(L kg^{-1}) as a function of cosolvent fraction for sample 1 (triangles) and sample 2 (squares).

and calculated K_d^w were found when Eq. (8) was used for the samples 4, 5, and 6, and for the samples 1 and 2 when Eq. (7) was applied.

Figure 5 shows that samples 1 and 2 give a good correlation of log K_d^m with f_c, as predicted by application of the solvophobic theory, while, as shown in Figure 6, samples 4 and 5 exhibit a linear correlation. Although the logarithmic equation, Eq. (7), could strictly be replaced by the linear approximation, Eq. (8), when the product $\alpha\sigma$ is very small (usually <0.1), in the present case Eq. (7) correlates the experimental data better for all cases in which $\alpha\sigma < 1$ (the differ-

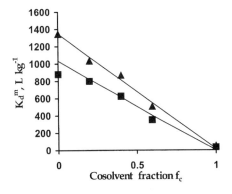

FIG. 6 K_d^m(L kg^{-1}) as a function of cosolvent fraction for sample 4 (triangles) and sample 5 (squares).

ence between $e^{\alpha\sigma}$ and $1 + \alpha\sigma$ is under 0.6 for samples 3–6, while it is 6.7 and 13.1 for samples 1 and 2, respectively).

An approximate estimation of the individual α- and σ-values can be carried out as follows: The cosolvency power σ depends on the solute–solvent interactions and can be estimated by applying Eq. (6) to the oils extracted from samples 1–6, respectively. Table 7 shows the σ_{app}-values thus determined (the squares of the calculated linear regression coefficients for this equation were $0.956 < r^2 < 0.999$). Taking into account that for the present work $a = 1$, from the calculated K_d values the product $\alpha\sigma$ can be estimated and, therefore, the values of α_{app} calculated. The values of α_{app} (which account for the solvent–sorbate (soil) interactions) are smaller than unity, as shown in Table 7.

According to theory, a value of α near 1 would indicate that the properties of the sorbent are independent of the changes in the composition of the water:cosolvent phase. In studies on sorption of hydrophobic organic compounds by soils from solutions containing varying fractions of organic cosolvent, $\alpha < 1$ has usually been obtained, which indicates that sorption from solvent mixtures was greater than that predicted from increased solubility alone. This behavior has been attributed to the swelling of soil organic matter in solvents. The gel swelling of sorbent organic matter results in enhanced permeation of compounds, leading to greater sorption. In our case, according to the available information a similar conclusion cannot be validated, since Patagonian soils are poor in organic matter. More investigations are necessary to draw sound conclusions with respect to α_{app}-values.

B. Effects of Spill Age

The equilibrium aqueous concentration, C_e^w, of the contaminated soil samples normalized by hydrocarbon percentage are shown in Figure 7 as a function of the age of the spills. The values are between 130 mg L^{-1} and 1,100 mg L^{-1}. When the rate of the degradative processes decreases with time, the concentrations of the nonpolar components tend to become constant while those of the polar ones decrease due to solubilization. Therefore, a decrease in aqueous solubility is expected with age, as observed in Figure 7. Those residuals that are the most aged have a superior hydrophobic behavior due to the loss of polar components.

The distribution coefficients K_d^w(L kg^{-1}) are shown in Figure 8 as a function of the age of the spills. The values are between 900 L kg^{-1} and 10,000 L kg^{-1}. Figure 8 shows that aged oil residuals exhibit a similar behavior, an increase of sorption with time, since they are from different crude oil sources and environmental exposure conditions [13,14]. The oil residual could be formed by recalcitrant original components, particularly the resins and the asphaltic fraction [15,16] and by the products of their successive transformations.

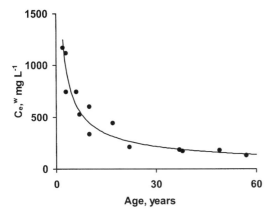

FIG. 7 C_e^w, equilibrium aqueous concentration (mg L^{-1}) as a function of the age of the spill (years).

For the interpretation of the hydrosolubility time dependence, the ratio (Aliph + Aro)/(Pol + Asph) could be used. The ratio (Aliph + Aro)/(Pol + Asph) for the case of regional crude oils are 4.59 ± 1.08 and for the degraded environmental samples are 1.03 ± 0.31 (age, 2–3 years) and 2.31 ± 0.48 (age, 6–57 years). It can be observed that the ratio (Aliph + Aro)/(Pol + Asph) for crude oils indicates a high content of aliphatic and aromatic components. In the

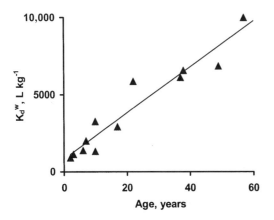

FIG. 8 K_d^w, distribution coefficients (L kg^{-1}) as a function of the age of the spill (years).

case of the degraded environmental samples, Aliph, Aro, Pol, and Asph represent component fractions (wt%) of the extractable hydrocarbons (EH).

As it is exposed to the environment, the aliphatic fraction decreases due to loss by volatilization and biodegradation, while the polar fraction decreases, too, due to loss by solubilization [13,14,17]. But polar compounds could additionally be formed through aliphatic biodegradation and photooxidative processes of aromatics [2,18]. This is consistent with the ratio (Aliph + Aro)/(Pol + Asph) for the youngest degraded environmental samples, which contain a high proportion of polar components and would exhibit important hydrophilic characteristics.

It is known that the chemical extractability and bioavailability of hydrophobic organic compounds (HOCs) from soil decrease with increasing contact time. The decrease in extractability may be controlled by physical sequestration of HOCs and limited mass transfer from soil to solvent or by the action of a soil's microbial community. This decrease in extractability and bioavailability has important implications for the risk assessment of HOCs in historically contaminated soil. The process of HOC sequestration in soil is thought to be driven by partitioning into the soil organic matter (SOM) and sequestration into soil micropores [19].

For example, the amount of PAHs extractable by butanol and dichloromethane decreased with compound aging in soil. The decrease in PAH extractability with aging, and the formation of nonextractable bound residues, increased with compound molecular weight, K_{ow} and K_{oc}. Calculated half-lives for the apparent loss of PAHs by sequestration were dependent on the method used to extract them from soil. Sorbed compounds are less available for partitioning and leaching in groundwater and exhibit reduced bioavailability, toxicity, and genotoxicity compared to dissolved counterparts.

Organic compounds that persist in soil exhibit declining extractability and bioavailability with increasing contact time, or "aging." In the past it was assumed that these observations were due to the degradation of contaminants by microbial processes in the soil. However, studies utilizing isotopically labeled compounds have demonstrated that significant amounts of compound are retained in the soil as nonavailable and nonextractable sequestered residues increase with increasing soil contact time, or aging. Aging is associated with a continuous diffusion and retention of compound molecules into remote and inaccessible regions within the soil matrix, thereby occluding the compounds from abiotic and biotic loss processes [20].

Since the rate of degradation decreases with time, the concentration of the aliphatic components tends to become constant, while that of the polar ones decreases due to its high solubility. Therefore, an increase in the ratio (Aliph + Aro)/(Pol + Asph) is expected with age, and the ratio (Aliph + Aro)/(Pol + Asph) tends to become constant, as observed. This could be interpreted as a probable indication of EH compositional stabilization. Then the increase in K_d-values with age not only could be attributed to the loss of the polar components, but it also suggests that sequestration may be an important process. Because the index reflects the actual EH composition obtained via organic solvent extraction, the K_d-value gives an idea of the water solubility of the components that are available to interact with the aqueous phase only.

C. Effects of Soil-Phase and Aqueous-Phase Ionic Strengths

A factor complicating bioremediation of crude oil spills is salinity. The oil residuals in exploration and production areas are generally accompanied by water spill that is extracted together with the oil and that frequently has a similar salinity to seawater. These salts stay on the soils for long times and they become part of the soil. The changes in ionic strength in the aqueous phase affect the partitioning of PAHs to surfactant micelles and sorbed surfactants, thus conditioning their remediation [21]. The aqueous solubility of organic compounds in the presence of dissolved salts can be expressed by the Setschenow relationship [22]:

$$\log S_{w,\text{salt}} = \log S_w - K_s[\text{salt}] \tag{9}$$

where $S_{w,\text{salt}}$ is the molar solubility in the presence of salts, S_w is the molar aqueous solubility, $K_s(M^{-1})$ is a function of the hydrophobic surface area of the compound, and [salt] (molar) is the concentration of dissolved salts. This relationship has been used, for example, to calculate the aqueous solubility of such organic pollutants as chloroform, lindane, and vinyl chloride in seawater [22].

Equation (9) is strictly valid for only a single solute; however, the applicability of the equation was tested considering the oil residual as only one solute. The scope of the equation to evaluate the variation of K_d with ionic strength was also examined. The aqueous concentration and the distribution coefficients in this case are global values, and therefore they account for the interactions among the components in the mixture and for the overall interactions of each of them with the mineral matrix [22]. The electrical conductivity of the aqueous phase, C, is a good measure of total ionic strength (the ionic content characteristic of the soil plus the added salt, calcium chloride in this work), and a relationship like that of Eq. (9) can be formulated between C and K_d^w:

$$\ln\left(\frac{K_d^w}{K_d^{w0}}\right) = a(C - C^0) \tag{10}$$

where "a" (μS^{-1} cm) is the slope of the straight line, C is the electrical conductivity of the aqueous phase (μS cm^{-1}), C^0 is the electrical conductivity of the aqueous phase without CaCl$_2$, K_d^w(L kg^{-1}) is the distribution coefficient observed with C, and K_d^{w0} is the distribution coefficient observed with C^0. The slope of the straight line, "a", is $(1.33 \pm 0.05) \times 10^{-2}\mu S^{-1}$ cm for the oldest degraded environmental samples and $(1.92 \pm 0.26) \times 10^{-2} \mu S^{-1}$ cm for the youngest degraded environmental samples. The regression values were $r^2 \geq 0.923$ in both cases. According to the model, an increase of sorption is observed when the ionic strength of the aqueous phase increases. The increase of slope "a" for the youngest degraded environmental samples implies a higher salinity effect on K_d, in agreement with the relative increase of polar compounds when age decreases.

A semiempirical model was developed that allows prediction of K_d as a function of exposure time, the salinity of the aqueous phase, and the soil's clay content. The last variable was included because previous studies show an important dependence of K_d on the soil's clay content [3]. The linear relationship between the calculated and measured values of K_d has a slope equal to 0.994 ($r^2 = 0.884$); this value indicates that $\ln K_d$ can be estimated with an error of less than 6%. Although the correlation coefficient is relatively poor, it can be considered a good fit, taking into account the diversity in the environmental conditions and in the sources and history of the residuals.

To evaluate the sensitivity of the model to variations in the main factors involved in the prediction of K_d, Monte Carlo simulation was applied. Data of soil electrical conductivity C_s, soil clay content (wt/wt%), and initial electrical conductivity of the aqueous phase C_i were generated, according to the distributions in Table 8 (five different simulations). C, K_d^0, and K_d were calculated for oil residuals with spill age equal to 2, 10, and 20 years.

Simulation 1.   It is assumed that the aqueous salinity is less than the soil salinity, a situation that could correspond to rainwater that has increased its salinity during its superficial runoff. Mean values of electrical conductivity have been assumed for soil salinity, according to regional data. The results are shown in Figure 9. The values of K_d(L kg^{-1}) are equal to or less than 1000 for 2-year-old residuals (95%), while only 42% and 15% present these values for 10-year-old

TABLE 8   Assumed Distributions of C_i, Soil Clay Content, and C_s for the Monte Carlo Simulations

Variable, distribution	Simulation				
	1	2	3	4	5
C_i, normal	$X = 300$,	$X = 300$,	$X = 300$,	$X = 300$,	$X = 500$,
	$\sigma = 60$	$\sigma = 60$	$\sigma = 60$	$\sigma = 60$	$\sigma = 50$
Clay, normal	$X = 50$,	$X = 50$,	$X = 10$,	$X = 85$,	$X = 50$,
	$\sigma = 15$	$\sigma = 15$	$\sigma = 5$	$\sigma = 5$	$\sigma = 15$
C_s, normal	$X = 600$,	$X = 2500$,	$X = 600$,	$X = 600$,	$X = 600$,
	$\sigma = 100$	$\sigma = 800$	$\sigma = 100$	$\sigma = 100$	$\sigma = 100$

X = mean, σ = standard deviation.

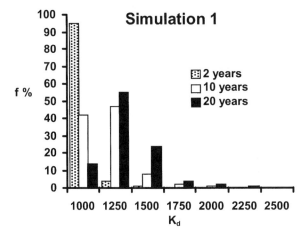

FIG. 9 Histograms showing results for Simulation 1 ($f\%$: percent frequency).

and 20-year-old residuals, respectively. When the age of the spill increases, the maximum frequencies shift to higher values of K_d.

Simulation 2.   A higher electrical conductivity for the soil has been assumed; the results are shown in Figure 10. When soil salinity is greater than aqueous salinity, K_d(L kg^{-1}) increases and the maximum frequencies appear at $1500 \leq K_d \leq 3000$, for all samples. Therefore, the age of the spill is a secondary factor, and the values of K_d would be affected mainly by soil salinity.

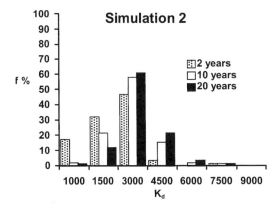

FIG. 10 Histograms showing results for Simulation 2 ($f\%$: percent frequency).

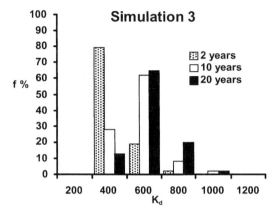

FIG. 11   Histograms showing results for Simulation 3 ($f\%$: percent frequency).

Simulation 3.   The results are shown in Figure 11. In this case, the assumed mean value and standard deviation for C_s correspond to regional sand-clay soils. The results given in Figure 11 show a decrease of K_d, due to the small soil clay content, and a marked effect of age.

Simulation 4.   We have assumed a C_s mean value and standard deviation corresponding to regional clay soils. The results given in Figure 12 show an increase in K_d, due to the high soil clay content: A bigger dispersion of the distribution values as a function of age is observed. These results, together with those

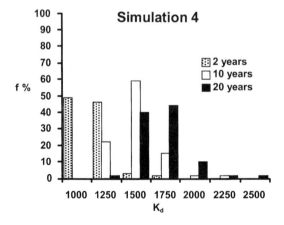

FIG. 12   Histograms showing results for Simulation 4 ($f\%$: percent frequency).

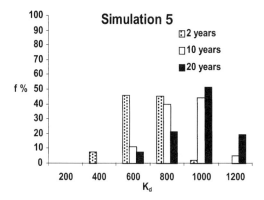

FIG. 13 Histograms showing results for Simulation 5 ($f\%$: percent frequency).

of Simulation 3, are consistent with the new sorption model proposed involving the oil–clay interaction.

Simulation 5. For the initial aqueous-phase salinity, a high mean value and standard deviation of C_i have been assumed, a situation that could correspond to oil residuals that are accompanied by water spills, which are extracted together with the oil and, frequently, have salinity similar to seawater. The results are shown in Figure 13. When the initial aqueous salinity is greater than the soil salinity, a decrease in K_d is observed: $300 \leq K_d \leq 1200$ for all samples.

The increase in K_d with increasing soil salinity (Simulation 2) would imply a high degree of oil sorption under these conditions. This would agree with the observation that, when soil salinity increases, the salinity of the equilibrium aqueous phase also increases and therefore that oil solubility decreases. On the other hand, this effect is more important than age. This same conclusion arises from the observation of a decrease in K_d when the initial aqueous-phase salinity increases (Simulation 5). Under these conditions, equilibrium aqueous-phase salinity decreases (due to the adsorption of ions by soil), which would imply an increase in oil solubility in relation to Simulation 1.

An increase in K_d has been observed when increasing the age of the residual in all of the simulations. However, the equilibrium aqueous-phase salinity minimizes this effect, while the clay content makes the differences more evident (Simulations 3 and 4). This is in agreement with our recent observations that the increase in K_d with the clay content could, in principle, be attributed to an increase in the sorption area. Nevertheless, since a differential uptake is observed for the different fractions, this is an indication of strong specific interactions between polar components of the sorbate and the clay, consistent with the sorption model proposed [3].

III. PHOTODEGRADATION OF OIL RESIDUALS UNDER ADVANCED OXIDATIVE PROCESSES

Photooxidation and biodegradation are among the two most important processes involved in the transformation of crude oil or its products that are released into a marine environment. Photooxidation affects mainly the aromatic compounds in crude oil and converts them to polar species; and the susceptibility of crude oil to biodegradation is increased by its photooxidation [23].

The phenomenon of photodegradation of crude oil via natural sunlight is less well understood in soil, but may provide an opportunity for the introduction of novel procedures for the remediation of oil spills. Due to the presence of strong chromophores and a variety of indigenous reactants in soil, photochemical processes can alter both soil surfaces and the chemicals sorbed to those surfaces. The heterogeneity of surfaces, however, has not allowed successful modeling of the photolysis process, as compared to water or air, which offer greater homogeneity. Recent efforts have sought to understand how various factors affect photochemical processes in soil. These include the depth of photolysis, photochemical quenching-sensitization reactions, and transport processes [15,24,25].

Advanced oxidative processes (AOPs) is the generic name given to a series of different processes in which OH radicals are the major oxidizing agent. The most common AOPs are: hydrogen peroxide, ozone, UV/H_2O_2, UV/O_3, ferrioxalate/H_2O_2, TiO_2, TiO_2/H_2O_2, $TiO_2/O_3/UV$, and Fenton's [24–26]. The photodegradation of oil residuals in Patagonian soils was examined along with the catalytic effect of some added oxidants. The oil residuals are of different ages, crude oil sources, and environmental exposure conditions. An artificial sample was also prepared, with crude oil and typical soil (50% clay content), and it was exposed to the same conditions as the other residuals. The experimental approach was to expose a series of thin, spiked soil layers (thickness typically between 0.25 and 2 mm) to a solar light source. The overall disappearance rate coefficient of the oil, which is generally reported as the photodegradation rate coefficient, is then determined by measuring the total loss of oil from the soil layers as a function of time. The selected AOPs in the present work were: H_2O_2, TiO_2, Fenton's, TiO_2/H_2O_2, and TiO_2/Fenton's.

All these photodegradation catalysts exhibit a similar pattern: a relatively rapid decrease in part of the contaminants (fast kinetic), followed by a much slower decrease in the remainder (slow kinetic). The data could be fitted by a nonlinear equation (11), with first-order constants for both kinetics, where C_t/C_0 is the fraction of oil remained at t days of the exposure time, C_t is the oil concentration at t, C_0 is the initial concentration, f is the fraction of the oil that is fast degraded, and k_1 and k_2 are the kinetic constants, for global first-order processes. A similar model was recently applied to a kinetic desorption of the contaminated soils [10].

TABLE 9 Experimental Parameters for Eq. (11)

Sample	Oil spill age (years)	Without catalyst			With catalyst		
		f	$k_F(\text{day}^{-1})$	$10^3 k_s(\text{day}^{-1})$	f	$k_F(\text{day}^{-1})$	$10^3 k_s(\text{day}^{-1})$
1	>10	0.063	0.09	3.9	0.086	0.11	3.2
2	10	0.159	0.043	0.1	0.135	0.06	1.0
3	6	0.156	0.10	0.8	0.086	0.08	1.9
4	3	0.135	0.09	0.4	0.159	0.04	1.4
5	3	0.198	0.13	0.02	0.246	0.07	0.2
6	2	0.051	0.13	0.4	0.136	0.14	0.9
7	—	0.170	0.08	0.9	0.166	0.19	1.6

$$\frac{C_t}{C_0} = f \exp\left(-k_F t\right) + (1 - f) \exp\left(-k_S t\right) \qquad (11)$$

The experimental parameters (f, k_F, and k_S) are summarized in Table 9 for the degraded environmental samples (1–6) and for an artificial sample (7), without catalyst and with catalyst. Figure 14 shows the experimental data for oil residual in soil with 10 years of exposure time. The results indicate that only the slow kinetics could correspond to a photodegradative process, because only it is affected by the AOP (TiO$_2$/Fenton's) catalysis. Figure 15 shows that, in the case of oil residual with two years of exposure time, it is probable that both kinetics could be affected by AOP (TiO$_2$/H$_2$O$_2$) catalysis. This is in agreement with our

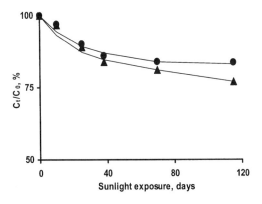

FIG. 14 C_t/C_0 (%), the fraction percentage of oil remaining as a function of sunlight exposure time (days) for sample 2: without catalyst (circles) and with TiO$_2$/Fenton's catalyst (triangles).

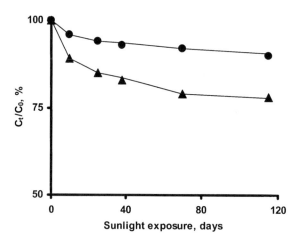

FIG. 15 C_t/C_0 (%), the fraction percentage of oil remaining as a function of sunlight exposure time (days) for sample 6: without catalyst (circles) and with TiO_2/H_2O_2 catalyst (triangles).

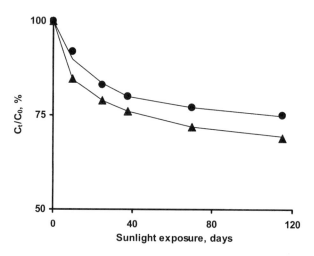

FIG. 16 C_t/C_0 (%), the fraction percentage of oil remaining as a function of sunlight exposure time (days) for the artificial sample: without catalyst (circles) and with TiO_2/H_2O_2 catalyst (triangles).

observation of similar behavior for an artificial sample. Figure 16 shows an important reduction in the concentration (probably by partial evaporation of the volatile fraction) and similar initial AOP (TiO_2/H_2O_2) catalytic effects from the beginning of the curve, as shown previously.

Because light penetration into soils is very limited (i.e., 0.1 to maximal 0.5 mm) and wavelength dependent, the fraction of total compounds actually exposed to light depends on the type of soil, on the thickness of the soil layer, and on the light absorption spectrum of the compounds. Thus, the rate of transport of the compounds from dark to irradiated zones influences the observed overall elimination rate. Because transport depends on the gas/solid partitioning behavior of the compounds, and since sorption is strongly influenced by humidity and other factors [25], the reported rates may have a comparative value. Transport-diffusion problems to the irradiated zone were excluded in the evaluation of the rates in the slow kinetics, because the slow kinetics is clearly affected by catalytic effects, Figures 14–16. The possibility of important catalytic surface effects on crude oil adsorption can be excluded in the present study, since the solid catalysts were no better than the liquids (i.e., H_2O_2).

IV. CONCLUSIONS

The determination of physical chemical parameters in natural field samples can be an important mechanistic tool for understanding the fate of oil residues, its significance to bioavailability, and the remediation of organic pollutants and a guide to the right choice of the cleanup technology. Studies with crude oil and aged oil residues were preferred to artificial, mock mixtures of few known components, since field studies are more realistic and the parameters and empirical equations determined can be used straightforwardly in the environmental models designed to evaluate likely remediation techniques.

Because of the exceptionally low organic matter content of Patagonian soils, an alternative model for sorption of oils in soils was proposed, involving interactions with clays (dipole–dipole, ion–dipole, and van der Waals types of interactions), based on the finding of biparametric relationships between K_d and the clay and water content of the soils. The model was confirmed by other measurements, which showed that the sorption and desorption of the oil residues depend on the age of the spill, the clay and water content of the soil, the salinity of the aqueous phase in contact with the residue, and the salinity of the soil. A characteristic compositional index could give the degree of oil residual stabilization.

The influence of AOP catalysts on oil residue photodegradation was shown to be important, especially in the slow kinetic steps, and catalytic photodegradation should be considered as a possible remediation treatment of the contaminated soils, together with other technologies. A numerical model was developed capable of handling complex and long time-dependence systems, one that could make sound contributions to the management of oil residues in the petroleum industry.

REFERENCES

1. CJ White, JJ Pignatello. Environ Sci Technol 33:4292–4298, 1999.
2. ME Balmer, K Goss, R Schwarzenbach. Environ Sci Technol 34:1246–1253, 2000.
3. N Nudelman, SM Ríos, O Katusich. Environ Technol 21:437–445, 2000.
4. MD Johnson, TM Keinath II, WJ Weber Jr. Environ Sci Technol 35:1688–1695, 2001.
5. D Unger, C Lam, C Schaefer, D Kosson. Environ Sci Technol 30:1081–1091, 1996.
6. TM Yong, A Mohamed, B Warkentin. Principles of Contaminant Transport in Soils. Amsterdam: Elsevier, 1992, p 39.
7. J Dragun. The Soil Chemistry of Hazardous Materials. Amherst, MA: Amherst Scientific, 1998, p 44.
8. S Karimi-Lotfabad, M Pickard, M Gray. Environ Sci Technol 30:1145–1151, 1996.
9. KU Goss. Environ Sci Technol 26:2287–2293, 1992.
10. D Opdyke, CRC Loehr. Environ Sci Technol 33:1193–1199, 1999.
11. WF Lane, RC Loehr. Environ Sci Technol 26:983–990, 1992.
12. N Nudelman, SM Ríos, O Katusich. Environ Technol. In press, 2001.
13. O Pucci, M Bak, S Peressutti. Proceedings of the 2das Jornadas de Preservación de Agua, Aire y Suelo en la Industria Petrolera, Instituto Argentino del Petróleo y del Gas, 1996, pp 291–297.
14. Z Wang, M Fingas, S Blenkinsopp, G Sergy, M Landriault, L Sigouin, P Lambert. Environ Sci Technol 32:2222–2232, 1998.
15. DA Wolfe, MJ Hameedi, JA Galt, G Watabayashi, J Short, CO Claire, S Rice, J Michel, JR Payne, J Braddock, S Hanna, D Sale. Environ Sci Technol 28:561–569, 1994.
16. K Venkateswaran, T Hoaki, M Kato, T Maruyama. Can J Microbiol 41:418–424, 1995.
17. R Garrett, I Pickering, C Haith, R Prince. Environ Sci Technol 32:3719–3723, 1998.
18. T Dutta, S Harayama. Environ Sci Technol 34:1500–1505, 1996.
19. CJA Macleod, K Semple. Environ Sci Technol 34:4952–4957, 2000.
20. G Xia, J Pignatello. Environ Sci Technol 35:1103–1110, 2001.
21. S Ko, M Schlautman. Environ Sci Technol 32:2776–2781, 1998.
22. RP Schwarzenbach, PM Gschwend, DM Imboden. Environmental Organic Chemistry. New York: Wiley, 1995, pp B16–B18.
23. TK Dutta, S Harayama. Environ Sci Technol 34:1500–1505, 2000.
24. GC Miller, SG Donaldson. Aquatic and Surface Photochemistry. Boca Raton, FL: Lewis, 1998, pp 97–109.
25. ME Balmer, K Goss, R Schwarzenbach. Environ Sci Technol 34:1240–1245, 2000.
26. M Fukushima, K Tatsumi, K Morimoto. Environ Sci Technol 34:2006–2013, 2000.

10

Effectiveness of Carbon Nanofibers in the Removal of Phenol-Based Organics from Aqueous Media

COLIN PARK   Synetix, Billingham, United Kingdom

MARK A. KEANE   University of Kentucky, Lexington, Kentucky, U.S.A.

I. BACKGROUND: THE ENVIRONMENTAL DIMENSION

A significant increase in public awareness and concern over global and local pollution has been prompted, at least in part, by the ever-growing evidence of environmental degradation. Air and water pollution constitute the two most prevalent forms, and volatile organic compounds (VOCs) have been identified as major contributors to the decline in air and water quality [1,2]. Volatile organic compounds enter the environment as a result of vehicle exhaust and industrial process emissions (oil refining, solvent usage in painting and printing, etc.) [3]. Phenol and chlorophenol(s) epitomize a class of particularly hazardous chemicals that are commonly found in industrial wastewater, notably from herbicide and biocide plants [3]. The proliferation of phenolic waste has meant that the responsible handling/treatment of such toxic material is now of high priority. Chemical spills may be much smaller than oil spills, but they can still be devastating in their impact. Such was the case in June 2001 with a phenol spill in Singapore's Jahor Strait, both one of the busiest seaways in the world and home to many commercial fish farms. An Indonesian-registered ship, the *Endah Lestari*, capsized in the strait between Malaysia and Singapore, releasing its cargo of 630 tons of phenol. While salvage activities took effect immediately to pump phenol from the damaged vessel, the phenol that had been leaked killed most marine life within 2 km of the ship. Phenol, a corrosive and severe skin irritant on land, also attacks gill tissues of fish when dispersed in water.

There are numerous methodologies in operation at this time to combat the problem of VOC pollution. The most frequently applied techniques are centered on incineration, absorption/adsorption, condensation, and biological treatment [1–7]. Incineration, which is the most widespread strategy for waste disposal (as opposed to treatment) has been heavily criticized in terms of cost and dioxin/ furan formation downstream of the oxidation zone. Combustion, as a destructive methodology, does not demonstrate an efficient management of resources and, even if fully effective, releases unwanted carbon dioxide into the environment. Although biological oxidation can be effective when dealing with biodegradable organics, chloroarenes are used in the production of herbicides and pesticides and, as such, are very resistant to biodegradation. Conversion of halogenated feedstock, where feasible, is in any case very slow, necessitating the construction of oversized and expensive bioreactors.

II. POLLUTANT ABATEMENT USING CARBON ADSORBENTS

Adsorption is perhaps the most widely employed nondestructive strategy, offering the possibility of VOC recovery. The adsorption of phenol, and chlorophenol(s) to a lesser extent, from aqueous media on various forms of amorphous carbon has been the focus of a number of studies published in the open literature [8–13]. Regeneration of the adsorbent, i.e., desorption of the organic pollutant, is usually carried out either by heating the adsorbent or by stripping with steam [6,14–17]. The uptake of VOCs, in general, from gas or liquid streams can, however, call on a variety of solid adsorbents, ranging from macroporous polymeric resins [18–22], mesoporous silica–based MCM-41 materials [23–25], and microporous zeolites [20,26,27] to carbons [28–35]. Currently, carbon is by far the preferred adsorbent, and it is generally derived from either a selection of natural products, e.g., coal, wood, peanut shells, and fruit stones or can be generated from a catalytic decomposition of a range of organics [10,36–41]. Carbon adsorbents find widespread use because they can be readily and precisely functionalized, often by simple yet effective chemical treatments, to meet various demands, e.g., surface oxidation by a gentle thermal oxygen treatment to aid mixing in aqueous media [42–45]. The importance of parameters such as solute concentration, solution pH, and adsorbent porosity/surface area in governing ultimate VOC uptake has been established [9,10,28,32,33,35,46]. The standard activated (amorphous) carbons do not perform well under "wet" conditions or when treating aqueous streams, and they exhibit indiscriminate adsorption. The uptake of both the contaminant and water molecules decreases the available volume for adsorption, limiting uptake effectiveness [47–57]. The adsorption of water on the surface is driven mainly by hydrogen binding interactions, e.g., the presence of certain surface functionalities: O, OH, and Cl can act as nucleation sites and/ or adsorption sites, resulting in the formation of adsorbed water clusters. Phillips and co-workers, in a series of studies [47,58–60], highlighted the complex relationship between the nature of the adsorbent surface and the uptake capacity and mechanism of adsorption. These authors, using a combination of microcalorimetry and adsorption techniques, demonstrated that hydrophobic carbon surfaces adsorb very small amounts of water, primarily by physisorption. In contrast, oxygenated carbon surfaces exhibit a significant capacity for water uptake [52–54,58–60]. The adsorption of methanol/water mixtures in activated carbon pores was studied using Monte Carlo simulations by Shevade and co-workers at ambient temperature [51]. The findings of this work suggest that water is preferentially adsorbed over methanol in the pores of a carbon surface functionalized by carboxyl groups. The hydrophilic nature of the carbon results in a complexation of both the water and methanol and a nonselective uptake [47–55]. Nevskaia and co-workers, using a commercially available activated carbon, found that an indiscriminate adsorption capacity could be inhibited somewhat by a HNO_3 treatment [61].

Moreover, recovery of the "loaded" carbon from the treated water can be problematic. Activated carbon is typically supplied in the form of a powder, and loss of fine particulates is often unavoidable but can be circumvented by additional (membrane) filtration. The major advantage of the activated/amorphous carbon that overrides such drawbacks is the high overall uptake that is synonymous with this material [62]. Indeed, a fibrous form of activated carbon has been manufactured that exhibits a greater adsorption capacity than the granulated form for the removal of liquid pollutants [39,63,64]. It has been claimed that the fibrous material is particularly selective for the adsorption of low-molecular-weight compounds, a feature that is linked to the molecular size of the organic adsorbate [32]. Graphite, on the other hand, the highly uniform and ordered form of carbon possesses delocalised π-electrons on the basal planes. This property imparts a weakly basic character that, in consort with its hydrophobic nature, allows selective VOC adsorption, but the characteristic low surface area/mass ratios (<20 m^2g^{-1}) results in lower overall uptake values [47,65–68]. One significant disadvantage of using activated carbon (or graphite) is the difficulty associated with separation from the solute; the fine carbon particles require a prolonged settling period to facilitate phase separation. Conversely, operation of a continuous-flow separation process, employing a fixed bed of activated carbon, although highly effective, is hampered by the associated high back-pressures. Maintenance of a constant flow is energy demanding, and flow disruptions/plugging can impair an effective processing of contaminant streams. A significant improvement in existing activated carbon–based VOC treatments would result from the development of an adsorbent that: (1) is readily separated from the solute, (2) exhibits high mechanical strength, (3) is resistant to crushing/attrition, and (4) delivers uptake values comparable with those of activated carbon.

III. APPLICATION OF CARBON NANOFIBERS

An ideal carbon adsorbent is one that encompasses the favorable aspects of both graphite (selective adsorption) and amorphous carbon (high uptake) combined with a facile separation from the treated phase. One possible material that may fall into this category is the catalytically generated carbon nanofiber. Carbon is unique in that it can bond in different ways to create structures with quite dissimilar properties. Carbon fibers are generally classified as graphitic structures, characterized by a series of ordered parallel graphene layers arranged in specific conformations with an interlayer distance of ca. 0.34 nm [69]. The direct synthesis of graphitic carbon fibers/filaments is possible by arc discharge and plasma decomposition, but such methodologies also yield polyhedron carbon nanoparticles (low aspect ratio) and an appreciable amorphous carbon component [70,71]. The latter necessitates an additional involved, cumbersome, and costly purification stage in order to extract the desired structured carbon. The generation of ordered carbon structures with different mechanical/chemical/electrical properties under milder conditions by catalytic means is now emerging as a viable lower-cost route [72]. The carbon product can be tailor-made to desired specifications by the judicious choice of both catalyst and reaction conditions. The pioneering studies by Baker, Rodriguez and co-workers [73–80] and Geus et al. [81–86] have established conditions and catalysts by which structured carbon with specific lattice orientations and properties can be prepared with a high degree of control. Much of the pertinent literature on the catalytic growth of carbon nanofibers, from its beginnings to the present day, has been the subject of five detailed review articles [73,77,87–89] that summarize the various aspects associated with the growth phenomena.

The applicability of these novel carbon materials as VOC adsorbents has yet to be established. In this chapter, we present the results of an evaluation of the performance of highly ordered carbon nanofibers to remove phenol and chlorophenol(s), as established VOC pollutants, from water. We adopted the decomposition of ethylene over supported and unsupported nickel catalysts as the synthesis route to generate carbon nanofibers of varying overall dimension and lattice orientations. The uptake measurements on commercially available activated carbon and graphite serve as a basis against which to assess the adequacy of the various forms of catalytically generated carbon nanofibers.

IV. EXPERIMENTAL PROCEDURES

A. Catalytic Production of Carbon Nanofibers

The catalytic growth of fibrous carbon adsorbents was carried out using both unsupported and supported Ni and Cu/Ni catalysts. The unsupported Ni and Cu/

Ni catalysts were prepared by standard precipitation/deposition [90], where the precipitate was thoroughly washed with deionized water and oven-dried at 383 K overnight. The precursor was calcined in air at 673 K for 4 h, reduced at 723 K in 20% v/v H_2/He for 20 h, cooled to ambient temperature, and passivated in a 2% v/v O_2/He mixture for 1 h. The supported Ni catalysts were prepared by impregnating a range of supports to incipient wetness with a 2-butanolic solution of $Ni(NO_3)_2$ to realize a 10% w/w Ni loading; the catalyst precursor was dried, activated and passivated as described previously. The substrates employed in this study include commercially available SiO_2, Ta_2O_5, and activated carbon. The range of metal carriers used provides a range of Ni/support interaction(s) that generate a variety of uniquely structured carbon materials. The Ni content was determined to within $\pm 2\%$ by atomic absorption spectrophotometry (VarianSpectra AA-10), where the samples were digested in HF (37% conc.) overnight at ambient temperature prior to analysis.

The procedure for the catalytic growth of carbon fibers has been discussed in some detail elsewhere [38,91], but specific features that are pertinent to this study are given here. Samples of the passivated catalysts were reduced in flowing 20% v/v H_2/He (100 cm^3 min^{-1}) in a fixed-bed vertically mounted silica reactor to the reaction temperature (798–873 K) and flushed in dry He before introducing the C_2H_4/H_2 mixture (1/4 to 4/1 v/v mixtures). The production of fibers with the desired dimensions/morphology and a particular predominant lattice orientation is strongly dependent on the nature of the catalyst and reaction conditions, as identified in Table 1. The catalyst/carbon was cooled to ambient temperature and passivated in 2% v/v O_2/He, and the gravimetric carbon yield was determined. Graphite (Sigma-Aldrich, synthetic powder) and activated carbon (Darco G-60, 100 mesh) were used as benchmarks with which to assess the performance of the catalytically generated carbon nanofibers. The carbonaceous adsorbents were subjected to acid washing (HCl and HNO_3) in order to remove the residual Ni

TABLE 1 Compilation of Catalysts and Reaction Conditions Used to Generate Carbon Nanofibers of Varying Conformation and Average Diameter

Catalyst	Nanofiber conformation	C_2H_4/H_2 v/v	Reaction temperature (K)	Carbon yield (g_c g_{cat}^{-1})	Nanofiber diameter (nm)
Ni/SiO_2	Ribbon	1/4	848	1.8	15.8
Cu-Ni/SiO_2	Fishbone	1/4	798	2.8	13.2
Ni/Ta_2O_5	Spiral	4/1	823	5.1	23.4
Ni/activated carbon	Branched	1/1	823	3.7	38.3
Unsupported Ni	Platelet/ribbon	1/1	873	7.3	114
Unsupported Cu/Ni	Fishbone	1/1	823	9.8	121

content. This acid treatment also served to introduce functional groups to the carbon surface. Oda and Yokokawa reported that the adsorption capacity of an activated carbon was intimately linked to the surface acidity of the adsorbent [92]. Carbon materials in their pristine form are hydrophobic in nature but, following oxidative treatment, can develop some hydrophilic character [92–94]. The carbonaceous materials (treated with HNO_3) were also subjected to a gentle oxidative treatment by heating in 5% v/v O_2/He (5 K min^{-1} to 723–973 K); up to 5% w/w carbon was oxidized/gasified in this step. In the case of the carbon nanofibers, an amorphous layer deposited during the cool-down stage of the reaction, and this was removed in the secondary oxidation step. The latter should allow greater access of the phenolic solutes to the ordered carbon layers/edge sites.

B.   Characterization of Adsorbent Materials

The pertinent characteristics of the carbon adsorbents used in this study (fibrous, graphite and activated carbon) were established using a variety of complementary techniques. Tap bulk densities of the carbonaceous materials (as supplied/grown) were calculated by weighing a known volume of gently compacted samples. Nitrogen BET surface area measurements (Omnisorb 100) were carried out at 77 K. Temperature-programmed oxidation (TPO) profiles were obtained from thoroughly washed, demineralized samples to avoid any possible catalyzed gasification of carbon by residual metals. A known quantity (ca. 100 mg) of a demineralized sample was ramped (25 K min^{-1}) from room temperature to 1233 K in a 5% v/v O_2/He mixture with on-line TCD analysis of the exhaust gas; the sample temperature was independently monitored using a TC-08 data logger. The associated T_{max} values corresponding to the major oxidation peaks are given in Table 2. High-resolution transmission electron microscopy (HRTEM) analysis was carried out using a Philips CM200 FEGTEM microscope operated at an accelerating voltage of 200 keV. The specimens were prepared by ultrasonic dispersion in butan-2-ol, evaporating a drop of the resultant suspension onto a holey carbon support grid. All gases [He (99.99%), C_2H_4 (99.95%), H_2 (99.99%), and 5% v/v O_2/He (99.9%)] were dried by passage through activated molecular sieves before use.

C.   Uptake of Volatile Organic Compounds

1.   Batch Adsorption Studies

Phenol and chlorophenol adsorption studies were conducted batchwise (298 K ± 3 K) in 100-cm^3-capacity polyethylene bottles, kept under constant agitation (Gallenkamp gyratory shaker) at 100 rpm. The solutes were of high purity (Sigma-Aldrich, 99+%), and stock solutions were used to prepare the test samples by

TABLE 2 Tap Densities, N_2 BET Surface Areas, and Characteristic TPO T_{max} Values Associated With "As-Grown"/Supplied (Catalytically Generated/Commercial) Carbon Adsorbents

Adsorbent (catalyst)	Density (g cm^{-3})	N_2 BET Surface area (m^2 g^{-1})	TPO T_{max} (K)
Activated carbon	0.35	625	848
Graphite	0.42	10	1233
Fishbone fibers (Cu-Ni/SiO$_2$)	0.09	160	889, 1048
Fishbone fibers (unsupported Cu/Ni)	0.17	140	916
Platelet/ribbon (unsupported Ni)	0.25	95	982, 1025
Ribbon fibers (Ni/SiO$_2$)	0.38	110	1040, 1233
Spiral fibers (Ni/Ta$_2$O$_5$)	0.39	80	838, 1064, 1126, 1233
Branched carbon (Ni/activated carbon)	0.49	230	872, 920, 1078, 1233

dilution in triply distilled deionized water. Uptake data were obtained at a constant adsorbate-to-adsorbent ratio of 100 cm^3 g^{-1}, in the absence of any buffered pH control; maximum uptake was generally realized within 3–4 days. The solute was routinely sampled (30 μL) and analyzed by HPLC (Jones chromatography) using a mobile phase (1/1 v/v acetonitrile/water, HPLC grade, Sigma-Aldrich) delivered at a constant rate (1 cm^3 min^{-1}). Sample injection via a 20-μL-sample loop onto a Genesis CII8 (7.5 × 300 mm) column ensured that the presence of any impurities in the feed was detected. Solute detection was by UV (Hitachi Model L-4700 UV detector), with the optimum wavelength set at 280 nm. Data acquisition and analysis were performed using the JCL 6000 (for Windows) chromatography data package. Peak area was converted to concentration using detailed calibration plots, with standards spanning the concentration range employed in this investigation. To ensure that adsorption on the polyethylene bottle walls or adsorbate volatilization did not contribute to the overall uptake, solutions of phenol and chlorophenol (in the absence of any adsorbent) were employed as blanks under the same adsorption conditions. Solutions pH was monitored continually for selected adsorbate/adsorbent systems by means of a data-logging pH probe (Hanna Instrument programmable pH meter). The pH probe was calibrated in the pH range 4–11 before the adsorption run and checked for reproducibility after the analysis period. A blank run was employed that involved pH monitoring of the carbon in deionized water.

2. Semibatch Operation

Phenol removal as a function of time was investigated using a differential column reactor. A stainless steel tube ($^1/_4$ inch o.d.) was packed with adsorbent, and the phenol solution (1.2 mmol dm^{-3}) was fed from a reservoir (1 L) using a Hitachi Model L-7100 pump operating in the constant-flow mode; the pump delivered a flow of 10 cm^3 min^{-1}, regardless of the back-pressure. The adsorbent bed was initially packed using compressed air to minimize the voidage and to facilitate packing: adsorbent bed length = 80 mm, bed volume = 1.83 cm^3, adsorbent weight = 0.2–0.9 g. Deionized water was first passed through the system and the packed adsorbent bed to wet the adsorbent before the aqueous solution of phenol was introduced. The exit stream was regularly sampled, using an on-line sampling valve, to monitor phenol concentration as a function of time; analysis was by HPLC, as described earlier.

V. RESULTS AND DISCUSSION

A. Characteristic Features of the Carbon Adsorbents

Representative transmission electron microscopy (TEM) images that illustrate the structural characteristics of the catalytically generated carbon nanofibers are shown in Figures 1 (unsupported catalyst) and 2 (supported catalysts). A simple schematic representation of the "ribbon" and "fishbone" fiber structure is shown in Figure 3 as a visual aide. In the fishbone (also termed "herringbone") configuration, the carbon platelets are parallel and oriented at an angle to the fiber axis [75,83,86]. This particular arrangement can lead to deviations in the interlayer spacing toward the outer edges of the graphitic platelets, making this particular structure a strong candidate as an effective adsorbent. The fishbone fiber can possess a narrow hollow channel that runs between the series of angled carbon platelets [86]. The so-called "ribbon" form is quite distinct, in that the carbon platelets are oriented solely in an arrangement that is parallel to the fiber axis [95]. The observed variations in carbon morphology and lattice structure are due to the differences in the nature of the catalytic metal site. The choice of both catalyst and reactant is critical when generating carbon nanofibers, because the metal particles can adopt well-defined geometries during the hydrocarbon decomposition step, thereby influencing the nature of the carbon precipitated and deposited at the rear face of the particle. For example, platelet nanofibers are generated from metal particles that are typically "rectangular" in shape, while rhombohedral/diamond-shaped particles produce nanofibers with a fishbone type

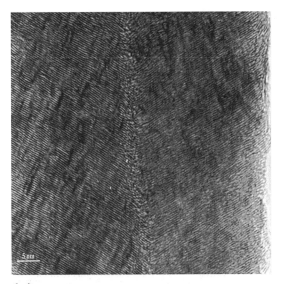

(a)

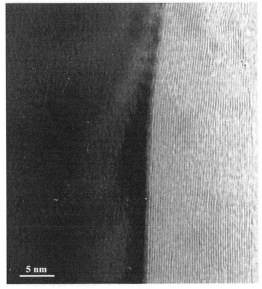

(b)

FIG. 1 Representative TEM images of (a) a fishbone and (b) ribbon nanofibers grown from unsupported (a) Ni/Cu and (b) Ni catalysts.

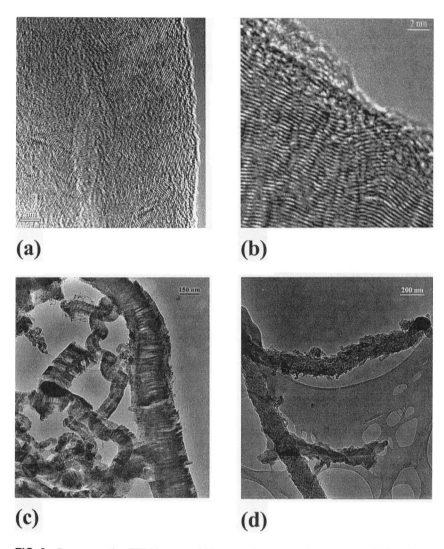

FIG. 2 Representative TEM images of fibrous carbon grown from supported Ni catalysts (details given in Table 1): (a) fishbone structures with platelets arrayed at an angle to the filament axis; (b) ribbon structures with platelets aligned parallel to the filament axis; (c) spiral structures with platelets oriented parallel to the filament axis; (d) "branched" fibers generated from Ni/activated carbon.

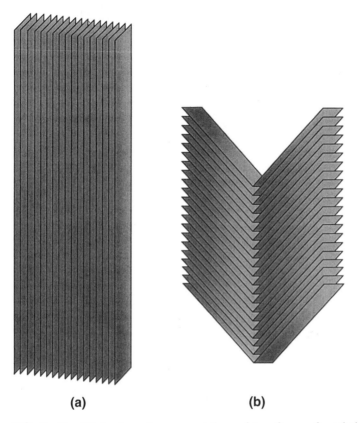

(a) **(b)**

FIG. 3 Simplified schematic representations of two forms of catalytically generated nanofibers employed as adsorbents in the current studies: (a) ribbon form, (b) fishbone form.

of configuration. As a means of aiding a visualization of this phenomenon, TEM images of an assortment of carbon nanofiber structures are given in Figure 4, where the relationship between the metal particle shape and the nanofilament structural characteristics can be seen. Three distinct growths are represented in Figures 4a–4d. The first (Fig. 4a) is monodirectional in nature, where the carbon is precipitated at the rear edge of the metal particle in a whiskerlike mode. The second is a bidirectional growth (Figs. 4b and 4c), where the carbon is precipitated at two opposite faces of the particle; the metal component remains entrapped within the body of the nanofiber during the growth process. The entrapped particle depicted in Figure 4b has assumed a diamond-like morphology, and the bidirectional growth of carbon platelets are arrayed around what appears to be a

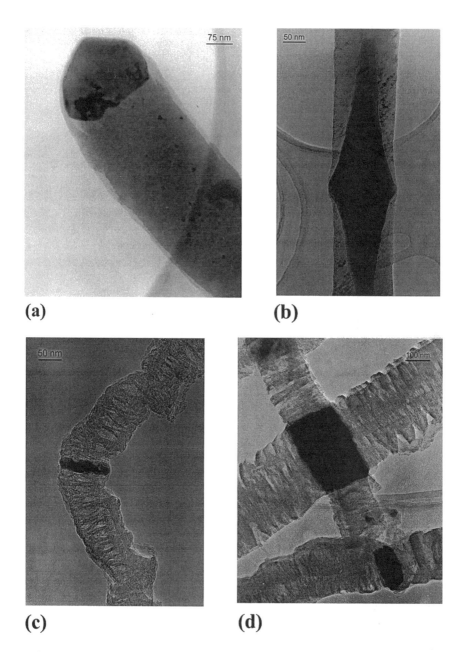

FIG. 4 TEM images illustrating the relationship between Ni particle shape and the nature of the associated carbon nanofiber growth: (a) pentagonal-shaped particle, monodirectional fiber growth; (b) diamond-shaped particle, bidirectional fiber growth; (c) rectangular-shaped particle, bidirectional spiral fiber growth; (d) rectangular-shaped particle, multidirectional spiral fiber growth.

hollow central core. On closer examination by HRTEM, distinct parallel platelets were found in this central core and aligned parallel to the fiber axis. Diffusion/ precipitation in this core region differs from that associated with the adjacent faces of the restructured Ni particle. Finally, a relatively uncommon type, a multi- directional growth, can be seen in Figure 4d, where two fibers are associated with two distinct sets of metal faces: The metal particle is locked at the hub of the four filamentous arms. From a consideration of these TEM images it becomes clear that the characteristics of the nanofiber are largely determined by the struc- ture adopted by the metal particle. The dimensions of the metal face at which the carbon is precipitated govern the fiber width. This effect is particularly evident in Figure 4d, where two distinct fiber diameters are generated that match the dimensions of the two sets of metal faces from which these fibers have been grown. By use of controlled-atmosphere electron microscopy, Baker and co- workers (96) demonstrated that the growth of each fibrous arm was identical and that the fiber grew in a symmetrical manner.

The commonly accepted fibrous carbon growth mechanism [73] involves re- actant (carbon source) decomposition on the top surface of a metal particle, fol- lowed by a diffusion of carbon atoms into the metal, with precipitation at other facets of the particle to yield the fiber, which continues to grow until the metal particle becomes poisoned or completely encapsulated by carbon. The growth of carbon nanofibers with a spiral (sometimes denoted helical) structure occurs due to an unequal diffusion of carbon through the metal particle, leading to the aniso- tropic growth; see Figures 4c and 4d. Zaikovskii and co-workers [97], using an MgO-supported bimetallic Ni-Cu catalyst, generated symmetrical spiral nanofi- bers. These authors proposed that a carbide mechanism was in operation, where Ni_3C, metastable at 723 K, exists during the hydrocarbon transformation before decomposing to metal and carbon. It was proposed that the different diffusional pathways taken by the carbon atoms through the carbide phase led to different rates of carbon growth, resulting in a "twisted," or spiral, growth. The generation of fibrous carbon with a spiral structure was also noted by Park and Keane [38,98] using alkali bromide–doped Ni/SiO_2 catalysts to generate substantial quantities of carbon with relatively small diameters. It was observed that the choice of alkali metal (from Li to Cs) had a direct impact on the degree of fiber curvature. The spiral growth was again assigned to an anisotropic diffusion of carbon atoms through the metal, generating a helical fiber. Moreover, doping the catalyst with alkali bromide enhanced both the carbon yield and overall structural order [99– 102].

The diameters of the individual carbon nanofibers generated from unsupported catalysts are appreciably greater than those grown from supported systems; see Table 1 for the details. This is a direct consequence of the much smaller metal particle size that can be stabilized on the support [74–76,80,86,91,95]. The degree of crystalline order of the carbon product is controlled by various factors, including the wetting properties of the metal with graphite and the crystallographic orientation of the metal faces that are in contact with the carbon deposit, features that are ultimately reliant upon the choice of catalyst [75,86]. The arrangement of the metal atoms at the face where the carbon is deposited ultimately regulates the nature of the precipitated carbon. If the atoms are arranged in such a manner that they are consistent with those of the basal plane structure of graphite, then the carbon that dissolves in and diffuses through the particle will be precipitated as an ordered structure. Conversely, if there is little or no match between the atomic arrangements of the depositing face and graphite, a more disordered carbon will be generated. The bulk densities of the carbon materials used as adsorbents are given in Table 2. There is a significant variation (fourfold) in the densities of the catalytically generated carbon. Those fibers that display a fishbone structure exhibit the lowest densities but possess the highest surface areas due to the large number of accessible edge sites in this more open structure. By comparison, the fibers that display a predominant ribbon or spiral shape are significantly denser, with a lower BET surface area. The nature of the carbon nanofibers grown from Ni supported on activated carbon (which also serves in this study as a model adsorbent) is shown in the micrograph given in Figure 2d. There is no discernible structural order, and the nanofibers exhibit a roughened (or "branched") exterior. The latter feature can be of benefit in terms of enhanced sites for solute attachment. Indeed, it is to be expected that carbon nanofibers grown from an activated carbon substrate should exhibit uptake characteristics that draw on the action of both carbonaceous species, i.e., original amorphous Ni support and catalytically grown fibers. Indeed, the associated surface area measurement (Table 2) is intermediate between the highly oriented nanofibers and the amorphous carbon.

High-resolution TEM (HRTEM) proved to be an invaluable aid in screening carbon nanofibers as potential adsorbents and linking uptake data with structural characteristics. The presence of an amorphous carbon layer on the filament edges (Fig. 2) is an artifact of the cooling stage, upon completion of the catalytic step. This layer may hinder uptake by blocking filament edge sites as potential points of solute attachment. A careful oxidation treatment was employed to remove this amorphous carbon overlayer, allowing access to the underlying adsorption sites, without disrupting the overall lattice structural order; a weight loss of ca. 5% was typically associated with this mild oxidative step. Similar oxidative treatments have been used by Baker and co-workers [78] to enhance the surface area of nanofibers, but it should be noted that a gasification of a significant filamentous component accompanied any substantial increase in area. Surface areas of up to 700 m^2 g^{-1} have, however, been quoted (with a 40% w/w burn-off), with no apparent damage to the overall structural integrity of the remaining carbon species [78].

Temperature-programmed oxidation (TPO) is a technique that has been put to good use in probing the degree of order in carbon structures where a move from an amorphous to a graphitic structure is accompanied by an elevation of the temperature at which gasification is induced [103]. The TPO profiles associated with selected demineralized carbon samples are shown in Figure 5, and the T_{max} values are recorded in Table 2. The oxidation of the model amorphous activated carbon takes place at a significantly lower temperature than that of the highly ordered model graphite (Figs. 5a and 5e). It can be readily seen that the oxidation characteristics of the carbon nanofibers fall somewhere between these two boundary cases. The ordered structure associated with the fibers elevates the onset of gasification relative to activated carbon, but the greater presence of edge sites means that fibers gasify at a lower temperature than the model graphite. Carbon generated from the Ni/activated-carbon catalyst exhibits a TPO peak (872 K) that roughly corresponds to the parent substrate (Figs. 5a and 5d) in addition to a higher-temperature response that can be linked to the structured fibers. The peak profile is very broad, indicative of the presence of a range of different carbon species. The TPO profile of the carbon generated with a spiral conformation is

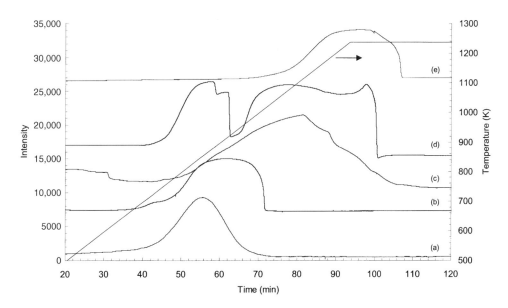

FIG. 5 TPO profiles of demineralized samples of (a) activated carbon, (b) fishbone fibers grown from unsupported Cu/Ni, (c) spiral fibers grown from Ni/Ta$_2$O$_5$, (d) fibers generated from Ni/activated carbon, and (e) graphite.

also broad, diagnostic of a diverse structure; a number of T_{max} values characterize this sample, as shown in Table 2. The spiral fibers, although highly ordered, do not exhibit quite the same regularity as the platelet or ribbon form. The fishbone nanofibers realize a sharper oxidation profile with one predominant characteristic T_{max}. The TPO characteristics of carbon grown from supported catalysts suggests a marginally greater degree of order than that associated with unsupported metal counterparts. The support material can alter the characteristics of the deposited metal and so impose changes to the carbon deposit.

B. Phenol Adsorption

The equilibrium phenol uptake values for a representative solute concentration are given in Table 3. The highest uptake was achieved using the model activated carbon, the lowest (by a factor of almost 4) with the model graphite, while the catalytically generated nanofibers delivered a range of values that fall within these two extremes. It should be noted that the carbon grown from unsupported metallic Ni took a predominantly platelet form (graphene layers are oriented perpendicular to the growing fiber axis), with a minor component of ribbon nanofibers. The extent of phenol adsorption matches, to a greater degree, the surface area associ-

TABLE 3 Effect of Acid Treatment and Partial Oxidation on Phenol Uptake Values for Model (As-Supplied) and Catalytically Generated (As-Grown) Carbon Adsorbents

Adsorbent	Phenol uptake (mmol g^{-1})			
	As grown/ as supplied	Demineralized with HCl	Demineralized with HNO$_3$	Partially oxidized
Activated carbon	1.63	1.98	1.84	2.13
Graphite	0.45	0.60	0.78	0.82
Fishbone fibers (Cu-Ni/SiO$_2$)	0.78	1.29	1.34	2.03
Fishbone fibers (unsupported Cu/Ni)	0.66	0.75	0.95	1.45
Platelet/ribbon fibers (unsupported Ni)	0.63	0.91	0.78	1.07
Ribbon fibers (Ni/SiO$_2$)	0.61	0.95	1.08	1.46
Spiral fibers (Ni/Ta$_2$O$_5$)	0.75	1.22	1.36	1.67
Branched carbon (Ni/activated carbon)	1.28	1.48	1.51	1.59

Initial phenol concentration = 30 mmol dm^{-3}.

ated with these carbon materials; i.e., uptake is dependent on the surface available for attachment. Within the range of nanofiber structures under investigation, the greatest phenol adsorption was achieved on the fishbone and spiral configurations generated from the supported catalysts. Treatment with the carbon fibers grown from an activated-carbon substrate resulted in a phenol removal that reflects a combined contribution from both carbon components. The commercial activated carbon and graphite as well as the catalytically generated nanofibers contain a residual metal component that is left over from the synthetic step(s). This metal can be removed by an acid washing; two mineral acids (HCl and HNO_3) were employed in this study. The demineralization agent can also influence the adsorption characteristics of the carbon by functionalizing the surface. Park and co-workers [39], studying the removal of low-molecular-weight alcohols from aqueous solution, demonstrated that nanofiber treatment with HCl resulted in enhanced adsorption. Demineralization with both acids raised the uptake of phenol by each carbon considered in this study (Table 3). The enhancement of uptake was greater in the case of the nanofibers; the fishbone and spiral structures continued to provide the highest uptakes among the catalytically generated carbons. In contrast, Pradhan and Sandle [21] reported that a treatment of activated carbons (granular and charcoal cloth) with HNO_3 and H_2O_2 under much harsher conditions than employed in this study resulted in a substantial decrease in adsorption capacity. This was ascribed to an increase in the concentration of oxygenated functional groups on the carbon surface (in particular at the entrance to the micropores), as was supported by the studies of Nevskaia and co-workers [61]. However, it has been established [104–106] that the adsorption of phenolic compounds on carbon involves the formation of electron donor–acceptor complexes, where basic surface oxygen– and/or surface electron–rich regions act as donors and the aromatic ring of the adsorbate serves as the acceptor. A surface functionalization by acid treatment is accordingly beneficial for phenol uptake, as was uniformly the case. A surface oxidation can be achieved by heat treatment in an oxidizing gas stream that also serves to remove any amorphous carbon overlayer from the nanofibers. The effect of this additional treatment on phenol adsorption characteristics can be assessed from the data presented in Table 3. The removal of the amorphous cover facilitates a more meaningful assessment of the influence of the carbon platelet orientation on phenol uptake where the treated fibers present an essentially clean surface. Once again, the most significant increases in uptake were recorded for the nanofibers grown from the supported catalyst. The extent of adsorption on the demineralized/oxidized fishbone nanofibers is equivalent to that achieved with the treated activated carbon. The variation in solution pH (as an important measure of water quality) during phenol uptake is shown in Figure 6 for the model activated carbon and two representative nanofibers. Agitating the carbon samples in water, as a blank, resulted in a slight shift in pH to more acidic conditions. The latter can be ascribed to a release of residual SO_x or NO_x spe-

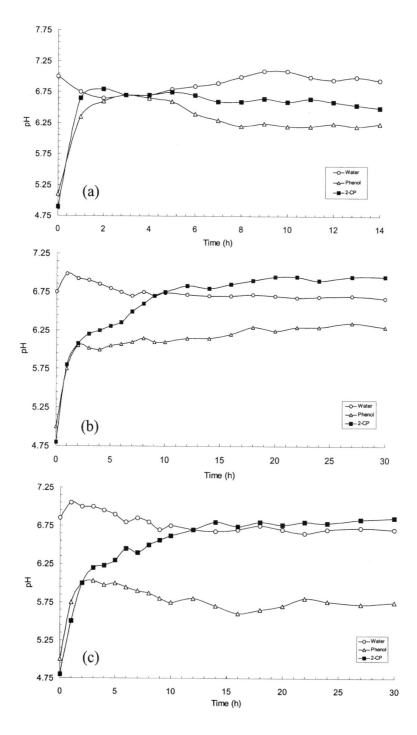

cies arising from the demineralization step. The initial phenol solution was acidic (pH < 5), but upon contact with each carbon, the pH was significantly raised as a result of the removal of the organic from solution. Phenol acts as a weak acid that dissociates to a small extent in aqueous solutions to give H_3O^+ and a phenoxide anion. The increase in pH may also be ascribed to an attachment of hydronium ions to the carbon surface, as proposed by Daifullah and Girgis [10].

One important aspect of separation processes involving carbon-based adsorbents is the ease of separation of the solid from the treated solution. The recovery of the carbon nanofibers from aqueous media was observed to be far more facile than phase separation involving the granular activated carbon powder. The nanofibers are extremely robust in nature and do not disintegrate or exhibit any appreciable damage during vigorous agitation, unlike the activated carbon, which shows signs of attrition with prolonged use. Indeed, the time taken for the separation of roughly the same weight of activated carbon from the treated solution was greater by a factor of up to 10. The intrinsic hydrophobicity of the carbon nanofibers may also serve to aid filtration by repelling water molecules. Moreover, unlike the activated carbon, separation of the fibers from solution was not accompanied by any significant loss of fine carbon particulates, and adsorbent reuse is greatly facilitated. Indeed, carbon fibers are known to exhibit high structural strength that is maintained over many cycles of adsorption/desorption and enhanced transport effects when compared with either graphite or activated carbon [75]. A TPO analysis of the carbon nanofibers before and after use demonstrated little change in the oxidation characteristics, a feature reinforced by HRTEM studies. This is a good indication that the highly ordered graphitic structure remains essentially unchanged. By comparison, the TPO profiles of the activated carbon revealed a small but significant shift in T_{max} to lower values, suggesting a loss of structural integrity. De Jong and Geus [86] have noted an improved mechanical strength due to filament interweaving associated with fibers wider than ca. 12 nm, dimensions that match the majority of carbon structures generated in this study. The latter feature would certainly be of importance in a fixed-bed adsorption configuration, where a high crushing resistance is required. Given the equivalency of solute uptake observed for both the oxidized activated carbon and the fishbone fibers, the greater ease of sorbent recovery associated with the latter warrants further study as part of an overall (financial and technical) assessment, perhaps in a pilot-plant phenol adsorption/recovery unit.

FIG. 6 Time dependence of solution pH values for the blank run (adsorbent in water) and uptake of phenol (initial concentration = 30 mmol dm^{-3}) and 2-chlorophenol (initial concentration = 53 mmol dm^{-3}) on demineralized adsorbents: (a) activated carbon; (b) fishbone fibers ($Cu-Ni/SiO_2$); (c) spiral fibers (Ni/Ta_2O_5).

C. Chlorophenol Adsorption

The results of the uptake of 2-chlorophenol (as a representative isomer) on the same carbonaceous materials are given in Table 4. As in the case of phenol, the performance of the as-grown nanofibers falls between that of the model activated carbon and graphite. One significant observation is the appearance of phenol in the solution treated with the fishbone nanofibers grown from the unsupported catalyst. Phenol in solution must arise from a dechlorination on the carbon surface with a subsequent release of the aromatic. The effects of a demineralization and gas-phase oxidation on 2-chlorophenol uptake are also presented in Table 4. Both pretreatments raised the level of adsorption, which is to be expected, since the presence of a strongly electron-withdrawing group (Cl) on the aromatic ring will favor the formation of sorbate/sorbent electron donor–acceptor complexes. The adsorption of phenol and the three chlorophenol isomers under the same conditions is compared in Table 5, taking the "as-received" activated carbon as a representative adsorbent. The adsorption capacity of a given activated carbon for a range of phenolic compounds has been related to the solute solubility in water, where the lower the solubility, the greater should be the uptake [28,37,104]. Comparing the solute solubilities in Table 5 with the uptake values, there is no obvious link between these two parameters. Uptake of the *meta-* and *para-*chloro-isomers

TABLE 4 Effect of Acid Treatment and Partial Oxidation on 2-Chlorophenol Uptake Values for Model (As-Supplied) and Catalytically Generated (As-Grown) Carbon Adsorbents

	2-Chlorophenol uptake (mmol g^{-1})		
Adsorbent	As grown/ as supplied	Demineralized with HNO_3	Partially oxidized
Activated carbon	2.47	3.29	4.17
Graphite	1.27	1.52	1.43
Fishbone fibers (Cu-Ni/SiO$_2$)	1.83	2.76 $(+0.12 \text{ phenol})^a$	4.09 $(+0.47 \text{ phenol})^a$
Fishbone fibers (unsupported Cu/Ni)	1.41 $(+0.15 \text{ phenol})^a$	19.8 $(+0.55 \text{ phenol})^a$	3.37 $(+0.64 \text{ phenol})^a$
Platelet/ribbon fibers (unsupported Ni)	1.74	2.32	3.15
Ribbon fibers (Ni/SiO$_2$)	1.92	2.45	3.87 $(+0.18 \text{ phenol})^a$

[a] Phenol concentration in solution (mmol g^{-1}).
Initial 2-chlorophenol concentration = 53 mmol dm^{-3}.

TABLE 5 Solubility and Uptake Data for Phenol and Three Chlorophenol Isomers on "As-Supplied" Activated Carbon at the Same Initial Solute Concentration (48 mmol dm^{-3})

Adsorbate	Solubility in water at 303 K (mmol dm^{-3})	Uptake (mmol g^{-1})
Phenol	871	2.4
2-Chlorophenol	222	2.5
3-Chlorophenol	202	3.0
4-Chlorophenol	211	3.1

was significantly greater than that recorded for phenol, which was, in turn, roughly equivalent to the *ortho*-substituted chlorophenol. The latter suggests the involvement of steric hindrance, in that the further the Cl atom is from the –OH group, the greater the ultimate uptake, and this points to a direct interaction of Cl with the carbon adsorbent. Yonge and co-workers [36] likewise concluded that substituent positioning influenced adsorption, whereas Singer and Yen [107] obtained equivalent uptakes for each isomer.

The occurrence of phenol in solution was even more significant over the treated samples, where the acid treatment induced dechlorination over the fishbone structure from the supported catalyst. Carbon oxidation further elevated dechlorination over both fishbone fibers and was responsible for the onset of dechlorination over the ribbon structure. The removal of the amorphous carbon overlayer combined with the oxidation/functionalization of the underlying surface enhanced chlorophenol interactions to such an extent that C—Cl bond scission results. The dehalogenation of arene derivatives mediated by activated carbon alone has been noted elsewhere in gas-phase [108–110] and liquid-phase [111] operation. In each case, dechlorination was promoted in the presence of hydrogen (hydrodehalogenation to aromatic and HCl) at temperatures in excess of 473 K. The observed dechlorination of 2-chlorophenol over the treated carbon fibers in the liquid phase at room temperature is indicative of a remarkably strong interaction/chemisorption that leads to C—Cl bond dissociation. The variation in solution pH (increasingly less acidic) shown in Figure 6, reflects 2-chlorophenol uptake, and there is no evidence of HCl release into solution. The highly reactive uptake sites on the treated filament surfaces must promote a dissociative adsorption of chlorophenol with both the aryl moiety and Cl attached to the surface. The resultant Cl–filament (sp^2) bonding is sufficiently strong that the extracted Cl remains on the surface while the dechlorinated phenol can re-enter the liquid phase. The presence of delocalized π-electrons situated between adjacent graphite layers is known to impart weakly basic character to the material in its pristine state and, in conjunction with the uniformly ordered, small-diameter carbon nanofibers, contributes to the high directional conductivity [75,112]. The high conductivity and greater availability of delocalized π-electrons, relative to conventional graphite, must be the source of the stronger sorbate/fiber interaction(s) that lead(s) to the observed apparent dechlorination activity. Indeed, it has been proposed that individual nanotubes exhibit unique conductivity properties, both metallic and nonmetallic, due to the variations in geometries and degree of graphitization [112]. One feature of the fishbone nanofibers that can have some bearing on the interactions is the variability of the d-spacings, especially at the edge regions. This feature may allow a stronger interaction with the delocalized electrons between adjacent layers that contributes to the dechlorination. The predominantly platelet form of nanofibers grown from unsupported Ni did not exhibit any significant dechlorination behavior. Platelet nanofibers are structurally similar to graphite, in that they possess two edges of similar dimension, are highly ordered structures, but possess an appreciably higher aspect ratio. This high degree of crystalline perfection does not appear to promote the same degree of chlorophenol interaction as that observed with the fishbone nanofibers, where variations in the interlayer spacing must be critical in promoting dechlorination.

The treated fishbone fibers grown from the supported catalysts again delivered equivalent solute uptake to the model activated carbon. It should be stressed that there was no detectable phenol in the solutions treated by both model carbons. It is instructive to note that uptake on the treated ribbon structures (grown from Ni/SiO$_2$) approached that of the treated activated carbon but that the same fibrous material acted as an indifferent adsorbent for phenol (Table 3). These structures are arranged in such a manner that only the edge regions are exposed; these nanofibers are characterized by a relatively large basal plane, bounded by two long and two short edges, perpendicular to each other. The carbon atoms at the edge sites can be arranged into two distinct conformations, "armchair" and "zigzag," and these can have quite different adsorption capacities. A preponderance of one particular face may have a significant influence on adsorption characteristics when compared with a nanofiber that has an equivalent number of exposed faces, e.g., the fishbone structure [113,114]. Park and Baker [114] illustrated that the nanofiber structure impacted strongly on the catalytic behavior of supported metal particles. This variation in behavior was attributed to the ability of the supported metal to adopt specific orientations, following deposition and nucleation on either the "zigzag" or the "armchair" faces of the fiber. From the results generated in this investigation, it is tentatively suggested that chlorophenol exhibits a higher affinity than phenol for adsorption at the longer edge sites. The benefits of the catalytically generated carbon in 2-chlorophenol treatment are twofold: (1) ease of recovery/enhanced mechanical strength and (2) dechlorination capability. Indeed, the treated fibers are obvious candidates as transition metal catalyst supports to promote chloroarene hydrodechlorination, which is now accepted as a viable means of chemical transformation/recycle [115–117].

D. Semibatch Phenol Uptake

The results generated for phenol uptake in a closed loop system using three representative carbon adsorbents are shown in Figure 7. In the earlier batch adsorption experiments, the fishbone nanofibers grown from the supported catalyst generated the highest uptakes of all the catalytically generated carbons. This form of structured carbon was accordingly chosen to test against standard and amorphous and graphitic samples. In semibatch operation, the activated carbon removed all traces of phenol from solution after 62 hours on-stream. At this point the reservoir concentration of phenol treated with graphite or nanofiber had been lowered by no more than 65%. The ultimate phenol uptakes agreed well with those determined in a solely batch operation. Phenol adsorption per gram of adsorbent was

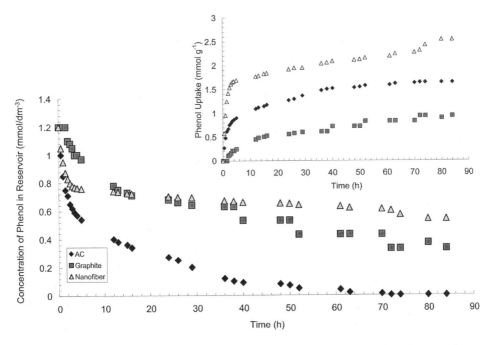

FIG. 7 Variation in phenol reservoir concentration with time in the semibatch-operated removal of phenol by activated carbon (◆), graphite (□), and fishbone filaments grown from Cu-Ni/SiO$_2$ (△). *Inset*: Time-dependent phenol uptake, symbols as previously: adsorbent bed volume = 1.83 cm^3; initial phenol reservoir concentration = 1.2 mmol dm^{-3}.

TABLE 6   Phenol Uptakes from a
Semibatch Operation

Adsorbent	Phenol uptake (mmol g^{-1})
Activated carbon	1.6
Graphite	0.9
Fishbone fiber (Cu-Ni/SiO$_2$)	2.5

Adsorbent bed volume $=$ 1.83 cm^3; initial
phenol concentration $=$ 1.2 mmol dm^{-3}.

nonetheless appreciably higher for the carbon nanofiber bed; see inset to Figure 7 and Table 6. The latter is a direct result of the differences in the density of the carbonaceous materials, where the maintenance of a fixed bed volume/space velocity required the use of quite different adsorbent weights. Nevertheless, these preliminary screening tests are positive in terms of flagging the potential of catalytically generated carbon nanofibers for application in continuous-flow water treatment. The pressure required to maintain a constant flow (10 cm^3 min^{-1}) of phenol solution through the bed of activated carbon and graphite (3800–3900 psig) was substantially higher than that recorded for the same nanofiber bed volume (2400–2700 psig). This pressure difference has significant ramifications in terms of energy usage/costs, in that operation of an activated carbon bed (to deliver equivalent levels of cleanup) would necessitate the design and operation of equipment rated for higher pressures.

VI. CONCLUSIONS

The removal of toxic phenolic pollutants from wastewater is an area of growing concerns as governmental legislation focuses on a substantial reduction in the emission of a broad range of compounds. Adsorption represents the most widely applied nondestructive control technology, offering the possibility of recovery/ recycle. Activated carbons, while effective under "dry" conditions, generally underachieve in aqueous media due to an indiscriminate adsorption of both pollutant and water. Adsorption on hydrophobic graphite is more selective, but the inherent low surface area/mass ratio mitigates against high specific uptakes. The use of catalytically generated highly ordered carbon nanofibers is a viable option for the uptake of phenolic compounds. A judicious choice of catalyst/synthesis conditions allows for a high degree of control in terms of the morphology and lattice structure of the carbon fibers that are produced. Phenolic adsorption on the "as-grown" nanofibers was less than that associated with model activated carbon and greater than that recorded for model graphite. Demineralization in acid and partial oxidation realized higher uptakes on all the carbon adsorbents, with comparable values recorded for the activated carbon and fibers bearing a fishbone lattice arrangement. The ordered fibers have the decided advantage of exhibiting a greater ease of separation from solution/operation in semibatch mode when compared with amorphous carbon, allied to higher mechanical strength and retention of structural integrity. Nanofibers with a ribbon structure exhibit an appreciably higher affinity for chlorophenol when compared with phenol. Moreover, the treated (fishbone/ribbon) nanofibers not only act to adsorb chlorophenol from water but promote a dechlorination, the degree of which is enhanced with acid treatment and partial oxidation. The emergence of a novel carbon that both adsorbs and dechlorinated chlorphenols at room temperature is a significant finding that has far-reaching implications in water treatment technologies based on separation and catalytic transformation processes.

REFERENCES

1. RE Hester, RM Harrison. Volatile Organic Compounds in the Atmosphere: Issues in Environmental Science and Technology. Cambridge, UK: Royal Society of Chemistry, 1995.
2. RM Harrison. Pollution Causes, Effects and Control. Cambridge, UK: Royal Society of Chemistry, 1990.
3. AJ Buonicore, W Davis, eds. Air Pollution Engineering Manual. New York: Van Nostrand Reinhold, 1992.
4. YJ Li, JN Armor. Appl Catal B: Environmental 1:L21–L29, 1992.
5. GJ Wilson, AP Khodadoust, MT Suidan, RC Brenner. Water Sci Technol 36:107–115, 1997.
6. B Hunger, S Matysik, M Heuchel, W-D Einicke. Langmuir 23:6249–6254, 1997.
7. S Pallerla, RP Chambers. Catal Today 40:103–111, 1998.
8. M Streat, JW Patrick, MJ Camporro Perez. Water Res 29:467–472, 1995.
9. AR Khan, A Al-Bahri, A Al-Haddad. Water Res 31:2102–2112, 1997.
10. AAM Daifullah, BS Girgis. Water Res 32(4):1169–1177, 1998.
11. W Fritz, W Merk, E Schluender. Chem Eng Sci 36:731–741, 1981.
12. F Caturla, JM Martin-Martinez, M Molina-Sobio, F Rodriguez-Reinoso, R Torregrosa. J Colloid Interface Sci 124:528–534, 1988.
13. W Fritz, W Merk, E Schluender. Chem Eng Sci 36:743–757, 1981.
14. S Susarla, GV Bhaskar, SMR Bhamidimarri. Environ Technol 14:159–166, 1993.
15. RL Gutafson, RL Albright, J Heisler, JA Lirio, OT Reid. Ind Eng Chem Fund 7: 107–115, 1968.
16. PC Chiang, EE Chang, JS Wu. Wat Sci Technol 35:279–285, 1997.
17. MA Ferro-Garcia, J Rivera-Utrilla, I Bautista-Toledo, C Moreno-Castilla. J Chem Tech Biotechnol 67:183–189, 1996.
18. T Kawai, K Tsutsumi. Colloid Polymer Sci 273:787–792, 1995.

19. DS Prakash, KV Athota, HL Greene, CA Vogel. AIChE Symp Ser 91:1–17, 1995.
20. B Okolo, C Park, MA Keane. J Colloid Interface Sci 226:308–317, 2000.
21. BK Pradhan, NK Sadle. Carbon 37:1323–1332, 1999.
22. R Kunin. Pure Appl Chem 46:205–211, 1976.
23. M Rozwadowski, M Lezanska, J Wloch, K Erdmann, R Golembiewski, J Kornatowski. Phys Chem Chem Phys 2:5510–5516, 2000.
24. M Kruk, M Jaroniec, R Ryoo, SH Joo. J Phys Chem B 104:7960–7968, 2000.
25. R Ryoo, SH Joo, S Jun. J Phys Chem B 103:7743–7746, 1999.
26. HT Shu, DY Li, AA Scala, YH Ma. Sep Purif Technol 11:27–36, 1997.
27. DM Ruthven. Chem Eng Prog 84:42–50, 1988.
28. C Moreno-Castilla, J Rivera-Utrilla, MV Lopez-Ramon, F Carrasco-Marin. Carbon 33:845–851, 1995.
29. PD Paulsen, BC Moore, FS Cannon. Carbon 37:1843–1853, 1999.
30. NA Eltekova, D Berek, I Novak, F Belliardo. Carbon 38:373–377, 2000.
31. B Singh, S Madhusudhanan, V Dubey, R Nath, NBSN Rao. Carbon 34:327–330, 1996.
32. C Brasquet, J Roussy, E Subrenat, P Le Cloirec. Environ Technol 17:1245–1252, 1996.
33. C Brasquet, E Subrenat, P Le Cloirec. Water Sci Technol 35(7):251–259, 1997.
34. A Seidel, E Tzscheutschler, KH Radeke, D Gelbin. Chem Eng Sci 40:215–222, 1985.
35. KS Kim, HC Choi. Water Sci Technol 38:95–101, 1988.
36. DR Yonge, TM Keinath, K Poznanska, ZP Jiang. Environ Sci Technol 19:690–694, 1985.
37. A Seidel, E Tzschutscler, K Radeke, D Gelbin. Chem Eng Sci 40:215–222, 1985.
38. C Park, MA Keane. Catal Commun 2:171–177, 2001.
39. C Park, ES Engel, A Crowe, TR Gilbert, NM Rodriguez. Langmuir 16:8050–8056, 2000.
40. BM van Vliet, WJ Weber Jr, H Hozumi. Water Res 14:1719–1728, 1980.
41. CD Chriswell, RL Ericson, GA Junk, KW Lee, JS Fritz, H Svec. J Am Wat Wks 69:669–674, 1977.
42. JA Mendendez, J Phillips, B Xia, LR Radovic. Langmuir 12: 4404–4410, 1996.
43. MV Lopez-Ramon, F Stoeckli, C Moreno-Castilla, F Carrasco-Marin. Carbon 37: 1215–1221, 1999.
44. G Calleja, J Serna, J Rodriguez. Carbon 31: 691–697, 1993.
45. SS Barton, MJB Evans, E Halliop, JAF MacDonald. Carbon 35:1361–1366, 1997.
46. F Haghseresht, GQ Lu, AK Whittaker. Carbon 37:1491–1497, 1999.
47. J Phillips, D Kelly, L Radovic, F Xie. J Phys Chem B 104:8170–8176, 2000.
48. Y Ueno, Y Muramatsu, MM Grush, RCC Perera. J Phys Chem B 104:7154–7162, 2000.
49. CT Hsieh, HS Teng. J Colloid Interface Sci 230:171–175, 2000.
50. RC Wang, SC Chang. J Chem Technol Biotechnol 74:647–654, 1999.
51. AV Shevade, SY Jiang, KE Gubbins. J Chem Phys 113:6933–6942, 2000.
52. G Calleja, J Serna, J Rodriguez. Carbon 31:691–697, 1993.
53. D Chatzopoulos, A Varma. Chem Eng Sci 50:127–141, 1995.
54. TM Grant, CJ King. Ind Eng Chem Res 29:264–271, 1990.

55. A Das, DK Sharma. Energ Source 20:821–830, 1998.
56. D Chatzopoulos, A Varma, RL Irvine. Environ Prog 13:21–25, 1994.
57. EA Muller, LF Rull, LF Vega, KE Gubbins. J Phys Chem 100:1189–1196, 1996.
58. JA Menendez, J Phillips, B Xia, LR Radovic. Langmuir 12:4404–4410, 1996.
59. JA Menendez, J Phillips, B Xia, LR Radovic. J Phys Chem 100:17243–17248, 1996.
60. JA Menendez, J Phillips, B Xia, LR Radovic. Langmuir 13:3414–3421, 1997.
61. DM Nevskaia, A Santianes, V Munoz, A Guerrero-Ruiz. Carbon 37:1065–1074, 1999.
62. GO Wood. Carbon 30:593–599, 1992.
63. MW Thwaites, ML Stewart, BE McNeese, MB Sumner. Fuel Process Technol 34: 137–145, 1993.
64. MP Cal, SM Larson, MJ Rood. Environ Prog 13:26–30, 1994.
65. E Papier, E Brendle, F Ozil, H Balard. Carbon 37:1265–1274, 1999.
66. AO Eremina, VV Golovina, ML Shchipko, EV Burmakina. Russ J Appl Chem 73: 266–268, 2000.
67. C Lepont, AD Gunatillaka, CF Poole. Analyst 126:1318–1325, 2001.
68. NK Shoniya, EV Vlasenko, GN Filatova, VV Avdeed, IV Nikol'skaya, IY Martynov. 73:2009–2013, 1999.
69. T Baird, JR Fryer, B Grant. Nature 233:329–330, 1971.
70. N Hatta, K Murata. Chem Phys Lett 217:398–402, 1994.
71. TW Ebbesen, PM Ajayan. Nature 358:220–222, 1992.
72. V Ivanov, A Fonseca, JB Nagy, A Lucas, P Lambin, D Bernaerts, XB Yang. Carbon 33:1727–1738, 1995.
73. RTK Baker, PS Harris. In: PE Thrower, PL Walker Jr, eds. Chemistry and Physics of Carbons. New York: Marcel Dekker, 1978, Chap 2, pp 83–165.
74. RTK Baker. Carbon 27:315–323, 1989.
75. NM Rodriguez. J Mater Res 8:3233–3250, 1993.
76. RTK Baker. J Adhesion 52:13–40, 1995.
77. NM Rodriguez, A Chambers, RTK Baker. Langmuir 11:3862–3866, 1995.
78. M-S Kim, NM Rodriguez, RTK Baker. Synthesis and Properties of Advanced Catalytic Materials, Mat Res Soc Symp Proc, Pittsburgh, Materials Research Society, Vol 368, 1995, pp 99–104.
79. A Chambers, RTK Baker. J Phys Chem B 101:1621–1630, 1997.
80. RTK Baker, MS Kim, A Chambers, C Park, NM Rodriguez. Studies Surf Sci Catal 111:99–109, 1997.
81. PK de Bokx, AJHM Kock, E Boellard, W Klop, JW Geus. J Catal 96:454–467, 1985.
82. AJHM Kock, PK de Bokx, E Boellard, W Klop, JW Geus. J Catal 96:468–480, 1985.
83. E Boellard, PK de Bokx, AJHM Lock, JW Geus. J Catal 96:481–490, 1985.
84. MS Hoogenraad, RAGMM van Leeuwarden, GJB van Breda Vriesman, A Broersma, AJ van Dillen, JW Geus. Studies Surf Sci Catal 91:263–271, 1995.
85. MS Hoogenraad, MF Onwezen, AJ van Dillen, JW Geus. Studies Surf Sci Catal 101:1331–1339, 1996.
86. KP de Jong, JW Geus. Catal Rev Sci Eng 42:481–510, 2000.

87. DL Trimm. Catal Rev Sci Eng 16:155–189, 1977.
88. JR Rostrup-Nielsen, DL Trimm. J Catal 48:155–165, 1977.
89. CH Bartholomew. Catal Rev Sci Eng 24:67–114, 1982.
90. RJ Best, WW Russell. J Am Chem Soc 76:838–842, 1954.
91. C Park, MA Keane. Chem Phys Chem 2:101–109, 2001.
92. H Oda, C Yokokawa. Carbon 21:485–489, 1983.
93. E Esumi, Y Kimura, Y Nayada, K Meguro, H Honda. Carbon 27:301–303, 1989.
94. K Esumi, M Suglura, T Mori, K Meguro, H Honda. Colloid Surf 19:331–336, 1986.
95. C Park, NM Rodriguez, RTK Baker. J Catal 169:212–227, 1997.
96. MS Kim, NM Rodriguez, RTK Baker. J Catal 134:253–268, 1992.
97. VI Zaikovskii, VV Chesnokov, RA Buyanov. Kinet Catal 40:552–555, 1999.
98. C Park, MA Keane. Solid State Ionics 141–142:191–195, 2001.
99. M Kawaguchi, K Nozaki, S Motojima, H Iwanaga. J Cryst Growth 118:309–313, 1992.
100. S Motojima, I Hasegawa, S Kagiya, M Momiyama, M Kawaguchi, H Iwanaga. Appl Phys Lett 62:2322–2323, 1993.
101. C Park, RTK Baker. J Catal 190:104–117, 2000.
102. S Motojima, M Kawaguchi, K Nozaki, H Iwanaga. Carbon 29:379–385, 1991.
103. PL Walker Jr, M Shelef, RA Anderson. In: PL Walker Jr, ed. Chemistry and Physics of Carbon. Vol 1. New York: Marcel Dekker, 1968, pp 287–383.
104. JC Liu, CP Huang. J Colloid Interf Sci 153:167–176, 1992.
105. CA Léon, JM Solar, V Calemma, LR Radovic. Carbon 30:797–811, 1992.
106. JS Mattson, HB Mark Jr. Activated Carbon, Surface Chemistry and Adsorption from Solution. New York: Marcel Dekker, 1971.
107. PC Singer, CY Yen. In: MJ McGuirie, IH Suffet, eds. Activated Carbon Adsorption from Aqueous Phase. Ann Arbor, MI: Ann Harbor Science, Vol 1, 1980.
108. D Santoro, R Louw. Carbon 39:2091–2099, 2001.
109. R Louw, P Mulder, J Environ Sci Health A25:555–569, 1990.
110. IWCE Arends, WR Ophurst, R Louw, P Mulder. Carbon 34:581–588, 1996.
111. M Farcasiu, SC Petronius, EP Lardner. J Catal 146:313–316, 1994.
112. M Terrones, WK Hsu, HW Kroto, DRM Walton. Topics Curr Chem 199:189–234, 1999.
113. C Park, RTK Baker. J Phys Chem B 102:5168–5177, 1998.
114. C Park, RTK Baker. J Phys Chem B 104:4418–4424, 2000.
115. EJ Shin, MA Keane. J Hazard Mater B 66:265–278, 1999.
116. C Menini, C Park, EJ Shin, G Tavoularis, MA Keane. Catal Today 62:355–366, 2000.
117. EJ Shin, MA Keane. Chem Eng Sci 54:1109–1120, 1999.

11

Effective Acidity-Constant Behavior Near Zero-Charge Conditions

NICHOLAS T. LOUX   U.S. Environmental Protection Agency, Athens, Georgia, U.S.A.

I. INTRODUCTION

Current geochemical paradigms for modeling the solid/water partitioning behavior of trace toxic ionic species at subsaturation mineral solubility porewater concentrations rely on two fundamental mechanisms: (1) solid solution formation with the major element solid phases present in the environment, and (2) adsorption reactions on environmental surfaces. Solid solution formation is the process leading to the substitution of a trace ion for a major ion in a natural solid phase (e.g., Ref. 1). For example, solid solution formation between Cr^{3+} and $Fe(OH)_3$ has been reported in the literature as a possible porewater solubility–limiting mechanism for dissolved Cr^{3+}. This reaction can be described by

$$n Cr^{3+} + Fe(OH)_3 \Leftrightarrow n Fe^{3+} + Fe_{(1-n)} Cr_n (OH)_3$$

where $n < 1$ [2].

The second mechanism, the topic of this chapter, is generally believed to be more widespread in environmental systems and is frequently described as the result of surface complexation reactions between ionizable species (Me^{z+}) and reactive surface sites ($>SOH$) present on environmental solids, including iron oxides, manganese oxides, aluminum oxides, silicon oxides, aluminosilicates, and particulate organic carbon. For example, a reaction of the form

$$Me^{z+} + {>}SOH \Leftrightarrow {>}SOMe^{(z-1)+} + H^+$$

can be described by the following generic mass action expression (e.g., see Ref. 3 and applications in Ref. 4):

$$K_{rxn} = \frac{[{>}SOMe^{(z-1)+}] a_{(H+)} e^{-\Delta G(excess)/RT}}{a_{Me(z+)} [{>}SOH]} \tag{1}$$

where

K_{rxn}	= formation constant for the rxn
$a_{(H+)}$	= bulk solution H^+ chemical activity
z	= valence of cation
R	= gas constant
$a_{Me(z+)}$	= bulk solution metal ion activity
$[>SOMe^{(z-1)+}]$	= concentration of complexed sites
e	= base of natural logarithm
$\Delta G(excess)$	= excess free energy
T	= absolute temperature
$[>SOH]$	= concentration of unbound sites

Equation (1) differs from a solution counterpart in two ways: (1) Analogous to surface protonation reactions, Eq. (1) is a mixed concentration/chemical activity expression. Most practitioners make the assumption(s) that was (were) originally applied to surface protonation reactions that the activity coefficients for bound sites are equal and hence cancel out in the mass action quotient. And (2), the presence of the exponential Boltzmann expression ($e^{-\Delta G(excess)/RT}$). The Boltzmann expression as commonly used is generally predicated on the assumption that any excess energy is primarily electrostatic in nature (i.e., $\Delta G_{excess} = \Delta G_{electrostatic}$) and that this energy results from moving mobile ions between bulk solution (where $\Delta G_{electrostatic} = 0$) and the interfacial region (where $\Delta G_{electrostatic} \neq 0$) (e.g., see Ref. 5).

By inspection of Eq. (1), one can observe that there is an inherent competition for reactive bound sites between metal ions and the hydrated proton. Pragmatically speaking, an inspection of Eq. (1) leads to a predicted "release" of bound (i.e., surface-complexed) metal ions when a solid/liquid system is acidified. Due to recognition of the inherent competition for bound sites by the hydrated proton and fundamental uncertainties in our ability to describe surface acidity reactions, two publications [6,7] concluded that the majority of uncertainty in our ability to model ionic contaminant adsorption behavior was due to limitations in our understanding of surface acidity behavior. Hence, a fundamental understanding of the protonation behavior of reactive sites on environmental surfaces is a prerequisite to a better understanding of the partitioning behavior of the ionizable species of toxicological interest.

Most researchers use the two-pK surface complexation model for describing the protonation behavior of environmental hydrous oxide adsorbents. They generally assume that bound surface sites can exist in one of three protonation conditions: $>SOH_2^+$, $>SOH$, and $>SO^-$. Mass action expressions commonly used for quantifying the equilibration among protonated surface sites in response to the chemical activity of the hydrated proton are:

$$K_{a1} = \frac{[>SOH]a_{(H+)}e^{-\Delta G(\text{electrostatic})/RT}}{[>SOH_2^+]} \tag{2}$$

$$K_{a2} = \frac{[>SO^-]a_{(H+)}e^{-\Delta G(\text{electrostatic})/RT}}{[>SOH]} \tag{3}$$

where the symbols are as defined previously. Activity coefficients for bound sites are ignored based on one or more of three assumptions: (1) $\gamma_{>SOH(x+1)} = \gamma_{>SOH(x)}$ [8–9], (2) the activity coefficients for the bound sites are already incorporated into the Boltzmann expression [10], or (3) the bound surface sites display ideal behavior (i.e., the activity coefficients $\gamma_{>SOH(x+1)}$ and $\gamma_{>SOH(x)}$ are both equal to 1 [11]).

For both computational convenience and as a result of experimental difficulties in measuring $\Delta G_{\text{electrostatic}}$, a number of authors adapted procedures previously applied to polyelectrolytes/latex particles [12–18] and rearranged Eqs. (2) and (3) into forms that are more amenable to computation from experimental data:

$$Q_{a1} = K_{a1}e^{\Delta G(\text{electrostatic})/RT} = \frac{[>SOH]a_{(H+)}}{[>SOH_2^+]} \tag{4}$$

$$Q_{a2} = K_{a2}e^{\Delta G(\text{electrostatic})/RT} = \frac{[>SO^-]a_{(H+)}}{[>SOH]} \tag{5}$$

These Q_a terms represent "ionization quotients," "concentration quotients," or effective acidity constants. Previous authors utilized Eqs. (4) and (5) for the purpose of estimating the intrinsic acidity constants by extrapolating Q_{a1} and Q_{a2} to conditions where $\Delta G_{\text{electrostatic}} = 0$ (mathematically, $Q_{a1} = K_{a1}$ and $Q_{a2} = K_{a2}$ when $\Delta G_{\text{electrostatic}} = 0$). For the purposes of this document, this extrapolation methodology for estimating intrinsic acidity constants will be termed the pH$_{zpc}$ extrapolation procedure (the pH$_{zpc}$ is the pH zero point of charge, i.e., the pH where $[>SOH_2^+] = [>SO^-]$ or the pH estimated by pH $= \frac{1}{2}[pK_{a1} + pK_{a2}]$). Of significance to the present study is that variations of Q_{a1} and Q_{a2} as functions of charge density, pH, and ionic strength can lend insight into the nature of those energies contributing to ΔG_{excess}.

Equations (1) to (5) are generally utilized with the assumption that the excess electrostatic Gibbs free energies for these systems (ΔG_{excess}) are reasonably approximated by integer multiples of $F\Psi$ (where F equals Faraday's constant and Ψ is the electrostatic potential in the interfacial region). As will be demonstrated in the next section, there are theoretical reasons to question this assumption.

A. Origin of the Charging-Energy Term

Chan et al. [9] defined the electrochemical potentials (u) of the surface reacting species in Eqs. (2) and (3) in the following way:

$$u_{H(+)} = u_{H(+)}^o + kT \ln(a_{H(+)}) - e\Psi \tag{6a}$$

$$u_{>SOH} = u_{<SOH}^o + kT \ln([>SOH]) + kT \ln(\gamma_{>SOH}) \tag{6b}$$

$$u_{>SOH2(+)} = u_{>SOH2(+)}^o + kT \ln([>SOH_2^+]) + kT \ln(\gamma_{>SOH2+}) + e\Psi \tag{6c}$$

$$u_{>SO(-)} = u_{>SO(-)}^o + kT \ln([>SO^-]) + kT \ln(\gamma_{>SO-}) - e\Psi \tag{6d}$$

where $\gamma_{>SOHx}$ is the activity coefficient for surface site $>SOHx$, e is the charge of the electron, and k is the Boltzmann constant. The electrostatic component of the electrochemical potential of the interfacial hydrated proton ($e\Psi$) in Eq. (6a) has been discussed extensively in the literature and results from moving mobile ions between bulk solution (where $\Psi = 0$) and the charged interfacial region (where $\Psi \neq 0$; e.g., see Ref. 5). The electrostatic components of the electrochemical potentials of the ionized surface sites in Eqs. (6c) and (6d) can be viewed as being representative of the charging energies associated with creating a net charge of $\pm e$ in an environment of constant potential Ψ. If one defines $\Delta G^o = \Sigma(u_{products}^o) - \Sigma(u_{reactants}^o)$, $K = e^{-\Delta Go/RT}$, and one assumes that the bound site activity coefficients in Eqs. (6b) to (6d) equal one another, then the electrostatic component of ΔG in the Boltzmann expression in Eqs. (4) and (5) ($\Delta G_{electrostatic}$) as derived from these electrochemical potentials should be $2e\Psi$ (on a per-ion basis) or $2F\Psi$ (on a molar basis) rather than the traditional value of $e\Psi$ or $F\Psi$. Specifically, with this thermodynamic analysis of surface protonation/deprotonation reactions occurring in the absence of surface charge neutralization by counterelectrolyte ions, the estimated energy in the Boltzmann term of $2F\Psi$ results from one $F\Psi$ being attributable to moving a mobile ion between neutral bulk solution and the charged interfacial region and one $F\Psi$ resulting from the creation of a site with a unit charge of "$\pm e$" under conditions of constant potential Ψ.

The present author [19] further examined charging energies by integrating a spherical Coulombic charge/potential relationship: $\Psi = Q/4\pi\varepsilon\varepsilon_0 r$(where $Q =$ the particle charge, $\varepsilon =$ the aqueous dielectric constant, $\varepsilon_0 =$ the permittivity of free space, and $r =$ the particle radius) from Q to $Q \pm e$. Specifically, Ref. 19 integrated ΨdQ from Q to $Q \pm e$ and derived a charging energy term of:

$$\Delta G_{charging} = \frac{(Q \pm e)^2 - Q^2}{8\pi\varepsilon\varepsilon_0 r}$$

(assuming an integration constant of zero). It was also demonstrated that when $Q \gg e$, then $\Delta G_{charging} \approx \pm e\Psi$. This analysis was predicated on the assumption that the surface region where charged sites are located is impenetrable to counterelectrolyte ions. Based on this analysis, $\Delta G_{electrostatic}$ in Eqs. (2) to (5) also should equal $2e\Psi$ (on a per-ion basis) or $2F\Psi$ (on a molar basis) under constant-potential conditions.

The present author [19] also examined circumstances where electrolyte ions can penetrate the surface region and partially neutralize the charge associated with the created charged site. Given that the fraction of net surface charge neutralized by electrolyte ions is assigned a value of τ (where τ ranges from zero to 1 [5,19–20]), the author integrated ΨdQ from Q to $Q \pm (1 - \tau)e$ and derived an integral of

$$\Delta G_{\text{charging}} = \frac{(Q \pm [1 - \tau]e)^2 - Q^2}{8\pi\varepsilon\varepsilon_0 r}$$

When $Q \gg e$, the charging energy was found to be approximated by $\Delta G_{\text{charging}} \approx \pm(1 - \tau)e\Psi$. If one then derived a mass action expression from the chemical potentials of the reacting species, the total electrostatic expression in the Boltzmann term ($\Delta G_{\text{electrostatic}}$) of the respective mass action expressions given in Eqs. (2) to (5) was estimated to be $(2 - \tau)F\Psi$ (on a molar basis) or $(2 - \tau)e\Psi$ (on a per-ion basis). Finally, through extensive computer simulations, it was also observed that $(2 - \tau)$ approaches a value of 1 at high charge densities for all ionic strengths (thereby supporting the historical mass action formulations). However, it also was predicted that $(2 - \tau)$ would significantly deviate from a value of 1 at low-charge conditions. In essence, it was predicted that charging energies will lead to increased values of calculated pQ_{a1} and pQ_{a2} terms in the pH_{zpc} region that is inconsistent with conventional diffuse layer modeling.

B. Significance of Aggregation-Derived Neutral Size Sequestration

Traditional approaches for using the pH_{zpc} extrapolation procedure in biprotic systems have relied on the assumption of monoprotic behavior both above and below the pH_{zpc}. Specifically, below the pH_{zpc} the concentration of negatively charged sites is assumed to be insignificant, and above the pH_{zpc} the concentration of positively charged sites is assumed to be insignificant. The rigorous definitions for pQ_{a1} and pQ_{a2} are given by

$$pQ_{a1} = pH - \log\frac{[>SOH]}{[>SOH_2^+]} \quad \text{and} \quad pQ_{a2} = pH - \log\frac{[>SO^-]}{[>SOH]}$$

However, if one defines a charge density σ and a maximum charge density σ_{tot} by

$$\sigma = \frac{\{[>SOH_2^+] - [>SO^-]\}F}{\{SSA * SC\}} \tag{7}$$

$$\sigma_{\text{tot}} = \pm\frac{\{[>SOH_2^+] + [>SOH] + [>SO^-]\}F}{\{SSA * SC\}} \tag{8}$$

(where SSA = specific surface area $[m^2/g]$ and SC = solids concentration $[g/L]$), then approximations incorporating the monoprotic behavior assumptions for calculating pQ_{a1} and pQ_{a2} values for titrimetric data are given by

$$pQ_{a1} = pH - \log \frac{(\sigma_{tot} - \sigma)}{\sigma} \qquad \text{(below the pH}_{zpc})$$

and

$$pQ_{a2} = pH - \log \frac{\sigma}{-(-\sigma_{tot} - \sigma)} \qquad \text{(above the pH}_{zpc})$$

As will be demonstrated in Section III, the assumptions of monoprotic behavior below and above the pH_{zpc} with the pH_{zpc} extrapolation procedure leads to an underestimate of the true pQ_{a1} values and an overestimate of the true pQ_{a2} values in the pH_{zpc} region. As a first approximation, these errors are the result of assuming that [>SOH] is directly proportional to $(\sigma_{tot} - \sigma)$ (below the pH_{zpc}) and $-(-\sigma_{tot} - \sigma)$ (above the pH_{zpc}) and that [>SOH$_2^+$] is directly proportional to σ (below the pH_{zpc}) and that [>SO$^-$] is proportional to σ (above the pH_{zpc}). In summary, these approximations suffer from an error that increases with proximity to the pH_{zpc} and is the result of simultaneously overestimating [>SOH] and underestimating charged site concentrations in the pH_{zpc} region.

It is hypothesized here that there exists an experimental artifact that can have a similar effect. Specifically, it is not uncommon for an experimenter to observe substantial aggregation in titrations at pH conditions adjacent to the pH_{zpc}. This phenomenon may be responsible for the widely reported observed hysteresis in forward and backward titrations of hydrous oxide slurries. Secondly, it is not unreasonable to believe that aggregation will render some sites inaccessible to a given titrant (at least within the equilibration times commonly used in these experiments). Finally, given the local acid–base disequilibrium conditions that exist prior to complete mixing of a titrant addition to a slurry in an experimental vessel, it is hypothesized here that neutral and oppositely charged sites will tend to be preferentially "buried" during the aggregation process. Qualitatively, and in contrast to charging energy phenomena, aggregation-derived sequestration of titrable sites in the pH_{zpc} region is predicted to cause the same type of error observed with the pH_{zpc} extrapolation procedure. That is, this error is hypothesized to simultaneously decrease pQ_{a1} estimates and increase pQ_{a2} estimates in the pH_{zpc} region.

The remainder of this chapter will focus on: (1) developing a method to generate simulated titrimetric data of known accuracy (using 17-digit double-precision GW-BASIC [21]), (2) developing two alternative methods to the pH_{zpc} extrapolation procedure for extracting Q_a values from titrimetric data, (3) assessing all three methods with simulated data, and, finally, (4) applying these methods to titrimetric data published in the literature for the purpose of identifying possible charging energy and/or aggregation-derived titrable site sequestration contributions to effective acidity-constant behavior.

II. METHODS

A. A Method for Simulating Titrimetric Data

If one combines Eqs. (4), (5), (7), and (8), the following expression for a biprotic system can be derived:

$$a_{H(+)}^2(\sigma_{tot} - \sigma) - a_{H(+)}Q_{a1}\sigma - Q_{a1}Q_{a2}(\sigma_{tot} + \sigma) = 0 \tag{9}$$

Expression (9) is particularly useful; among other things, it may be used to simulate titrimetric data. For a given system with specified values for temperature, ionic strength, σ_{tot}, K_{a1}, and K_{a2} and assuming traditional diffuse layer model behavior, one can ultimately estimate the hydrogen ion activities required to yield a given value of σ with the quadratic solution. For example, for a given value of σ, one can first calculate a value of Ψ using the Gouy–Chapman 1-dimensional solution to the Poisson–Boltzmann equation; e.g., at 25°C,

$$\Psi = \sinh^{-1}\frac{\sigma/\{0.1174 * I^{1/2}\}}{19.46 * z}$$

Values for Q_{a1} and Q_{a2} can then be generated by

$$Q_{a1} = K_{a1}e^{F\Psi/RT} \qquad \text{and} \qquad Q_{a2} = K_{a2}e^{F\Psi/RT}$$

Finally, with the substitutions $a = (\sigma_{tot} - \sigma)$, $b = -Q_{a1}\sigma$, and $c = -Q_{a1}Q_{a2}(\sigma_{tot} + \sigma)$, the hydrogen ion activity required to achieve a given value of σ can be calculated by

$$a_{H(+)} = \frac{-b \pm (b^2 - 4ac)^{1/2}}{2a}$$

B. Alternate Methods for Estimating Effective Acidity Constants

For a monoprotic surface (e.g., a latex bead with one anionic functional group),

$$Q_a = \frac{[>SO^-]a_{H+}}{[>SOH]}$$

$$\sigma = -\frac{\{[>SO^-]\}F}{(SSA * SC)}$$

$$\sigma_{tot} = -\frac{\{[>SOH] + [>SO^-]\}F}{\{SSA * SC\}}$$

Hence, Q_a values can be extracted from titrimetric data for a monoprotic system with the expression $Q_a = \sigma a_{H+}/(\sigma_{tot} - \sigma)$. In contrast to biprotic systems, these values for Q_a can be obtained directly from titrimetric data without the approximation errors in relating $[>SO^-]$ and $[>SOH]$ to σ_{tot} and σ.

Equation (9) also may be used to extract Q_{a1} and Q_{a2} values from experimental data derived from a biprotic system. By inspection of Eq. (9), the reader can discern that for any given data point characterized by $a_{H(+)}$ and σ (and where σ_{tot} is known), one cannot solve explicitly for Q_{a1} and Q_{a2} because there exists only one equation [Eq. (9)] and two unknowns. In theory however, Eq. (9) can be solved for two unknowns by using two adjacent data points in a titration curve if the effective acidity constants can be assumed to remain nearly constant for these two points. Although only two data points are required with this procedure, the present author [20] found that solving for values of Q_{a1} and Q_{a2} twice using three consecutive data points and averaging the values tended to minimize extreme estimates of Q_a behavior. The procedure of solving Eq. (9) twice with three consecutive data points and averaging the results will be used in this work and will be termed the *direct substitution procedure*.

One may also take partial derivatives of Eq. (9) with respect to $a_{H(+)}$ and σ and obtain the following relationship:

$$\frac{\partial(\sigma)}{\partial(a_{H(+)})} = \frac{2\sigma_{tot}a_{H(+)} - 2a_{H(+)}\sigma - \sigma Q_{a1}}{a_{H(+)}^2 + a_{H(+)}Q_{a1} + Q_{a1}Q_{a2}} \tag{10}$$

As with the direct substitution method, differentials can be taken between a given data point and the two data points preceding and following the central point. Average effective pQ values can then be calculated by averaging the values obtained from twice solving two equations for two unknowns.

In summary, this work will involve using Eq. (9) to generate simulated titration curves at various ionic strengths for a biprotic surface (with intrinsic acidity constants of 10^{-6} and 10^{-8}) using the Gouy–Chapman charge/potential relationship. These computations will be performed using double-precision GWBASIC^R with an accuracy of 17 digits [21]. Data obtained from the simulated curves will then be subjected to the conventional pH_{zpc} extrapolation procedure and the substitution and differential methodologies described earlier for the purpose of assessing the accuracy of these methods for extracting Q_{a1} and Q_{a2} values from the simulated experimental data. Lastly, these extraction methodologies will be applied to experimental data obtained from the peer-reviewed literature for the purpose of interpreting anomalous pQ behavior in the pH_{zpc} region within the context of possible charging-energy and aggregation-derived site sequestration phenomena.

III. RESULTS

A. Results from Simulated Data

Figure 1 illustrates simulated pQ_{a1} values as a function of ionic strength derived for a Gouy–Chapman surface with intrinsic acidity constants of $K_{a1} = 1E-6$ and $K_{a2} = 1E-8$. The maximum site density for this surface was set at 0.32 C/m², and the temperature was held at 298 K with these simulations. The "fictional" 10^4 M ionic strength simulations were used to saturate the Gouy–Chapman electrostatic term (i.e., the maximum estimated surface potential at an "ionic strength" of 1E4 M was estimated to be ± 0.0004 V). The reader should note that the pQ_{a1} values for ionic strengths 1E-1, 1E-2, and 1E-3 M display logistic or S-shaped curves as functions of charge density; these shapes are more characteristic of a diffuse layer model of the interface. The pQ_{a1} values at an ionic strength of 1E-1 M generate a more "linear" curve and, hence, illustrate a possible situation

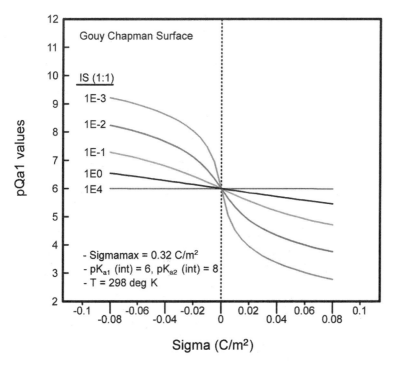

FIG. 1 Simulated pQ_{a1} values as functions of ionic strength and charge density for a Gouy–Chapman diffuse layer surface in aqueous solution. Maximum charge density = 0.32 C/m², $T = 298$ K, $pK_{a1} = 6$, and $pK_{a2} = 8$.

for justifiably using a constant-capacitance-charge/potential relationship [22]. Although not shown here, the pQ_{a2} values displayed identical curves that were offset from the pQ_{a1} values by 2 pK units.

Figure 2 displays simulated titration data for the biprotic Gouy–Chapman surface described in Figure 1. These data were generated by inserting the previously estimated pQ_{a1} and pQ_{a2} values used to construct Figure 1 into the quadratic equation [Eq. (9)] and solving for the required hydrogen ion activity.

Figure 3 depicts estimated values of pQ_{a1} and pQ_{a2} extracted from the simulated data at an ionic strength of 1E4 M displayed in Figure 2. It is gratifying to note that the substitution and differential procedures yielded effective acidity constants comparable to the "true" values for pQ_{a1} below the pH_{zpc} and for pQ_{a2} above the pH_{zpc}. The pH_{zpc} extrapolation methodology suffered from significant error in the pH_{zpc} region due to the assumption that the concentrations of oppositely charged sites both above and below the pH_{zpc} were insignificant. Estimated

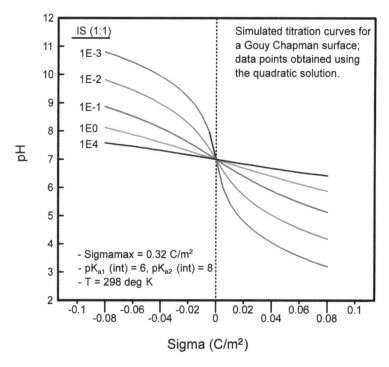

FIG. 2 Simulated titration curves for the Gouy–Chapman surface discussed in Figure 1 using the quadratic equation [Equation (9)]; the "fictional" ionic strength of 1E-4 molar was performed to minimize electrostatic effects.

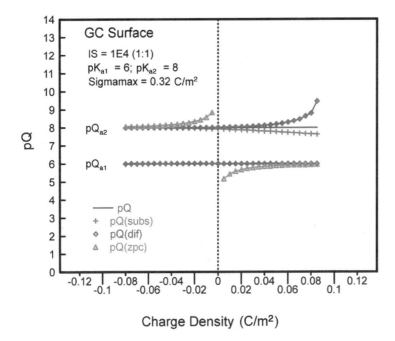

FIG. 3 Comparison of "true" pQ values with the pH_{zpc} extrapolation procedure, substitution, and differential methodologies for estimating effective acidity constants from the 1E4 M simulated data presented in Figure 2. Note the errors in the pH_{zpc} region using the pH_{azpc} extrapolation procedure. The substitution and differential methodologies yield significant deviations for pQ_{a2} below the pH_{zpc} (possibly due to round-off errors).

values for pQ_{a2} below the pH_{zpc} with the substitution and differential procedures displayed diminished accuracy, possibly due to round-off errors in the algorithm used to estimate these numbers.

Figure 4 compares the results from the three extraction methodologies as applied to the 1E-3 M simulated data displayed in Figure 2. In contrast to the results displayed in Figure 3, the pH_{zpc} methodology yields a slightly superior accuracy (when compared to its relative performance in Figure 3). The marginally improved performance of the pH_{zpc} extrapolation procedure at an ionic strength of 1E-3 M (when compared with the substitution and differential methodologies) can be ascribed to the fact that the performance of the two other methodologies degrade, possibly due to nonequivalence of pQ_a values between adjacent data points.

The results displayed in Figures 3 and 4 tend to support a contention that there is no perfect methodology for extracting pQ_a values from titrimetric data for

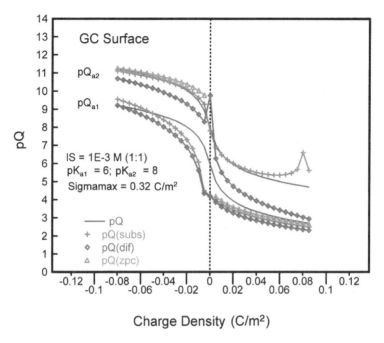

FIG. 4 Comparison of "true" pQ values with the pH_{zpc} extrapolation procedure, substitution, and differential methodologies for estimating effective acidity constants from the 1E-3 M simulated data presented in Figure 2. Note the diminished errors in the pH_{zpc} region using the pH_{zpc} extrapolation procedure. The substitution and differential methodologies yield significant deviations from the "true" $pQ_{a1,a2}$ values in the vicinity of the pH_{zpc}, presumably because of violations of the assumption of equivalence of pQ values between adjacent data points. The differential methodology yields excessive errors in pQ_{a2} estimates with this data.

biprotic systems. These analyses were performed on computer-simulated data of 17-digit accuracy that is unachievable with current experimental methodologies.

Figure 5 is a pictorial representation of the predicted generic effects expected near the pH_{zpc} in the event that either charging energies or site-sequestration phenomena become significant in titration datasets derived from biprotic systems. Both site-sequestration and charging-energy phenomena are predicted to increase the relative pQ_{a2} values in the vicinity of the pH_{zpc}. However, significant site sequestration is expected to decrease calculated pQ_{a1} values, and charging energies are predicted to increase pQ_{a1} estimates in the pH_{zpc} region. This difference in behavior can then presumably be used to distinguish between these two phenomena.

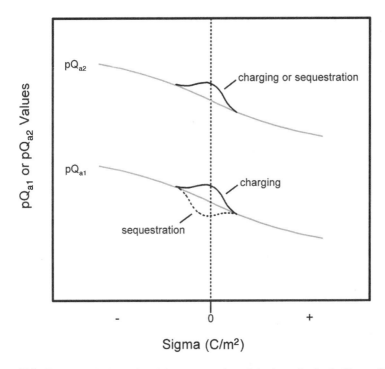

FIG. 5 A qualitative, pictorial representation of the hypothesized effects of both charging-energy and aggregation-induced site-sequestration effects on pQ behavior with effective acidity-constant estimates derived from potentiometric titration data. Both charging energies and site sequestration are predicted to increase pQ_{a2} values in the vicinity of zero-charge conditions; charging energies are predicted to increase pQ_{a1} values, and site sequestration is hypothesized to decrease pQ_{a1} estimates in the pH$_{zpc}$ region.

B. pQ Values Derived from Data in the Published Literature

Figure 6 illustrates another possible means of distinguishing between possible charging-energy and site-sequestration phenomena in experimental potentiometric titration data. These pQ values were obtained from data published for a monoprotic latex (sigmamax = 0.091 C/m^2 [23]). The pQ values at an ionic strength of 1E-4 M display a significant upward trend near zero-charge conditions; this behavior would be consistent with either a charging-energy or site-sequestration phenomenon; i.e., these pQ values are equivalent to a pQ_{a2} formulation with a biprotic system. In contrast, the pQ values derived from data at an ionic strength of 1E-1 M display a downward trend near zero-charge conditions. Given that

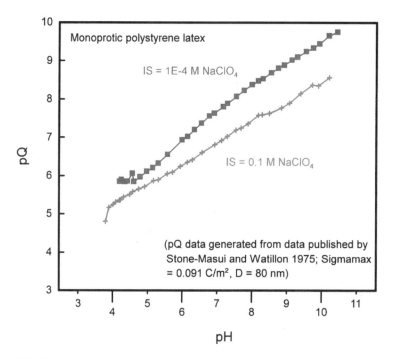

FIG. 6 pQ estimates from titrimetric data at ionic strengths of 1E-1 and 1E-4 M NaClO$_4$ M for a monoprotic latex. Given that pQ estimates for a monoprotic system do not require the assumptions required for analyzing data from biprotic substrates, the upward trend in estimated pQ values at low-ionic-strength and low-pH conditions is consistent with a charging-energy interpretation. (From Ref. 23.)

aggregation is well known to be enhanced at higher ionic strengths and that an upward curve in the pQ values is not observed with the higher-ionic-strength data, the upward curve observed with the low-ionic-strength, low-charge-density data can plausibly be attributed to a charging-energy phenomenon. The reader may recall that the methodology for estimating pQ values from potentiometric titration data derived from monoprotic systems does not require any of the assumptions utilized in analyzing data obtained from biprotic systems.

Figure 7 depicts estimated pQ_{a1} and pQ_{a2} values derived from titrimetric data for spherical anatase particles [24] using the two-pK model (sigmamax = 2.08 C/m^2; IS = 0.1 M KCl). These pQ_{a1} and pQ_{a2} profiles are inconsistent with a traditional diffuse layer model of the interface; specifically, the upward trends near zero-charge conditions would be consistent with a charging-energy phenom-

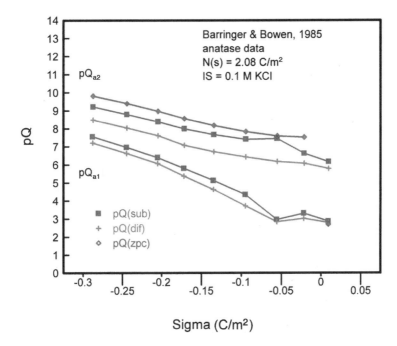

FIG. 7 pQ_{a1} and pQ_{a2} estimates from potentiometric titration data for anatase at an ionic strength of 0.1 M KCl. Estimates from the pH_{zpc} extrapolation procedure agree well with most estimates from the substitution methodology. Calculated differential estimates of pQ_{a2} do not agree as well with values estimated using the other two methodologies. The "spoon"-shaped curves near zero-charge conditions have been observed with numerous other datasets and are inconsistent with traditional diffuse layer theory. (Raw data from Ref. 24.)

enon. The performance of the differential methodology for estimating pQ_{a2} values is significantly degraded with these data.

Figure 8 illustrates estimated pQ_{a1} and pQ_{a2} values for the same spherical anatase samples depicted in Figure 7; the sole difference is that these data were derived at an aqueous ionic strength of 0.001 M KCl. As in Figure 7, the differential methodology yields results that differ significantly from the results obtained with the other methodologies when pQ_{a2} estimates are compared.

Figures 9–11 compare estimated pQ_{a1} and pQ_{a2} values derived from potentiometric titration data for corundum (Ref. 25, cited in Ref. 26). As with Ref. 26, the maximum site density for corundum is assumed to equal 22 sites/nm² (or 3.52 C/m²). The ionic strengths used to derive these data were 0.139 M, 0.03

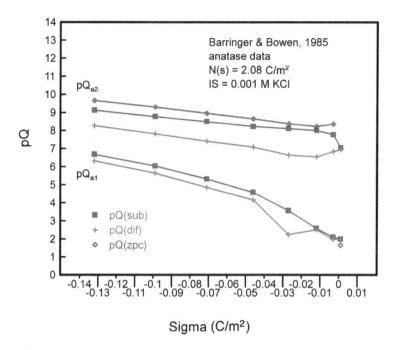

FIG. 8 pQ_{a1} and pQ_{a2} estimates from potentiometric titration data for anatase at an ionic strength of 0.001 M KCl. Estimates from the pH_{zpc} extrapolation procedure agree well with most estimates from the substitution methodology. Calculated differential estimates of pQ_{a2} do not agree as well with values estimated using the other two methodologies. A "fragment" of a traditional logistic S-shaped curve is observed with only one profile in Figure 8. (Raw data from Ref. 24.)

M, and 0.005 M $NaNO_3$. As in Figures 7 and 8, the differential methodology yields significantly different pQ_{a2} values than is observed with the substitution and pH_{zpc} extrapolation procedures. One also can observe significant deviations in the vicinity of the pH_{zpc} for the pH_{zpc} extrapolation procedure (when compared with the other methodologies). Generally speaking, the substitution and pH_{zpc} extrapolation methodologies yield comparable pQ estimates in regions of the curve distant from the pH_{zpc}.

Figure 12 displays estimated pQ_{a1} and pQ_{a2} values derived from forward and backward potentiometric titration data for rutile (Ref. 27, cited in Ref. 26). As with Refs. 26 and 28, the maximum site density for rutile is given a value of 12.5 sites/nm² (2 C/m²). In contrast to the previous figures, only pQ estimates from the substitution and pH_{zpc} extrapolation methodologies are presented. The solid lines designated pQ estimates from the forward titrations, and the dashed

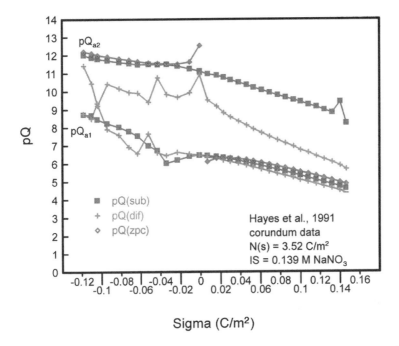

FIG. 9 pQ_{a1} and pQ_{a2} estimates from potentiometric titration data for corundum at an ionic strength of 0.139 M NaNO$_3$. Estimates from the pH$_{zpc}$ extrapolation procedure agree well with most estimates from the substitution methodology (at least for data distant from zero-charge conditions). Errors due to ignoring oppositely charged site concentrations may be operative in pH$_{zpc}$ extrapolation procedure estimates. Calculated differential estimates of pQ_{a2} fare poorly with values estimated using the other two methodologies. (Raw data from Refs. 25 and 26.)

lines represent pQ estimates from the backward titrations. Although the pQ estimates from the backward titration data tend to be "noisier" than the results from the forward titrations, both datasets tend to yield comparable pQ estimates. The pH$_{zpc}$ extrapolation methodology appears to yield significant error in the vicinity of the pH$_{zpc}$ with these data. The findings depicted in Figure 12 tend to support a contention that reproducibility in forward and backward titrations can be experimentally achieved.

IV. CONCLUSIONS

A major conclusion of this work is that there exists no perfect methodology for extracting effective acidity constants from titrimetric data for biprotic systems.

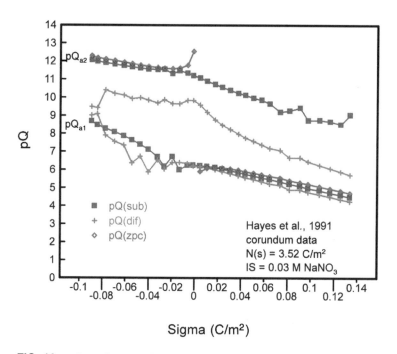

FIG. 10 pQ_{a1} and pQ_{a2} estimates from potentiometric titration data for corundum at an ionic strength of 0.03 M NaNO$_3$. Estimates from the pH$_{zpc}$ extrapolation procedure agree well with most estimates from the substitution methodology (at least for data distant from zero-charge conditions). As in Figure 9, errors due to ignoring oppositely charged site concentrations may be operative in pH$_{zpc}$ extrapolation procedure estimates. (Raw data from Refs. 25 and 26.)

The traditional pH$_{zpc}$ extrapolation procedure suffers from the assumption of monoprotic behavior on either side of the pH$_{zpc}$. Generally speaking, this assumption is most nearly correct with data points distant from the pH$_{zpc}$. The substitution and differential methodologies introduced in this work both suffer from the assumption of constant-pQ behavior between adjacent data points. Essentially, this approximation is best met in the absence of excess energies that can significantly alter pQ behavior (i.e., in systems that behave as soluble diprotic acids). In comparing all three methodologies, agreement was best between the pH$_{zpc}$ extrapolation and substitution procedures. This finding suggests that at a minimum, the substitution procedure may be useful for assessing the significance of ignoring oppositely charged sites with the pH$_{zpc}$ extrapolation methodology.

Although not presented here, the author has found that the accuracy of the substitution and differential methodologies is enhanced with a decrease in the size of surface charge intervals between points in the titration curve (this presumably

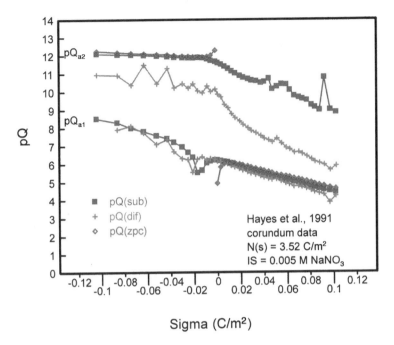

FIG. 11 pQ_{a1} and pQ_{a2} estimates from potentiometric titration data for corundum at an ionic strength of 0.005 M NaNO$_3$. Estimates from the pH$_{zpc}$ extrapolation procedure agree well with most estimates from the substitution methodology (at least for data distant from zero-charge conditions). Errors due to ignoring oppositely charged site concentrations would be significant in estimating intrinsic pK values near zero-charge conditions with the pH$_{zpc}$ extrapolation procedure with these data. (Raw data from Refs. 25 and 26.)

improves the assumption of the equivalence of pQ values for adjacent data points). This suggests two possible methods for improving the accuracy in estimating pQ values with these methodologies: (1) increasing the number of data points by decreasing the quantity of titrant used in each titrant addition, and (2) statistically fitting titration curves for the purpose of generating additional data points through interpolation. Clearly, the first methodology is likely to be preferable.

Given the difficulties in estimating pQ values from titrimetric data derived from biprotic systems, pQ values from data obtained with a monoprotic latex were presented in Figure 6. The pQ values obtained from data at an ionic strength of 1E-4 M suggested that charging energies may have contributed to pQ behavior near zero-charge conditions. The data from the biprotic systems, although more likely to be influenced by computational errors, also tend to support charging-energy contributions. Specifically, charging energies should increase both pQ_{a1}

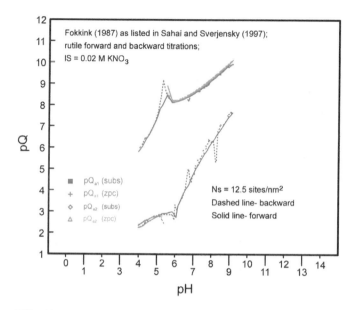

FIG. 12 pQ_{a1} and pQ_{a2} estimates from forward and backward titration potentiometric titration data for rutile at an ionic strength of 0.02 M KNO$_3$. Estimates from the differential procedure were not presented in this figure. Although pQ estimates from the backward titration data were "noisier," both datasets tended to yield comparable curves. The shapes of these curves are not consistent with the traditional logistic S-shaped curve expected from diffuse layer model theory. (Raw data from Refs. 26 and 27.)

and pQ_{a2} values in the vicinity of the pH_{zpc}; this trend is generally observed in the data displayed in this chapter.

The literature is rife with evidence of anomalous behavior in the pH_{zpc} region. For example, the titration data summarized in Ref. 26 contain several datasets illustrating hysteresis in the pH_{zpc} region of surface-charge/pH data when forward titration data is compared with data obtained from a back titration of the same sample. This hysteresis is consistent with the aggregation-derived site-sequestration phenomenon postulated earlier in this chapter. In contrast to purely electrostatic phenomena that are predicted to similarly offset pQ values from their "intrinsic" values, aggregation-derived site sequestration is predicted to decrease pQ_{a1} values and increase pQ_{a2} estimates. The pQ estimates derived from experimental data illustrated in Figures 7–12 also tend to support a contention that aggregation-derived phenomena may be influencing pQ estimates.

This work was conducted in an effort to demonstrate that effective acidity-constant behavior can be another means of probing interfacial excess free energies. Deviations from an ideal "logistic" or S-shaped curve are apparent in Figures 7–12; charging energies and/or site-sequestration phenomena can presumably account for at least some of these discrepancies.

More recent efforts at establishing databases of intrinsic adsorption/protonation constants for environmental surfaces have bypassed the use of graphical extrapolation methodologies and instead have focused on linear least squares analytical techniques (e.g., FITEQL [29–31]). However, an earlier publication in this area concluded that for a given experimental dataset, one can achieve comparable accuracy with any number of models [32]. The present findings may have significant implications for this conclusion. Specifically, based on the present work and the findings in Ref. 19, the error associated with statistically fitting an experimental potentiometric titration dataset to a diffuse layer model is likely to be decreased by excluding datapoints in the vicinity of the pH_{zpc}. Reference 19 suggested that the traditional Boltzmann expression for estimating the excess free energy is reasonably accurate with datapoints distant from the pH_{zpc} (i.e., charging energies are likely to be minimal under these conditions); findings derived from experimental data in the present study also support this suggestion.

ACKNOWLEDGMENTS

Appreciation is expressed to the reviewers of this document for their constructive comments that strengthened the final product. The author also acknowledges previous work on developing methodologies for estimating effective acidity constants from titrimetric data while a graduate student in the Department of Water Chemistry at the University of Wisconsin—Madison. Finally, the author wishes to thank the U.S. Environmental Protection Agency for providing the resources necessary to conduct this work.

DISCLAIMER

Mention of trade names or commercial products does not constitute endorsement or recommendation for use by the U.S. Environmental Protection Agency.

REFERENCES

1. H Gamsjager, E Konigsberger, W Preis. Aq. Geochem. 6:119–132, 1999.
2. D Rai, JM Zachara. Geochemical Behavior of Chromium Species. EPRI Report EA-4544, Electric Power Research Institute, Palto Alto, CA, 1986.
3. DA Dzombak. Toward a Uniform Model for the Sorption of Inorganic Ions on Hydrous Oxides. PhD dissertation, Massachusetts Institute of Technology, Cambridge, MA, 1986.

4. NT Loux, DS Brown, CR Chafin, JD Allison. J. Chem. Spec. Bioavail. 1:111–126, 1989.

5. NT Loux, MA Anderson. Coll. Surf. A. 177:123–131, 2001.

6. RW Smith, EA Jenne. Environ. Sci. Tech. 25:525–531, 1991.

7. EA Jenne. In: EA Jenne, ed. Adsorption of Metals by Geomedia: Variables, Mechanisms, and Model Applications. San Diego, CA, Academic Press, 1998.

8. GA Parks, PL deBruyn. J. Phys. Chem. 66:967–973, 1962.

9. D Chan, JW Perram, LR White, TW Healy. J. Chem. Soc. Faraday Trans. I 71: 1046–1057, 1975.

10. KF Hayes, JO Leckie. J. Coll. Interface Sci. 115:564–572, 1987.

11. N Sahai, DA Sverjensky. Computers Geosci. 24:853–873, 1998.

12. RH Ottewill, JH Shaw. Kolloid Zeit. 218:34–40, 1967.

13. PW Schindler, HR Kamber. Helv. Chim. Acta 51:1781–1786, 1972.

14. PW Schindler, H Gamsjager. Koll. Z. Polymere 250:759–763, 1972.

15. CP Huang, W Stumm. J. Coll. Interface Sci. 43:409–420, 1973.

16. JA Davis, RO James, JO Leckie. J. Coll. Interface Sci. 63:480–499, 1978.

17. RO James, JA Davis, JO Leckie. J. Coll. Interface Sci. 65:331–344, 1978.

18. RO James. In: MA Anderson, AJ Rubin, ed. Adsorption of Inorganics at Solid–Liquid Interfaces. Ann Arbor, MI: Ann Arbor Science, 1981, Chapter 6.

19. NT Loux. Variable bound-site charging contributions to surface complexation model mass action expressions Paper #114, pp 341–343; Preprints of Environmental, Computational and Geochemistry Divisions. 219th American Chemical Society National Meeting, San Francisco, March 26–30, 2000.

20. NT Loux. Energetics Associated with the Diffuse Layer Structure Surrounding Hydrous Oxide Substrates. PhD dissertation, University of Wisconsin, Madison, 1985.

21. Microsoft. Microsoft GW-BASIC User's Guide and User's Reference. Seattle, WA: Microsoft Corp., 1987.

22. J Lutzenkirchen. J. Coll. Interface Sci. 217:8–18, 1999.

23. J Stone-Masui, A Watillon. J. Coll. Interface Sci. 52:479–503, 1975.

24. EA Barringer, HK Bowen. Langmuir 1:421–428, 1985.

25. RK Hayes et al. 1991. Data cited and listed in Ref. 26.

26. N Sahai, DA Sverjensky. Geochim. Cosmochim. Acta 61:2801–2826, 1997.

27. LGJ Fokkink. 1987. Data cited and listed in Ref. 26.

28. DE Yates. The Structure of the Oxide/Aqueous Electrolyte Interface. PhD dissertation, University of Melbourne, Melbourne, Australia, 1975.

29. J Westall, F Morel. FITEQL: A General Algorithm for the Interpretation of Experimental Data. Technical Note #19. Ralph M. Parsons Laboratory, Massachusetts Institute of Technology, Cambridge, MA, 1977.

30. KF Hayes, G Redden, W Ela, JO Leckie. J. Coll. Interface Sci. 142:448–469, 1991.

31. A Heberlin, J Westall. FITEQL version 4.0. Dept. of Chemistry, Oregon State University, Corvallis, OR, 1999.

32. J Westall, H Hohl. Adv. Colloid Interface Sci. 12:265–294, 1980.

12
The Activity, Mechanism, and Effect of Water as a Promoter of Uranium Oxide Catalysts for Destruction of Volatile Organic Compounds

STUART H. TAYLOR, RICHARD H. HARRIS, and GRAHAM J. HUTCHINGS Cardiff University, Cardiff, United Kingdom

IAN D. HUDSON BNFL, Seascale, United Kingdom

I. INTRODUCTION

In recent years, concern for protection of the environment has increased, and environmental legislation has imposed increasingly stringent targets for permitted atmospheric emissions. In particular, the release of volatile organic compounds (VOCs) has received much attention. Such VOCs represent a wide-ranging class of chemicals derived from many sources and containing over 300 compounds as designated by the U.S. Environmental Protection Agency [1]. Their release has widespread environmental implications and has been linked to the increase in photochemical smog [2], the depletion in atmospheric ozone [3], and the production of ground-level ozone [4]. In addition, many VOCs are inherently toxic and/ or carcinogenic. The U.S. Clean Air Act (1990) called for a 90% reduction in emissions of 189 toxic chemicals by 1998; many of these chemicals are classed as VOCs. In 1994 it was estimated that 706,000 tons of organic pollutants were discharged to the atmosphere from the United States alone [5]. Approximately 70% of these compounds can be classed as VOCs, and, although it cannot be determined directly, it is estimated that discharges worldwide are at least twice that of the United States. In view of the scale of the problem presented to the chemical and processing industries, the major challenge they face is to reduce the emission of pollutants without stifling economic growth.

Abatement technologies to control the release of VOCs to the environment are therefore of paramount importance. Many technologies for the treatment of

215

VOC-contaminated effluent have been developed. The most widely adopted is adsorption, often using carbon or zeolite type of adsorbents. However, this process can generate further waste, because the adsorbent is usually buried in landfill sites. The most widely adopted technique is thermal combustion, or incineration, which requires temperatures in excess of 1000°C. Though this is a simple and often effective method of control, the high temperatures required culminate in a relatively fuel-intensive technique with little control over the ultimate products. The latter is particularly problematic and can result in incomplete oxidation of the waste stream and the formation of toxic byproducts such as dioxins, dibenzo-furans, and oxides of nitrogen if conditions are not carefully controlled. Alternatively, heterogeneous catalytic oxidation offers many potential advantages. The use of a catalyst in the oxidative destruction of VOCs significantly lowers the process operating temperature, which is typically in the range 300–600°C. This reduction in temperature is advantageous, for supplementary fuel requirements are reduced and legislatively the process is no longer regarded as an incineration process, eliminating certain regulatory requirements. In addition, catalytic oxidation offers a much greater degree of control over the reaction products and can operate with dilute effluent streams (<1% VOC) that cannot be treated easily by thermal combustion. Hence, catalytic oxidation may be considered a more appropriate method for end-of-pipe pollution control.

Two classes of catalyst are commonly used: noble metal–based and metal oxide–based systems. A prospective catalyst must be active at relatively low temperatures and show high selectivity to carbon oxides. Ideally, the catalyst must also be able to destroy effectively low concentrations of VOCs at very high flow rates with little or no deactivation. Supported noble metal systems, primarily platinum and palladium, show high activity for the oxidation of many VOCs, with high selectivity to carbon oxides. However, these tend to be relatively expensive and can be rapidly deactivated by the presence of chlorinated compounds, sulfur, or other metals in the waste stream [6]. The second class of catalysts are metal oxides, and some of the most active are based on copper [7], cobalt [8], chromium [9], and manganese [10]. Generally, these are less expensive than precious metals and show higher resistance to poisoning. However, for complete oxidation they are inherently less active. The development of oxide catalysts that may be used for the combustion of a wide range of volatile organic compounds presents a major challenge for future research.

The application of catalytic oxidation for VOC control is an end-of-pipe process, and performance of the catalysts is also dependent on the process conditions. Effluent streams often contain moisture, which is known to be detrimental to the deep oxidation performance of the most widely used precious metal catalysts. The presence of water in the effluent stream can have a dramatic effect, and it has been shown to inhibit activity over supported Pd [11,12] and Pt [13] catalysts. Attempts have been made to improve the tolerance of precious metal catalysts to water vapor, and supporting Pt on a hydrophobic support such as a fluorinated carbon can reduce the inhibiting effect of water, but inhibition is still observed [13]. In combination with improving the activity of oxide catalysts it is also important to investigate the effect of catalyst poisons such as water.

This work outlines the advances made in the development of uranium oxide–based catalysts for VOC oxidation. Uranium oxide was initially selected as a catalyst for several reasons; in particular, U_3O_8 has uranium present in mixed oxidation states, with a facile transition between states, and can also show a wide range of metal/oxygen stoichiometry. These are important features that are characteristic of other oxidation catalysts. Additionally, uranium oxides have shown relatively high activity for carbon monoxide oxidation [14].

The effect of water inhibition on catalyst performance has also been addressed, and results are presented that indicate that the addition of low concentrations of water to the effluent stream further enhance the catalytic activity of uranium oxide catalysts. This is in contrast to many metal oxide–based catalysts and to precious metal catalysts, which are less active when water is present. Studies to probe the reaction mechanism have been performed using a temporal analysis of products (TAP) reactor to unravel the reaction mechanism.

II. EXPERIMENTAL

A. Catalyst Preparation

The U_3O_8 catalyst was prepared by decomposition of $UO_2(NO_3)_2 \cdot 6H_2O$ (Strem 99.9%) by calcination in static air at 300°C for 1 h and then at 800°C for 3 h. A supported uranium catalyst was also prepared by impregnation of fumed silica (BDH, Cab-O-Sil M5) with 4.2 mL g^{-1} of uranyl nitrate solution (0.397 mol L^{-1}). The resulting material was dried at 100°C and subsequently calcined using the same conditions as the unsupported catalyst. The uranium loading for this catalyst was 10 mol% (U/SiO$_2$), approximately representing theoretical monolayer coverage. For comparison of catalyst performance, the oxidation catalyst Mn_2O_3 (Aldrich, 99.9%) was selected, because it is known to have high complete oxidation activity [15].

B. Catalyst Characterization

Ex situ powder X-ray diffraction patterns were collected using an Enraf FR590 instrument with a Cu source operated at an X-ray power of 1.2 kW (30 mA and 40 kV). A Ge (111) monochromator was used to select Cu K□ X-rays. The powdered samples were compressed into an aluminum sample holder, which was rotated during data collection to compensate for any crystallite ordering. The diffraction pattern was measured by means of a position-sensitive detector (Inel

PSD120), covering all 2θ values in the range 4.4–124.6θ. Raw data were corrected against a silicon standard, and phase identification was performed by matching the experimental pattern against standard entries in the JCPDS powder diffraction file.

In situ powder XRD studies were performed using a Phillips X-PERT diffractometer with a high-temperature Parr XRK reaction chamber and a position-sensitive detector. Copper Cu Kα X-rays (30 KeV, 40 mA) were used, and data in the 2Θ range 18°–60° were collected. The in situ reaction cell was designed so that gases flowed through the catalyst sample, which was heated from ambient to 600°C. Experiments were carried out with a flow of dry air and an air stream containing ca. 4% water. Typical analysis times at each temperature were in the region of 1.5 minutes.

C. Steady-State Catalytic Activity

The catalysts were tested for VOC destruction using a fixed-bed laboratory microreactor equipped with an on-line gas chromatograph analysis system using propane and benzene. Gas flow rates were regulated with electronic thermal mass-flow controllers. Catalyst performance was screened using a dry flow of gas, while a series of experiments also investigated the effect of cofeeding water. Water was introduced by passing the air flow through a set of two saturators. The concentration of water was controlled by oversaturating the gas stream at room temperature in the first saturator and then reduced by passing through the second saturator maintained in a thermostatically controlled bath. The water concentration was calculated using water vapor pressure data. The reactant gases were heated to 150°C prior to entering the reactor. Catalysts were tested in powdered form using a $1/4''$ o.d. stainless steel reactor using a gas hourly space velocity of either 35,000 or 70,000 h^{-1}. The VOC concentrations used were 1% propane and 600-ppm benzene in air. Conversion of VOCs was calculated from the difference of concentration at reaction temperature and a lower temperature at which the catalyst was inactive. Carbon balances were in the range 100 ± 10%.

D. Temporal Analysis of Products: Catalytic Studies

The TAP reactor was used in continuous-flow and TAP pulse modes to investigate the oxidation of a variety of VOCs, including benzene and butane. A detailed explanation of the design and capabilities of the TAP reactor are given elsewhere [16]. Prior to reactivity studies, detailed experiments were carried out to accurately determine the mass spectral fragmentation patterns of the VOCs and expected reaction products. This was achieved by preparing gas mixtures in a high-purity (99.99%) neon standard and the fragmentation patterns collected from a continuous flow of the mixture through a reactor packed with inert quartz particles sieved to a particle size distribution comparable to that of the catalyst. These data were also used to determine the total sensitivity of reactant and products relative to the m/e peak at 20 for neon.

III. RESULTS AND DISCUSSION

A. Catalyst Characterization

The powder X-ray diffraction patterns of the U_3O_8 and U_3O_8/SiO_2 catalysts are shown in Figure 1. The XRD patterns confirm that the preparative calcination procedure produced orthorhombic U_3O_8 from the nitrate precursor. The diffraction peaks from the silica-supported uranium oxide catalyst were centered at the same d spacing as U_3O_8 and confirm that the supported catalyst also contained U_3O_8. The diffraction peaks from U_3O_8/SiO_2 were significantly broader when compared to U_3O_8, indicating that the supported U_3O_8 crystallite size was considerably smaller. From X-ray line broadening, the supported U_3O_8 crystallite was estimated to be in the region of 150 Å.

B. Activity of Uranium Oxide Catalysts

The oxidation activity of uranium oxide catalysts has been determined for a wide range of typical VOCs that are chemically diverse in nature. The compounds investigated include benzene, propane, butane, butyl acetate, cyclohexanone,

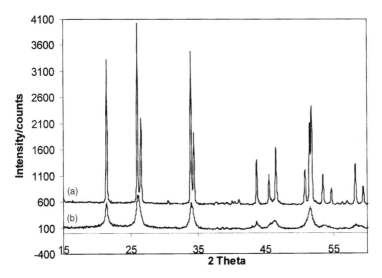

FIG. 1 Powder X-ray diffraction patterns of the U_3O_8 and U_3O_8/SiO_2 catalysts: (a) U_3O_8, (b) U_3O_8/SiO_2.

chlorobenzene, chlorobutane, acetylene, methanol, and toluene. Blank reactions in an empty reactor tube and using a catalyst bed of pelleted silica indicated that the blank activity at 70,000 h^{-1} was negligible. For example, benzene showed 1% conversion to CO_2 over SiO_2 at 500°C, while at 600°C 3% butane conversion to CO_2 was observed. Representative data for the oxidation of a range of VOCs are shown in Table 1.

The uranium oxide catalysts showed high activity for deep oxidation. The sole carbon reaction products were carbon oxides, and no partially oxygenated or other hydrocarbon byproducts were detected. In the case of the chlorinated VOCs, HCl, determined by mass spectroscopy, was the sole chlorine-containing product. HCl is preferred to Cl_2, because it can be readily removed by aqueous scrubbing.

TABLE 1 Catalytic Activity of Uranium Oxide Catalysts for Oxidation of a Range of VOCs

				Selectivity/%	
Catalyst	VOC type	Temperature/°C	Conversion/%	CO	CO_2
U_3O_8	Benzene	400	99.9	23	77
U_3O_8/SiO_2	Benzene	400	99.9	26	74
Co_3O_4	Benzene	400	83	—	100
U_3O_8	Butane	500	3	5	95
U_3O_8	Butane	600	81	14	86
U_3O_8/SiO_2	Butane	500	99.9	37	63
Co_3O_4	Butane	500	75	—	100
Co_3O_4	Butane	600	68	—	100
U_3O_8	Chlorobenzene	350	99.7	41	59
U_3O_8/SiO_2	Chlorobenzene	400	99.9	41	59
Co_3O_4	Chlorobenzene	400	0	—	—
Co_3O_4	Chlorobenzene	600	62	16	84
U_3O_8	Chlorobutane	350	>99.5	45	55
U_3O_8	Cyclohexanone	300	99.9	29	71
U_3O_8/SiO_2	Cyclohexanone	300	99.9	34	66
Co_3O_4	Cyclohexanone	300	0	—	—
Co_3O_4	Cyclohexanone	350	99.9	4	96
U_3O_8	Butylacetate	350	99.9	23	77
U_3O_8/SiO_2	Butylacetate	350	99.9	22	78
U_3O_8/SiO_2	Butylacetate	350	99.9	3	97
U_3O_8	Methanol	300	99.9	47	53
U_3O_8	Acetylene	400	97.4	35	65
U_3O_8/SiO_2	Toluene	400	99.9	10	90

1% VOC in air, GHSV = 70,000 h^{-1}.

A general comparison of catalytic activity was made with Co_3O_4, which, along with Mn_2O_3, is recognized as a highly active catalyst for deep oxidation [15]. In many cases, comparison with the activity of Co_3O_4 has been made, and it is evident that uranium oxide–based catalysts show superior deep oxidation activity. The comparison for cyclohexanone and chlorobenzene is striking, for the uranium oxide catalysts show high conversions at temperatures at which Co_3O_4 is inactive. Conversion of VOCs was generally greater over U_3O_8 than over Co_3O_4, and it must also be noted that U_3O_8 has a surface area of 0.8 m^2g^{-1} compared to 4.2 m^2g^{-1} for Co_3O_4. It is also evident that the uranium oxide catalysts are active at relatively low temperatures; generally high VOC conversion was achieved below 450°C, which compares very favorably with temperatures in excess of 1,000°C that are required for thermal combustion.

Supporting uranium oxide on silica had little effect on the conversion and product selectivity of the VOCs like benzene, cyclohexanone, and butylacetate. However, the supported catalyst showed increased activity for butane oxidation. At 500°C, butane conversion over U_3O_8/SiO_2 was 100%, compared to 3% over U_3O_8. The surface area for the silica-supported catalyst (110 m^2g^{-1}) is far greater than for U_3O_8. It is difficult to measure the active surface area of the supported catalyst; the uranium oxide loading was calculated to be in the region of that required for monolayer coverage. The identification of U_3O_8 crystallites by XRD from the U_3O_8/SiO_2 catalyst indicate the monolayer dispersion was not achieved, but the average crystallite size of 150 Å suggests that U_3O_8 is relatively highly dispersed on the support.

C. Temporal Analysis of Products: Mechanistic Studies

When TAP experiments were carried out with a range of VOCs, many similarities between different VOCs were observed. A typical TAP response for a pulse of butane in the absence of oxygen (19.8% butane, 80.2% neon) over U_3O_8/SiO_2 at 479°C is shown in Figure 2.

Several important observations can be made from the TAP pulse experiment. The first is that oxidation takes place in the absence of gas-phase oxygen, suggesting that oxygen species from the catalyst are active in the oxidation cycle. All the catalysts were vacuum treated in situ at >500°C prior to pulse experiments; consequently, the concentration of adsorbed oxygen is expected to be at a minimum. The oxidation activity was maintained after many thousands of pulses, indicating that lattice oxygen was the active species. Second, analysis of the TAP data indicated that the time for peak maxima of the products increased relative to neon and butane, and the peaks are all significantly broader than for neon. Because diffusion effects are controlled in the TAP reactor system, this would indicate that the products and reactants interact with the catalyst surface and are

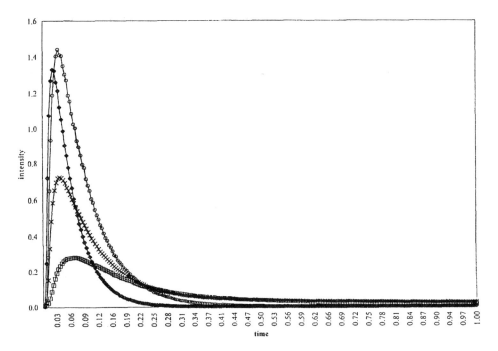

FIG. 2 TAP pulse response for butane oxidation at 479°C in the absence of gas-phase oxygen: ◆ neon; ○ butane; □ carbon dioxide; ✖ carbon monoxide.

adsorbed to varying degrees. This observation provides an indication that the reaction is occurring on the catalyst surface and not via gas-phase processes.

It is also evident that the only carbon-containing products were carbon monoxide and carbon dioxide, which is consistent with the steady-state reactor studies. The TAP reactor is particularly suited to the detection and identification of gas-phase intermediate species that cannot be detected readily in conventional steady-state studies. In the TAP system the relatively low number of molecules passing through the catalyst bed and the absence of a carrier results in molecular beam transport through the bed, thus minimizing collision between reactants and products. Consequently, highly reactive and short-lived intermediates that are not detected by conventional steady-state techniques are readily observed in the TAP reactor. The absence of any partially oxidized intermediates in these studies indicates that the fundamental reaction pathway for the oxidation of these VOCs by U_3O_8-based catalysts takes place on the catalyst surface. The surface reaction pathway is unclear, and it may be via a partially oxygenated intermediate; however, such intermediates do not desorb to the gas phase, for they would be de-

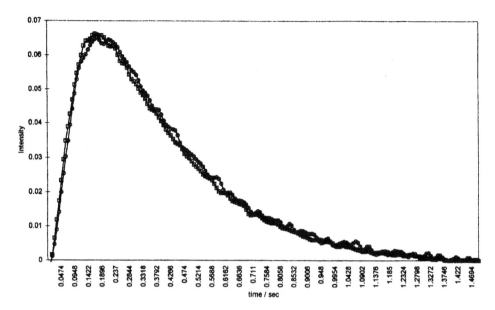

FIG. 3 Normalized TAP pulse response for carbon dioxide produced from benzene in the presence and absence of gas-phase oxygen at 366°C: □ gas-phase oxygen; ○ no gas-phase oxygen.

tected in these studies. This clearly contrasts with many other TAP studies, which have identified a series of reaction intermediates. Direct comparison can be made with other systems for the oxidation of butane, such as vanadium phosphate catalysts. Under similar conditions, TAP studies with VPO catalysts have identified a series of reaction intermediates, such as butadiene and maleic anhydride, during the oxidation of butane to the thermodynamically more stable carbon oxides [17].

Similar TAP pulse studies have also investigated oxidation with gas-phase oxygen, and the results with all VOCs are identical to experiments without gas-phase oxygen. The normalized response in the absence of gas-phase oxygen can be superimposed on the response in the presence of oxygen (Fig. 3). In the non-steady-state conditions of the TAP reactor, this indicates that there is no difference in conversion with gas-phase oxygen present and absent. It can therefore be concluded that the oxygen species that is utilized in total oxidation is derived from the uranium oxide catalyst.

In order to confirm the origin of the oxygen in the oxidation products, a model study has investigated the oxidation of CO with isotopically labeled oxygen. These studies were performed with a continuous flow of $C^{16}O/^{18}O_2$ (25% $C^{16}O$, 25% $^{18}O_2$, 50% neon). Data are shown for CO oxidation over U_3O_8/SiO_2 at 596°C

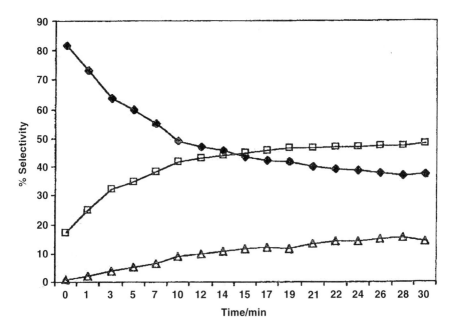

FIG. 4 Isotopic selectivity of CO_2 species during continuous flow of $C^{16}O/^{18}O_2$ over U_3O_8/SiO_2 at 596°C: ◆ $C^{16}O_2$; □ $C^{16}O^{18}O$; △ $C^{18}O_2$.

(Fig. 4). Carbon dioxide is the only reaction product, and initially only ^{16}O is observed. This is again consistent with oxidation by lattice oxygen. With time on-line, the concentration of $C^{16}O_2$ decreased and the concentration of the isotopically labeled product, $C^{16}O^{18}O$, increased. This type of behavior indicates that the catalyst is operating by a redox mechanism with reoxidation of the catalyst by the gas-phase $^{18}O_2$. Some $C^{18}O_2$ is also observed; however, in comparison, levels are relatively low. Either the $C^{18}O_2$ product may be derived from oxidation of a $C^{18}O$ species derived from exchange of oxygen in $C^{16}O$ once the surface is enriched with ^{18}O, or it may be derived from oxygen exchange of the CO_2 product with the surface. Identical behavior was also observed with the U_3O_8 catalyst, indicating that it also operated via a redox mechanism involving lattice oxygen.

D. Effect of Cofeeding Water

The oxidation of benzene over silica-supported U_3O_8 showed no activity for hydrocarbon oxidation below 300°C in the absence of water (Fig. 4). The conversion increased to 99.9% at 450°C and was maintained at higher temperatures. On the addition of 2.6% water, the catalyst became active at 250°C, which was 100°C lower than with dry conditions. With low levels of water added, 99.9% conversion was achieved at 400°C, 50°C lower than the dry experiment. Conversely, the addition of a greater concentration of water (12.1%) resulted in a 50°C increase in the temperature required for 99.9% conversion to 500°C, and the conversion was lower than for the dry conditions throughout the active temperature range. As we have observed previously, the only products formed under all conditions tested were CO_2 and CO. The CO_2/CO ratio was dependent on the concentration of water cofed. The product selectivities for benzene oxidation using U_3O_8/SiO_2 in the absence of cofed water and 2.6% water are shown in Figure 5. At 400°C without water cofeeding, selectivity was 70% toward CO_2; on the addition of 2.6% water, CO_2 selectivity was increased to 93%. At 500°C in the absence of co-fed water and with the addition of 2.6% water, the CO_2 selectivity increased to 75% and 96%, respectively. Similar results were also observed when the reaction temperature was increased to 600°C. The increase in the CO_2 selectivity with cofed water is consistent with the water–gas shift reaction.

To establish whether the effect of water was specific for benzene oxidation, and to probe further the effect of water, propane oxidation experiments were performed with the U_3O_8 catalyst; results are shown in Figure 7. It is clear that in the presence of 2.6% water, 99.9% propane was converted at 450°C over U_3O_8/SiO_2; however, in the absence of co-fed water, U_3O_8 produced only 70% conversion at 600°C. To establish whether this type of behavior was limited to uranium oxide catalysts, the oxidation of was investigated using highly active Mn_2O_3 [15]. The Mn_2O_3 catalyst produced 65% conversion at 600°C in the absence of co-fed water,

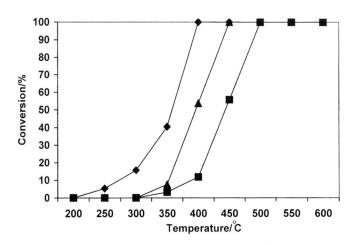

FIG. 5 Effect of co-feeding water on the oxidation of 600-ppm benzene in air over U_3O_8/SiO_2: ▲ no water; ◆ 2.6% water; ■ 12.1% water.

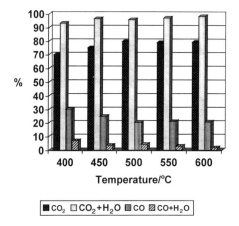

FIG. 6 Product selectivities for benzene oxidation using U_3O_8/SiO_2 in the absence of co-fed water and 2.6% water.

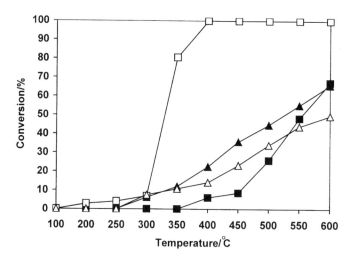

FIG. 7 Comparison of catalytic activity for 1% propane oxidation in air: $\triangle$ U_3O_8; $\square$ U_3O_8 + 2.6% water; $\blacktriangle$ Mn_2O_3; $\blacksquare$ Mn_2O_3 + 2.6% water.

and conversion was suppressed to 49% when 2.6% water was present. Over Mn_2O_3 the conversion of propane was lower at all temperatures when water was cofed. Determination of the BET surface areas showed that the surface area of Mn_2O_3 was 1.8 m^2g^{-1}, while for U_3O_8 it was 0.8 m^2g^{-1}. These data may account, in part, for the higher activity of Mn_2O_3 under dry reaction conditions, but it is clear that the effect of cofeeding water has opposing effects for the two catalysts.

In situ powder X-ray diffraction studies have investigated the catalyst structure under reaction conditions; the diffraction patterns of the U_3O_8/SiO_2 catalyst in flowing 4% water/air over the temperature range 100–600°C is shown in Figure 8. The results indicated that even at 600°C no change in the initial U_3O_8 phase occurred. Some thermal broadening and loss of peak resolution were observed; but even after approximately 7 h at elevated temperature, patterns were comparable with those obtained under ambient conditions. Similar studies in a dry flow of air over the same temperature range also showed that the structure of U_3O_8 was unchanged during the reaction.

Further TAP studies are required to elucidate more fully the role of water in this reaction, but it is interesting to comment on the mechanism of the promotional effect of water over U_3O_8. It has been shown that the complete oxidation of VOCs by uranium oxide catalysts takes place by a redox mechanism with lattice oxygen as oxidant. The rate-determining step over uranium oxide catalysts has not been unequivocally determined. However, it is feasible that the rate-

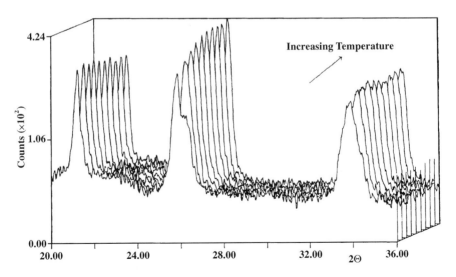

FIG. 8 In situ powder X-ray diffraction patterns of the U_3O_8/SiO_2 catalyst in flowing 4% water/air over the temperature range 100–600°C.

determining step is reoxidation of the reduced catalyst, and this has also been proposed for other redox oxidation catalysts [18]. The presence of water vapor is also known to promote the oxidation of uranium oxides when compared to oxidation rates under dry conditions [19]. Therefore it is possible that the oxidation rate for VOCs is promoted by water, since the catalyst reoxidation rate is enhanced. The reoxidation of the catalysts may also aid the desorption of adsorbed CO_2, although the higher rates of benzene oxidation compared to propane indicates that desorption of CO_2 may not be rate limiting. The role of water may also have an effect on the initial activation of the hydrocarbons; indeed, cofeeding 2.6% water reduced the light-off temperature for propane oxidation by 100°C. This effect may be due to modification of the uranium oxide surface by hydroxylation, which effectively aids hydrocarbon activation. Finally, the contribution from new reaction pathways when water is cofed, such as steam reforming, cannot be discounted. Uranium oxide catalysts have been identified as highly active steam-reforming catalysts [20]; however, at the active temperatures used in this study, steam-reforming activity is expected to be negligible.

At this stage no attempt has been made to optimize the concentration of water required for maximum effect, since this can be expected to vary with the nature of the VOC, but it is feasible that further activity enhancement could be observed at lower concentrations (<1%). However, it is clear that the addition of water is beneficial for the activity of uranium oxide catalysts for the oxidative destruction of VOCs. Since effluent streams often contain water, this is an additional advantage for this very active heterogeneous catalyst.

IV. CONCLUSIONS

Uranium oxide catalysts show high activity for the oxidative destruction of a range of VOCs at high space velocity. Catalysts consisting of U_3O_8 and U_3O_8 supported on silica demonstrated high rates of conversion to carbon oxides, with no traces of partially oxidized products. Studies using a TAP reactor have proved a valuable approach in starting to understand the mechanism of uranium oxide catalysts for the oxidation of VOCs. Investigations using a continuous-flow method have shown that benzene, butane, and chlorobenzene are combusted directly to carbon oxides with no partially oxidized species. Further studies, using a combination of TAP pulse experiments in the presence and absence of gas-phase oxygen and with isotopically labeled gas-phase oxygen, demonstrate that the active oxygen species are derived from the lattice of the oxide catalyst, which operates via a redox mechanism.

Steady-state reactor studies have been extended to investigate the effects of cofeeding water on catalytic activity for benzene and propane oxidation. The addition of low levels of water (2.6%) had a marked effect on oxidation activity, promoting the rate of complete oxidation. The addition of a higher concentration of water (12.1%) had a detrimental effect on oxidation activity. The addition of water also increased selectivity toward CO_2, and this may be due to the water–gas shift reaction. A comparison of catalytic activity has been made with highly active Mn_2O_3. In contrast to the U_3O_8 catalysts, the addition of 2.6% water suppressed the Mn_2O_3 oxidation activity. In situ powder XRD showed that the bulk U_3O_8 structure was stable under the reaction conditions used in this study. The origin of the increased activity is not clear, but it is most likely due to modification of the U_3O_8 catalyst surface by hydroxylation. Contribution from new reaction pathways, such as the steam-reforming reaction, cannot be discounted but appear less likely, considering the low reaction temperature. The use of depleted uranium oxide as a catalyst may be considered by some as controversial; however, uranium oxide-based catalysts have been widely used by the chemical industry for a considerable time [21]. Well-established procedures for the safe handling of these materials exist, and these are determined primarily by issues of chemical toxicity. The results presented in this chapter demonstrate that uranium oxides show high and stable activity for the destruction of a range of VOCs under industrially relevant flow rates and temperatures. Previously, uranium oxides have been considered by many as a burden on our environment; however, this work demonstrates that they can be effectively used to provide a solution to a major problem affecting the environment.

ACKNOWLEDGMENTS

We would like to thank Catherine Heneghan and Vicki Boyd (BNFL) for their contribution to this work.

REFERENCES

1. N Mukhopadhyay, EC Moretti. Current and Potential Future Industrial Practices for Controlling Volatile Organic Compounds. Center For Waste Control Management, 1993.
2. MS Jennings, MA Palazzolo, NE Krohn, RM Parks, RS Berry, KK Fidler. Catalytic incineration for the control of volatile organic compound emission. Pollution Technology Review (Noyes Ed.), No. 121, 1985.
3. MJ Molina, FS Rowland. Nature 249:810–812, 1974.
4. Chemistry in Britain. Royal Society of Chemistry, London, February 1997.
5. U.S. Environmental Protection Agency. Toxic Release Inventory Rep. 745-R-96-002. U.S. EPA, Washington, DC, June 1996.
6. JJ Spivey, JB Butt. Catal. Today 11:465–500, 1992.
7. YM Kang, B-Z Wan. Appl. Catal. A 114:35–49, 1994.
8. RS Drago, K Jurczyk, DL Singh, V Young. Appl. Catal. B 6:155–168, 1996.
9. SK Agarwal, JJ Spivey, JB Butt. Appl. Catal. A 82:259–275, 1992.

10. N Watanabe, H Yamashita, H Miyadera, S Tominaga. Appl. Catal. B 8:405–415, 1996.
11. JC van Giezan, ER van den Berg, JL Kleinen, AJ van Dillen, JW Geus. Catal. Today 47:287–293, 1999.
12. D Chiuparu, L Pfefferle. Appl. Catal. A, 209:115–128, 2000.
13. M Zhang, B Zhou, KT Chang. Appl. Catal. B 13:123–130, 1997.
14. E Nozaki, K Ohki. Bull. Chem. Soc. Japan 45:473–474, 1972.
15. GI Golodets. Stud. Surf. Sci. Catal. 15:657, 1983.
16. JT Gleaves, JR Ebner, TC Kuechler. Catal. Rev. Sci. Eng. 30:49–116, 1988.
17. JT Gleaves, GS Yablonskii, P Phananwadee, Y Schuuman. Appl. Catal. A 160:55–88, 1997.
18. MA Banares, JLG Fierro, JB Moffat. J. Catal. 142:406–417, 1993.
19. CA Colmenares. Prog. Solid State Chem. 15:257–312, 1984.
20. DG Gavin. U.S. Patent 3,974,098.
21. RK Grasselli, JL Callahan. J Catal. 14:93–103, 1969.

13

Detoxification of Concentrated Halogenated Gas Streams Using Solid Supported Nickel Catalysts

MARK A. KEANE   University of Kentucky, Lexington, Kentucky, U.S.A.

I. BACKGROUND

Chlorinated organic compounds are now an established source of environmental pollution [1–3]. The presence of these nonbiodegradable compounds in effluent discharges is of increasing concern due to the mounting evidence of adverse ecological and public health impacts [4,5]. As a direct consequence, ever more stringent legislation is being introduced to limit those chloro-emissions that lead to contamination of wastewater and trade effluent [6–8]. The control strategies that are currently favored involve some form of "end-of-pipe" control, entailing either phase transfer/physical separation (adsorption, air/steam stripping, and condensation) or chemical degradation/destruction (thermal incineration, catalytic oxidation, chemical oxidation, and wet-air oxidation) operations. A catalytic transformation of chlorinated waste represents an innovative "end-of-process" strategy that offers a means of recovering valuable raw material, something that would be very difficult if not impossible to achieve with end-of-pipe technologies.

The application of heterogeneous catalysis to environmental pollution control is a burgeoning area of research. This chapter will focus on one case study, the gas-phase hydrodechlorination of chlorinated aromatics (chlorobenzenes and chlorophenols) promoted using supported nickel catalysts. Chlorinated benzenes/phenols represent a class of commercially important (world market in tens of thousands of tons) but particularly toxic chemicals that enter the environment as industrial effluent from herbicide/biocide production plants, petrochemical units, and oil refineries [9,10]. Haloarenes have been listed for some time by the EPA as "priority pollutants" [11,12] and targeted in terms of emission control. In response to such issues as climate change, water protection, and air quality, the concept of a waste management hierarchy has emerged, embracing the "four Rs," i.e., reduction, reuse, recycling, and (energy) recovery [13]. The application of catalytic hydrodechlorination to the treatment of chlorinated waste fits well within this environmental remediation ethos. A concerted safety and environmental approach is now called for, one that incorporates advanced "green" processing technology as a means of canceling any negative environmental impact without stifling the commercial activities of the chemical industry.

II. STRATEGIES FOR HANDLING/DISPOSING OF CHLORO-ORGANICS

A reduction in organic pollutants can be achieved through a combination of resource management, product reformulation, and process modification. In choosing the best strategy, many considerations must be taken into account, such as recycling potential, the phase and character of the organic compound(s), the volume of the stream to be treated, and the treatment costs. The established technologies are based on incineration/oxidation, biological treatment, absorption, and adsorption processes. Incineration is a widely used, robust methodology for treating/destroying hazardous waste [14]. However, chlorinated organics fall under the category of *principal organic hazardous constituents*, compounds that are inherently difficult to combust. As a direct consequence of the thermal stability of these compounds, complete combustion occurs at such high temperatures (>1700 K) as to be economically prohibitive, while the formation of such hazardous by-products as polychlorodibenzodioxins (PCDD) and polychlorodibenzofurans (PCDF) (dioxins/furans) can result from incomplete incineration [15,16]. These severe conditions render the process very expensive and chloroaromatic incineration costs can amount to over US$2000 per metric ton [17]. Ever more stringent limiting values for PCDD/PCDF emissions (of the order of 0.1 mg m^3) from municipal and hazardous waste incinerators are being introduced worldwide [18]. At present, primary measures such as design and operation of the firing system to minimize the formation of products of incomplete combustion or boiler technology cannot guarantee compliance with the legislated emission levels [19].

Catalytic oxidation represents a more progressive approach, where conversion proceeds at a much lower temperature and fuel/air ratio, with an associated reduction in energy costs and NO_x emissions [20,21]. Oxidation of chlorinated VOCs has been reported using supported Pd and Pt catalysts over the temperature range 523–823 K [22–24]. By-products, however, include CO, Cl_2, and $COCl_2$, which are difficult to trap, while complete oxidation (the ultimate goal) generates unwanted CO_2. Catalyst deactivation is also an important consideration, given the expense involved in synthesizing noble metal–based catalyst systems. Less effective chromia-based oxidation catalysts, though also active in chlorohydrocarbon oxidation, are susceptible to attack by Cl, leading to loss of chromium content and catalyst deactivation [25]. The application of photolysis, ozonation, and su-percritical oxidation to the treatment of recalcitrant organic compounds falls un-der what is now regarded as advanced oxidation technologies [21,26–29]. Ultra-sonic irradiation as applied to the treatment of chloroarenes is also undergoing feasibility studies [30,31]. While these approaches show promise, especially at low contaminant concentrations [32], each is hampered by practical consider-ations in terms of high energy demands and cost [33]. Although biological oxida-tion can be effective when dealing with biodegradable organics, chlorophenols are used in the production of herbicides and pesticides and, as such, are very resistant to biodegradation [34,35]. Even the monochlorinated 2-chlorophenol isomer, as a priority pollutant, is poorly biodegradable, and waste streams con-taining concentrations above 200 ppm cannot be treated effectively by direct biological methods [36]. Conversion of chloro-organics, where feasible, is in any case very slow, necessitating the construction of oversized and expensive bioreac-tors [37]. Because the biological toxicity in polychlorinated organics is linked directly to the chlorine content, a feasible bioprocess would require a pretreatment (preferably catalytic) that served to remove some of the chlorine component in a controlled fashion, rendering the waste more susceptible to biodegradation.

Adsorption, as a separation process, is an established technology in chemical waste treatment [38]. Activated carbon, usually derived from natural materials (e.g., coal, wood, straw, fruit stones, and shells) and manufactured to precise surface properties, is widely used in water cleanup due to its high adsorption capacity coupled with cost effectiveness [39]. The uptake of chloroaromatics on carbon has been the subject of a number of reports [40–44] that have revealed the importance of such parameters as concentration, pH, carbon porosity, particle size, and surface area on the ultimate removal efficiency. However, adsorption in common with other separation processes involves only phase transfer of pollut-ants without a transformation or decomposition of the hazardous material and really serves to prolong the ultimate treatment step. Catalytic treatment under nonoxidizing conditions is now emerging as a viable nondestructive (low-energy) recycle strategy [45,46]. The possibility of achieving a dechlorination of various organochlorine compounds by electrochemical means has been addressed in the literature [47–49]. However, high dechlorination efficiency typically necessitates the use of nonaqueous (aprotic solvent) reaction media or environmentally de-structive cathode materials (Hg or Pb), which has mitigated against practical ap-plication. Catalytic steam reforming has been viewed as a feasible methodology [50] but is again destructive in nature, albeit the possibility of generating synthe-sis gas as product.

By and large, the existing treatment technologies involve a separation (or con-centration) step followed by a destruction step. Catalytic hydrodehalogenation, the focus of this chapter, represents an alternative approach where the hazardous material is transformed into recyclable products in a closed system with negligible toxic emissions. Hydrodehalogenation, the hydrogen cleavage of C–X (carbon–halogen) bonds can be represented by

R–X + reducing agent → R–H + HX

No dioxins are formed in a reducing environment, and any dioxin-containing waste can be detoxified, with recovery of valuable chemical feedstock. Such a strategy promotes an efficient use of resources, greatly reducing both direct and indirect waste/emissions costs, and fosters sustainable development. While separation methodologies offer a means of concentration, if the extracted materials are mixtures of chlorinated isomers, then these are not, without some difficulty, recovered for reuse. Mixed isomers arising from an uncontrolled chlorination process can readily be converted by hydrodechlorination back to the single parent raw material precursor from which they originated. The principal advantages of catalytic aromatic hydrodechlorination when compared with the approaches described earlier are: (a) low-temperature (<600 K) nonoxidative and nondestructive process with lower energy requirements and no directly associated NOx/SOx emissions; (b) absence of thermally induced free-radical reactions leading to toxic intermediates; (c) possibility of selective chlorine removal to generate a reusable/recyclable product; (d) operability in a closed system, with no toxic emissions; (e) gas-phase operation requires low residence times; (f) can be employed as a pretreatment step to detoxify concentrated chlorinated streams prior to biodegradation.

III. POTENTIAL IMPACT

A. Environmental Considerations

The increasing threat posed to the environment by hazardous halogenated waste has intensified the research efforts into safer methods of handling/disposal. The direct link between halogenated emissions into the environment and ozone depletion is now well established and widely recognized. Chloroarene production by direct chlorination is typically unselective and gives rise to a range of isomeric products where overchlorination is often unavoidable [51]. The overall chloroarene market has been in decline, in part due to the associated negative environmental impact, but still represents a significant commercial activity. The potential deleterious effect to human health associated with exposure to halogenated compounds is cause for grave concern. The U.S. EPA has recently posted an Advisory Document on the internet (http://www.epa.gov/oppt/24dcp.htm) that deals with the 2,4-dichlorophenol isomer, describing this "high-production-volume chemical feedstock" as being a significant occupational hazard risk and known to be responsible for a number of worker fatalities in the chemical industry. In Europe, the EC Framework Directive has catalogued 129 substances in a "Black List," among them a range of organohalogens, considered to be so toxic, persistent, or bioaccumulative in the environment that priority is given to eliminating such compounds as pollutants. In all cases the directive designates emission limits and quality objectives, and the use of the best available technology is strongly encouraged.

Incineration, as the present established and preferred method of disposal, is certainly not the best possible environmental option, even when taking into account the considerable precautions that can be employed to prevent emission of toxic by-products. Over the past five years, the EPA has imposed regulations on major dioxin emitters, including municipal waste combustors, medical waste incinerators, hazardous waste incinerators, and cement kilns that are used to burn hazardous waste. The permissible emission levels associated with treating chlorinated compounds will certainly be lowered in the future, and the potential costs involved in legal prosecution alone lend a high degree of urgency to the development of safe methods for the handling of such organics. While combustion does not demonstrate an efficient use of resources, chemical hydroprocessing of the hazardous waste can serve to both detoxify and transform the waste into recyclable products. In this chapter, the catalytic hydrodechlorination of polychlorinated aromatics is presented as following two possible strategies: (1) a complete removal of the chlorine component to generate the parent aromatic, (2) a selective partial hydrodechlorination to a less chlorinated target product. Both routes represent unique processes of chemical desynthesis and must be viewed as a progressive approach to environmental pollution control.

B. Economic Considerations

Taking incineration as the principal means of "disposal," a move to a catalytic hydrogen treatment represents immediate savings in terms of fuel consumption and/or chemical recovery. The actual conditions that must be employed for safe incineration of chlorinated compounds is still somewhat controversial, but a common rule of thumb is to limit the waste feed to a minimum heat of combustion content of 10,000 Btu/lb [52], which corresponds to a chlorine content of 20% to 50%. Effective combustion can require the use of auxiliary fuel, but an efficient heat recovery system will recoup a proportion of the heat that is liberated. The energy needed for the hydrogenolytic route is that required to generate the hydrogen that is consumed in the process, and this can be subtracted from the energy in the recycled fuel product to give a net energy production. Kalnes and James [53], in a pilot-scale study, clearly showed the appreciable economic advantages of hydrodechlorination over incineration. Incorporation of catalytic hydrodehalogenation units in distillation/separation lines is envisaged with a HCl recovery unit, where HCl absorption into an aqueous phase produces a dilute acid solution that can be concentrated downstream to any level desired. The HCl effluent can be further trapped in basic solution and the hydrogen gas scrubbed and washed to remove trace contaminants and recycled to the reactor.

IV. CATALYTIC HYDRODECHLORINATION: REVIEW OF RECENT LITERATURE

While there is a wealth of published data concerning hydrodenitrogenation, hydrodesulfurization, and hydrodeoxygenation reactions [54], catalytic hydrodechlorination is only now receiving a comprehensive consideration, and kinetic and mechanistic studies are urgently required to evaluate the potential of such an approach to environmental pollution control. The number of papers related to hydrodehalogenation has certainly mushroomed over the past two years, as even a cursory glance through any recent issue of *Applied Catalysis B: Environmental* will reveal. There are two comprehensive review articles that deal with dehalogenation reactions, dating from 1980 [55] and 1996 [56]. Both reviews are largely concerned with organic synthetic aspects of dehalogenation, and the environmental remediation aspect is only now truly emerging.

Thermal (noncatalytic) dehalogenation has been successfully applied to a range of halogenated compounds, but elevated temperatures (up to 1173 K) are required to achieve near-complete (ca. 99.95%) dehalogenation to HX [57,58]. A thermodynamic analysis of gas-phase hydrodechlorination reactions has shown that HCl formation is strongly favored [14,59], and the presence of a metal catalyst reduces considerably the operating temperature, providing a lower-energy pathway for the reaction to occur [60]. Catalytic hydrodehalogenation is established for homogenous systems, where the catalyst and reactants are in the same (liquid) phase [61,62]; while high turnovers have been achieved, this approach is not suitable for environmental remediation purposes, due to the involvement of additional chemicals (as solvents/hydrogen donors) and the often-difficult product/solvent/catalyst separation steps. Hydrodechlorination in heterogeneous systems has been viewed in terms of both nucleophilic [63,64] and electrophilic [65–67] attack. Surface science studies on Pd(111) suggest that homolytic cleavage predominates and is insensitive to any substituent inductive effect [68,69]. Chlorine removal from an aromatic reactant has been proposed to be both more [12,70,71] and less [55] facile than dechlorination of aliphatics. The nature of both the surface-reactive adsorbed species and catalytically active sites is still open to question. It is, however, accepted that hydrodechlorination, in common with most hydrogenolysis reactions, is strongly influenced by the electronic structure of the surface metal sites [72], where the nature of the catalyst support can influence catalytic activity/selectivity and stability [12,73].

Chlorobenzene has been the most widely adopted model reactant to assess catalytic aromatic hydrodechlorination activity in both the gas [63,67,74–85] and liquid [86–90] phases using Pd- [63,81,82,86–90], Pt- [84,87], Rh- [81,82,87], and Ni- [46,59,60,65,67,74–81,83,85] based catalysts. The hydrodechlorination of monochlorophenols has received less attention, but reaction rates have been reported in the liquid phase over Pd/C [91,92] and Ru/C [93] and in the gas phase over Ni-Mo/Al$_2$O$_3$ [83,94], Ni/Al$_2$O$_3$ [78], and Ni/SiO$_2$ [65,66,95]. The removal of multiple chlorine atoms from an aromatic host has also been studied to a lesser extent [59,60,67,79,80,96–99], while hydrodebromination reactions have received scant attention in the literature [74,100,101]. Urbano and Marinas [12] have noted that the ease of C–X bond scission decreases in the order, R–I > R–Br > R–Cl ≫ R–F, which matches the sequence of decreasing C–X bond dissociation energies. However, in gas-phase debromination and dechlorination promoted by Ni/SiO$_2$ [74], the relative rates of Cl and Br removal depend on the nature of the organic host, in that debromination rates are higher in the case of aliphatic reactants and lower for the conversion of aromatics. In the treatment of polychlorinated aromatics, a range of partially dechlorinated isomers has been isolated in the product stream where the product composition depends on the nature of the catalyst and process conditions, i.e., temperature, concentration, residence time, etc. [60,96].

Taking an overview of the reported data [12], it appears that Pd is the most active dechlorination metal, but Pd catalysts suffer from appreciable deactivation with time on-stream [101,102]. Halogens are known to act as strong poisons in the case of transition metal catalysts [103], and catalyst deactivation during hydrodechlorination has been reported for an array of catalyst/reactant systems [63,77,81,84,87,91,99,101,102,104]. Deactivation has been attributed to different causes, ranging from deposition of coke [84,105] to the formation of surface metal halides [77,106] to sintering [106–108], but no conclusive deactivation mechanism has yet emerged. Hydrodechlorination kinetics has been based on both pseudo-first-order approximations [59,60,79,80] and mechanistic models [63,67,81,109,110]. There is general agreement in the literature that the reactive hydrogen is adsorbed dissociatively [63,75,77,81,82], while the involvement of spillover species has also been proposed [109–111]. The mechanism of C–Cl bond hydrogenolysis is still open to question, and this must be established and combined with a robust kinetic model in order to inform reactor design and facilitate process optimization. An unambiguous link between catalyst structure and dechlorination activity/selectivity has yet to emerge. The latter is essential in order to develop the best strategy for both promoting and prolonging the hydrogenolysis activity of surface metal sites.

V. CASE STUDY: GAS-PHASE HYDRODECHLORINATION OF CHLOROARENES OVER SUPPORTED NICKEL

A. Nature of the Catalysts

Three standard synthetic routes were considered in anchoring Ni to a range of supports: impregnation (Imp); precipitation/deposition (P/D); ion exchange (IE).

The Ni content of the catalyst precursors, method of preparation, and average Ni particle diameter (and range of diameters) in the activated catalysts are given in Table 1, wherein the experimentally determined chlorobenzene hydrodechlorination rates over each catalyst under the same reaction conditions are identified. Supported Ni catalysts prepared by deposition/precipitation have been shown to exhibit a narrower distribution of smaller particles when compared with the less controlled impregnation route [112,114]. Nickel can be introduced into a microporous zeolite matrix by ion exchange with the charge-balancing sodium cations [115]. Reduction of Ni-exchanged Y zeolites under similar conditions is known [116,117] to generate a metal phase that exhibits a wide size distribution, with particle growth resulting in the formation of larger metal crystallites supported on the external surface. While metal dispersion is dependent on metal loading, the array of supported Ni catalysts (where %Ni w/w = 8 ± 2) included in Table 1 present a range of particle sizes. There is ample evidence in the literature linking the extent of the metal/support interaction(s) to the ultimate morphology and dimensions of the metal crystallites [72,118,119]: The stronger the interactions, the greater the metal dispersion. Weak interactions between metal and carbon-based supports have been reported elsewhere [119], leading to Ni particle growth. Enhanced dispersion on alumina has been attributed to the ionic character of the

TABLE 1   Physical Characteristics of a Range of Supported Ni Catalysts[a] and Associated Chlorobenzene Hydrodechlorination Rates (R)[b]

Support	Ni loading (% w/w)	Preparation	Ni diameter range (nm)	Average Ni diameter (nm)	$R(\text{mol g}^{-1}\text{h}^{-1})$
SiO$_2$	1.5	P/D	<1 to 3	1.4	2×10^{-5}
SiO$_2$	6.2	P/D	<1 to 5	1.9	6×10^{-5}
SiO$_2$	11.9	P/D	<1 to 6	2.5	10×10^{-5}
SiO$_2$	15.2	P/D	<1 to 8	3.1	12×10^{-5}
SiO$_2$	20.3	P/D	<1 to 8	3.8	15×10^{-5}
SiO$_2$	10.1	Imp	<1 to 40	12.9	4×10^{-5}
MgO	9.3	Imp	<1 to 25	10.5	4×10^{-5}
Al$_2$O$_3$	8.6	Imp	<1 to 15	5.7	2×10^{-5}
Activated carbon	9.0	Imp	<2 to 70	23.4	6×10^{-5}
Graphite	10.3	Imp	<2 to 80	27.1	2×10^{-6}
Zeolite Y	6.4	IE	<5 to 80	38.2	1×10^{-6}

[a] Prepared by precipitation/deposition (P/D), impregnation (Imp), and ion exchange (IE).
[b] $T = 523$ K.

support and the existence of partially electron-deficient metal species leading to strong interactions with the support [72,120].

B.   Hydrodechlorination and Catalyst Structure

The magnitude of the hydrodechlorination rates (related to catalyst weight) recorded in Table 1 cover a wide range, where the highest value is greater by over two orders of magnitude than the lowest. It has been demonstrated [59,60,65, 66,74–76,95–97] that nanodispersed nickel metal on amorphous silica in the presence of hydrogen is highly effective in the catalytic dehalogenation of concentrated halogenated gas streams. The performance of supported metal catalysts, in general, is governed by a number of interrelated factors, notably metal particle dispersion, morphology, and electronic properties. The observed diversity of hydrodechlorination activity can be related to variations in the nature of the supported Ni sites. The Ni crystallite sizes fall within the so-called mithohedrical region, wherein catalytic reactivity can show a critical dependence on morphology [121]. Taking the family of Ni/SiO$_2$ catalysts, the specific hydrodechlorination rates (per exposed nickel surface area) for chlorobenzene and 4-chlorophenol are plotted as a function of Ni particle size in Figure 1. An increase in the supported Ni particle size consistently generated, for both reactants, a higher specific chlorine removal rate. The reaction can then be classified as structure sensitive,

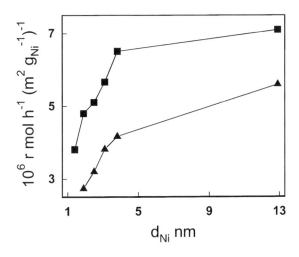

FIG. 1   Specific hydrodechlorination rate (r) as a function of nickel particle size (d_{Ni}) for the hydrodechlorination of chlorobenzene (▲) and 4-chlorophenol (■) over Ni/SiO$_2$ at 523 K.

where higher specific activities are associated with larger Ni particle sizes. There is no general consensus regarding structure sensitivity or insensitivity in hydrodechlorination systems. However, Karpinski and co-workers [122,123] have noted a higher turnover frequency of CF_3CFCl_2 and CCl_2F_2 for larger Pd particles supported on Al_2O_3 and attributed this to an ensemble effect. Marinas et al. [101,124] also found that the liquid-phase hydrodechlorination of chlorobenzene and bromobenzene over Pd/SiO_2-$AlPO_4$ was enhanced at lower Pd dispersions. Efremenko [125] has recently demonstrated the impact of metal particle geometry and electronic structure on the reactivity and mobility of adsorbed hydrogen. It is well established that different forms of hydrogen with different degrees of interaction are present on the surface of supported Ni catalysts, with reported adsorption enthalpies ranging from -110 to in excess of -400 kJ mol^{-1} [126]. The presence of chlorine is known to limit the degree of hydrogen chemisorption on supported nickel [108] and Ni (100) [104], disrupting interaction energetics. Moreover, the nature of the reactive hydrogen in hydrogenolysis and hydrogenation reactions has been shown [75,109] to be quite different, with spillover hydrogen on the support metal/support interface proposed as the reactive hydrodechlorination agent [109,111].

There are many instances in the literature [121] where reactivity is strongly influenced by the electron density of small supported metal particles. Hydrogenolysis reactions have been used as tests or probes for metal charge effects in catalysis, where the metal/support interface plays a significant role [72]. Variations in basicity/acidity of the support have been shown to have a dramatic effect on hydrogenolysis rate [127–129]. The effect of doping Ni/SiO_2 with KOH and CsOH on hydrodechlorination activity is shown in Table 2, where the incorpora-

TABLE 2 Effect of Doping Ni/SiO_2[a] with KOH and CsOH on Associated Chlorobenzene Hydrodechlorination Rates (R)[b]

% Ni w/w	Preparation	Alkali dopant	R(mol h^{-1}g$_{Ni}$$^{-1}$)
11.9	P/D	—	83×10^{-5}
11.9	P/D	KOH	6×10^{-5}
11.9	P/D	CsOH	2×10^{-5}
10.1	Imp	—	38×10^{-5}
10.1	Imp	KOH	4×10^{-5}
10.1	Imp	CsOH	1×10^{-5}

[a] Prepared by precipitation/deposition (P/D) and impregnation (*Imp*).
[b] $T = 523$ K; Ni/alkali metal mol ratio = 1.

tion of an alkali metal component lowered rates by a factor of up to 50. The possible formation of electron-rich Ni particles via electron donation from the K and Cs dopants resulted in a significant suppression of hydrodechlorination activity. The latter effect suggests an enhanced C–Cl scission activity associated with electron-deficient metal sites. The drop in activity may, however, be due to a spreading of K/Cs over the Ni surface that in effect occludes the active phase, as has been demonstrated elsewhere [130]. The effect of prolonged contact of the catalysts with concentrated chlorinated gas streams, in terms of alterations to Ni particle size and hydrodechlorination rates, can be assessed from the results presented in Table 3. The tabulated data represent continual operation in a single-pass dechlorination through a fixed catalyst bed for up to 800 h; this translates into a total Cl-to-Ni mol ratio of up to $(2 \times 10^4):1$. The nickel-dilute catalyst prepared by precipitation/deposition (P/D) largely retained its initial activity, while the higher-loaded P/D catalyst exhibited a decided loss of activity but was still appreciably more durable than the sample prepared by impregnation (Imp). Loss of activity was accompanied by a shift in the surface-weighted mean Ni metal particle size. The Ni particle diameter histograms shown in Figure 2 illustrate the overall shift in size to higher values after catalyst use. A halide-induced agglomeration of Ni particles (on activated carbon) has been reported by Othsuka [131], who attributed this effect to a surface mobility of Ni-Cl species. Vaporization of $NiCl_2$ crystals at temperatures as low as 573 K has been proposed to occur, leading to a deposition and growth of surface Ni particles [118]. There was no evidence of any significant metal particle growth in the lower-Ni-loaded P/D sample, which may be attributed to stronger metal–support interactions. The spent samples contained an appreciable residual Cl content, and it has been shown elsewhere [75] that the catalyst surface, under reaction conditions, is saturated with hydrogen halide. Moreover, STEM/EDX elemental maps of the used catalysts revealed an appreciable halogen concentration on the surface [132]. Nickel

TABLE 3 Effect of Total Amount of Chlorine That Contacted Ni/SiO_2 on Average Ni Particle Size (d_{Ni}), Cl/Ni Ratio in Spent Samples, and Ratio of Final (x_{Cl}) to Initial (x_0) Chlorobenzene Conversion

| % Ni w/w | Preparation | d_{Ni}(nm) | | Mol Cl processed | Cl/Ni | x_{Cl}/x_0 |
		Freshly activated	Used			
1.5	P/D	1.4	1.5	2.6	0.56	0.93
15.2	P/D	3.1	12.2	2.8	0.15	0.72
10.1	Imp	12.9	19.4	2.2	0.17	0.61

$T = 573$ K.

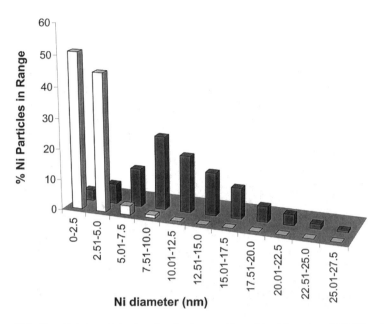

FIG. 2 Nickel particle size distribution profiles of freshly reduced 15.2% w/w Ni/SiO$_2$ (open bars) prepared by precipitation/deposition and the same catalyst after extended use in the hydrodechlorination of chlorobenzene (solid bars).

particle growth alone, on the basis of the structure sensitivity patterns shown in Figure 1, should serve to raise the dechlorination rate. Prolonged contact with the concentrated chlorinated gas stream must result in a restructuring of the metal particles, where the presence of the surface halogen has been shown to result in strong electronic perturbations of the Ni sites that can impact on the hydrogen activation step [109,118] with a consequent loss of hydrogenolysis activity. The deactivated samples contained a significant carbon content (up to 10% w/w), suggesting that coke formation may also contribute to catalyst deactivation. The nature of the carbon deposits in spent catalysts can be probed by means of temperature-programmed oxidation (TPO). A TPO profile for a representative used catalyst is shown in Figure 3, which also includes a profile generated from a commercial amorphous carbon sample. Both profiles are essentially superimposable, suggesting that the carbon deposit is essentially amorphous. The presence of residual Cl on a catalyst surface has been noted elsewhere to result in a greater degree of coke formation [133,134]. A displacement of charge density from the surface nickel sites can also occur through the surface carbon, where such carbonaceous deposits retain a halogenated character. The observed loss of activity

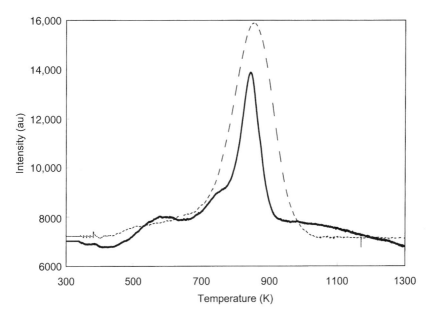

FIG. 3 TPO profiles generated from model amorphous carbon (dashed line) and the carbon deposit on a spent sample of 10.1% w/w Ni/SiO$_2$ prepared by impregnation after extended use in the hydrodechlorination of 4-chlorophenol (solid line): $T = 573$ K.

can be linked to a restructuring/electronic perturbation of the Ni crystallites and possible site blocking by residual chlorine/amorphous carbon deposit.

C. Hydrodechlorination Activity

The effect of secondary aromatic ring substituents has been shown to have a considerable effect on the catalytic hydrogenolysis of aromatic compounds [46,135,136]. Catalytic hydrodechlorination rates (expressed as moles Cl cleaved from the aromatic ring per gram of Ni), obtained under identical reaction conditions, for a range of chloroarenes are given in Table 4; the tabulated data represent steady-state values. The magnitude of the determined rates spans a wide range, where the hydrodechlorination rate for 3-chlorophenol (highest value) was greater by over two orders of magnitude than that recorded for hexachlorobenzene (lowest value). Hydrodechlorination activity associated with the different isomers is dependent on the halogen content and the nature of the cosubstituent. In substituted aromatic systems, reactivity is typically related to localized (inductive) and delocalized (resonance) effects [137]. Taking an overview of the tabulated data and lumping the isomers together, the families of chlorobased haloarenes exhib-

TABLE 4 Hydrodechlorination Rates (R) for a
Range of Chlorinated Benzenes and Phenols
Reacted at 523 K over a 15.2% w/w Ni/SiO$_2$[a]

Reactant	$R(\text{mol}_{Cl} g_{Ni}^{-1} h^{-1})$
Chlorobenzene	7.9×10^{-4}
2-Chlorotoluene	8.4×10^{-4}
3-Chlorotoluene	1.1×10^{-3}
4-Chlorotoluene	1.3×10^{-3}
2-Chlorophenol	8.2×10^{-4}
3-Chlorophenol	1.7×10^{-3}
4-Chlorophenol	1.4×10^{-3}
1,2-Dichlorobenzene	1.2×10^{-4}
1,4-Dichlorobenzene	2.1×10^{-4}
2,3-Dichlorophenol	4.4×10^{-4}
2,5-Dichlorophenol	2.3×10^{-4}
1,2,3-Trichlorobenzene	7.1×10^{-5}
1,3,5-Trichlorobenzene	4.9×10^{-5}
2,3,6-Trichlorophenol	3.8×10^{-4}
2,4,6-Trichlorophenol	4.7×10^{-4}
Hexachlorobenzene	4.8×10^{-6}
Pentachlorophenol	2.8×10^{-5}

[a] Prepared by precipitation/deposition (P/D).

ited the following trend of decreasing hydrodehalogenation rates: chlorophe-
nol(s) ~ chlorotoluene(s) > chlorobenzene > dichlorophenol(s) ~ trichlorophe-
nol(s) > dichlorobenzene(s) > trichlorobenzene(s) > pentachlorophenol >
hexachlorobenzene. The higher dechlorination rates associated with the chlorophenols and chlorotoluenes are indicative of an electrophilic mechanism, where
the presence of the hydroxyl or methyl group (as opposed to a hydrogen atom)
serves to activate the ring for electrophilic attack via an inductive effect that
increases the electron density of the aromatic ring, i.e., stabilizes the cationic
transition state. An electrophilic mechanism presumes the involvement of a hydronium ion as a reactive species, and there is ample evidence for the coexistence
of charged and uncharged hydrogen spillover species on catalyst surfaces [138].
As a direct corollary, the additional presence of a second [dichlorobenzene(s)
and -phenol(s), third (trichlorobenzene(s) and -phenol(s)], fifth (pentachlorophenol), and sixth (hexachlorobenzene) Cl on the ring has a deactivating effect. There
is some variation of reactivity among the different isomers, but the pattern that
emerges points to steric effects as the limiting feature. Resonance effects appear
to have a negligible role to play in determining reaction rate in that ortho/para
isomers cannot be linked in terms of reactivity when compared with the meta-

form. This is particularly marked in the case of the chlorophenols and -toluenes, where, if resonance effects governed reactivity alone, the dechlorination rates for 2-chlorotoluene/-phenol and 4-chlorotoluene/-phenol should be similar but different from 3-chlorotoluene/-phenol. Indeed, it has been demonstrated elsewhere [60] that the product composition resulting from the catalytic hydrodechlorination of polychlorinated aromatics is quite distinct from that predicted on the basis of resonance considerations.

The effect of switching from an aromatic to an aliphatic host in terms of dechlorination reactivity is examined in Figure 4, where the dechlorination rate for chlorobenzene over Ni/SiO_2 is compared with that of cyclohexyl chloride as a function of temperature. It is immediately evident that Cl removal is more facile from the aliphatic host, as has been reported previously [55,74]. Dechlorination of cyclohexyl chloride generated cyclohexene as the predominant organic product, with cyclohexane and benzene formed as secondary products. The formation of cyclohexene results from the internal elimination of HCl. Supported nickel catalysts promote the hydrogenolytic cleavage of the C–Cl bond in aromatic systems and a dehydrochlorination in the case of the chloroalkane. The latter has been viewed in terms of an E1 elimination mechanism, where the chlorine component interacts with the catalyst with electron withdrawal, weakening the C–Cl bond and inducing intermediate carbocation formation [74]. The adsorbed chlorine species may then serve as a base for the removal of the hydrogen atom in a fast step followed by C=C bond formation and desorption of cyclohexene

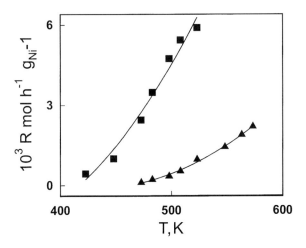

FIG. 4 Temperature dependence of the dechlorination rates (R) of chlorobenzene (▲) and cyclohexyl chloride (■) over 15.2% w/w Ni/SiO_2 prepared by precipitation/deposition.

from the surface. In gas-phase homogeneous systems operating under equilibrium conditions, thermodynamic analysis [74] predicts that cyclohexane is by far the preferred product. Cyclohexane is formed only in the catalytic process at temperatures in excess of 473 K, and its yield is elevated with further increases in temperature, because catalytic hydrogenolysis is favored. Mixtures of cyclohexene and cyclohexane are generated in an uncatalyzed dechlorination, whereas the Ni catalysts impart a high degree of process selectivity (particularly where T < 473 K) in favor of the alkene.

D. Hydrodechlorination Selectivity

The gas-phase hydrodechlorination of the range of chloroarenes listed in Table 4 over the range of nickel catalysts identified in Table 1 generated partially/fully dechlorinated aromatics and HCl as by far the predominant products (selectivity > 99%). The catalytic hydrogen treatment of each mono-and polychlorinated phenol yielded only HCl and phenol (as the ultimate dechlorinated organic), and there was no detectable formation of cyclohexanone or cyclohexanol as a result of a further hydrogenation of phenol. Likewise, there was no evidence of any ring reduction in the conversion of chlorobenzenes or chlorotoluenes. Moreover, there was no observable catalytic hydrodehydroxylation in the case of chlorophenol(s) transformation(s), i.e., C–OH bond scission. A selective hydrodechlorination is to be expected, given the reported [139] bond dissociation energies of aromatic C–Cl (406 kJ mol^{-1}) and C–OH (469 kJ mol^{-1}), which show that the hydrodeoxygenation step is the more energetically demanding. A minor degree of chlorophenol isomerization activity (<1 mol% conversion) was evident at T > 523 K. Qualitative analysis for the presence of chlorine gas was negative in every instance, confirming that hydrogenolytic cleavage of chlorine from an aromatic host yields HCl as the only inorganic product.

Hydrodechlorination selectivity is an important feature in the conversion of polychlorinated aromatics, i.e., the degree of partial vs. full dechlorination. Taking the family di- and trichlorobenzenes, the ratio of complete to partial dechlorination under the same reaction conditions is recorded in Table 5. The overall selectivity trend points to a more limited degree of dechlorination where the Cl substituents are spaced further apart on the aromatic ring. Partial dechlorination is more predominant in the case of the trichlorobenzenes, in keeping with the deactivating effect of the additional Cl substituent. The attractive feature of catalytic hydrodechlorination in terms of treating "halogenated waste" is that the waste can be transformed into a recyclable product. Adopting dichlorobenzenes as representative of unwanted chloro-products, judicious choice of process conditions can facilitate a conversion to the parent raw material (benzene) or to chlorobenzene as a target product. With the former goal in mind, benzene selectivity from the three dichlorobenzene isomers is plotted as a function of W/F_{DCB} in

TABLE 5   Ratio of Full to Partial
Hydrodechlorination in the Conversion of
Dichlorobenzene and Trichlorobenzene Isomers[a]

Reactant	Full dechlorination/ partial dechlorination
1,2-dichlorobenzene	8.33
1,3-dichlorobenzene	0.34
1,4-dichlorobenzene	0.29
1,2,3-trichlorobenzene	0.47
1,2,4-trichlorobenzene	0.28
1,3,5-trichlorobenzene	0.04

[a] At 573 K over a 6.1% w/w Ni/SiO_2 prepared by
precipitation/deposition (P/D).

Figure 5. The W/F_{DCB} parameter represents the ratio of catalyst weight to the inlet
dichlorobenzene molar feed rate and has the physical significance of representing
contact time. Benzene formation is clearly enhanced at higher contact times, and
the ultimate selectivity is dependent on the nature of the chloro-isomer. Complete
dechlorination is, however, possible at extended contact times or by recycling

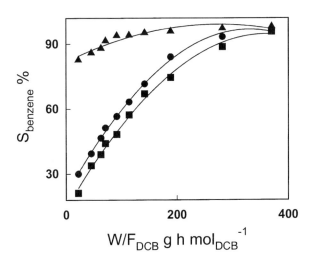

FIG. 5   Benzene selectivity ($S_{benzene}$) as a function of W/F_{DCB} for the hydrodechlorination
of 1,2-dichlorobenzene (▲), 1,3-dichlorobenzene (●), and 1,4-dichlorobenzene (■) at 573
K over 6.1% w/w Ni/SiO_2 prepared by precipitation/deposition.

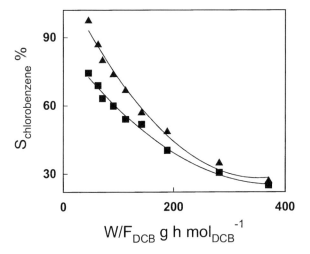

FIG. 6   Chlorobenzene selectivity ($S_{\text{chlorobenzene}}$) as a function of W/F_{DCB} for the hydrode-chlorination of 1,4-dichlorobenzene over 1.5% w/w Ni/SiO$_2$ prepared by precipitation/deposition at 523 K (▲) and 573 K (■).

the effluent (with HCl trapping) for additional dechlorination. The response of chlorobenzene selectivity (as the alternative goal) from a 1,4-dichlorobenzene feed over a less active (see Table 1) Ni-dilute catalyst is illustrated in Figure 6. A high selectivity (>97%) is certainly possible by operating the reactor at lower contact times and reduced temperatures. It is evident from the data presented that selectivity dependence on temperature is negligible at high contact times, where the two selectivity profiles converge.

VI.   CONCLUSIONS

The present preferred method of dealing with or disposal of chlorinated waste, i.e., incineration, not only is an expensive operation but clearly does not represent the best possible management of resources. Legislation governing the handling of chlorinated waste is certain to become increasingly more restrictive, as the censure of defaulters receives higher priority in Europe and the United States. Catalytic hydrodechlorination offers an alternative to disposal, a chemical processing of concentrated hazardous waste that serves to detoxify and transform it into a reusable feedstock. The treatment of chlorinated waste is an important issue in the handling of raw materials/products used for heat exchangers, dyes, herbicides, insecticides, and agricultural materials and, as such, applies to a broad industrial sector. Such a nondestructive treatment methodology represents an important step forward in reducing the negative environmental impact of chemical industries in general.

The gas-phase hydrodechlorination of chloroarenes over supported Ni catalysts is fully selective in the hydrogen scission of the C–Cl bond(s), leaving the aromatic nucleus and hydroxyl/alkyl substituents intact. Chlorine is cleaved from the aromatic host as HCl (which is easily trapped for reuse), and there is no evidence of any Cl_2 formation. Catalyst deactivation appears to be dependent on Ni loading and is accompanied by an increase in Ni particle size, disruption to the Ni electronic structure, and an appreciable level of coke deposits. As a general observation, the presence of electron-donating substituents on the aromatic ring serves to increase the rate of hydrodechlorination, while doping the catalyst with electron-donating atoms lowers C–Cl hydrogen scission activity. Dechlorination of chloroalkanes is more facile and proceeds via HCl elimination or dehydrochlorination. Hydrodechlorination of polychlorinated feed proceeds via stepwise and concerted routes, where steric hindrance impacts process selectivity. Preliminary studies reveal that a judicious choice of both catalyst and operating conditions will permit a control of the ultimate product composition.

The use of supported nickel catalysts to promote a selective dechlorination of chloroaromatics is a feasible progressive approach to minimizing chlorinated waste. High dechlorination rates can be achieved at temperatures less than 600 K. Low-temperature operation is desirable for economic reasons where the catalyst is operated downstream of wet scrubbers as part of an "end-of-process" recycle strategy. The conversion of overly chlorinated aromatic waste can follow two directions: (1) regenerating raw material that can be reused; (2) direct formation of a desirable chloro-aromatic product. Catalytic hydrodechlorination as a detoxification/recycle methodology can be extended further to consider the catalytic conversion of waste halogenated polymeric materials into fuel. The thermal degradation of chloropolymers is known to generate both inorganic and organic chlorine-containing products [140,141]. A complete removal of chlorine from the polymeric waste material is essential before the use of the product oil as a fuel is at all feasible; the application of catalytic dechlorination can facilitate this process.

REFERENCES

1. JK Fawell, S. Hunt. Environmental Toxicology, Organic Pollutants. Chichester, UK: Ellis Horwood, 1988.
2. Toxics Release Inventory, Public Data Release. Washington, DC: USEPA, Office of Pollution Prevention and Toxics, 1991.
3. A Kaune, D Lenoir, KW Schramm, R Zimmermann, A Kettrup, K Jaeger, HG Ruckee, F Frank. Environ Eng Sci 15:85–95, 1998.
4. C Rolf. Kyoto Protocol to the United Nations Framework Convention on Climate

Change: A Guide to the Protocol and Analysis of Its Effectiveness. West Coast Environmental Law Association, 1998.

5. C Denbesten, JJRM Vet, HT Besselink, GS Kiel, BJM Vanberkel, R Beems, PJ Vandladeren. Toxicol Appl Pharm 11:69–81, 1991.

6. RE Hester, RM Harrison. Volatile Organic Compounds in the Environment, Issues in Environmental Science and Technology. Cambridge, UK: Royal Society of Chemistry, 1995.

7. EN Ruddy, LA Carroll. Chem Eng Progr 89:28–81, 1993.

8. N Supranat, T Nunno, M Kravett, M Breton. Halogenated Organic Containing Waste: Treatment Technologies. Park Ridge, NJ: Noyes Data Corp., 1988.

9. AJ Buonicore, W Davis, eds. Air Pollution Engineering Manual. New York: Van Nostrand Reinhold, 1992.

10. RB Clark. Halogenated Hydrocarbons in Marine Pollution. 2nd ed. Oxford, UK: Oxford Science, 1989.

11. LH Keith. AIChE Symp Ser 77:249–263, 1980.

12. FJ Urbano, JM Marinas. J Mol Catal A Chemical 173:329–343, 2001.

13. JF McEldowney, S McEldowney. Environment and the Law. Essex, UK: Longman, 1996.

14. DR van der Vaart, EG Marchand, A Bagely-Pride. Crit Rev Environ Sci Technol 24:203–236, 1994.

15. A Converti, M Zilli, DM De Faveri, G Ferraiolo. J Hazard Mater 27:127–135, 1991.

16. BF Hagh, DT Allen. Innovative Hazardous Waste Treatment Technology. Lancaster, PA: H. M. Freemont, TECHNOMIC, 1990.

17. R Broadbank. Process Eng 72 (3a):41–43, 1991.

18. R Weber, T Sakurai, H Hagenmaier. Appl Catal B Environmental 20:249–256, 1999.

19. H Hagenmaier, K Horch, H Fahlenkamp, G Schetter. Chemosphere 23:1429–1437, 1991.

20. JJ Spivey. Ind Eng Chem Res 26:2165–2180, 1987.

21. KS Lin, HP Wang. J Phys Chem B 105:4956–4960, 2001.

22. JR González-Velasco, A Aranzabal, R López-Fonseca, R Ferret, JA González-Marcos. Appl Catal B Environmental 24:33–43, 2000.

23. T-Ch Yu, H Shaw, RJ Farrauto. ACS Symp Ser 495:141–152, 1992.

24. T Miyake, M Hanaya. Appl Catal A General 121:L13–17, 1995.

25. AM Padilla, J Corella, JM Toledo. Appl Catal B Environmental 22:107–121, 1999.

26. JM Tseng, CP Huang. Water Sci Technol 23:377–387, 1991.

27. E Piera, JC Calpe, E Brillas, X Domènech, J Peral. Appl Catal B Environmental 27:169–177, 2000.

28. MR Hoffmann, ST Martin, W Choi, D Bahnemann. Chem Rev 95:69–96, 1995.

29. DP Fernandez, ARH Goodwin, EW Lemmon, JMH Levelt Sangers, RC Williams. J Phys Chem Ref Data 26:1125–1166, 1997.

30. C Petrier, M Nicolle, G Merlin, JL Luche, G Reverdy. Environ Sci Technol 26: 1639–1642, 1992.

31. G Zhang, I Hua. Adv Env Res 4:219–224, 2000.

32. DF Ollis, H Al-Ekabi, eds. Photocatalytic Purification and Treatment of Water and Air. Amsterdam: Elsevier, 1993.
33. DD Dionysiou, AP Khodadoust, AM Kern, MT Suidan, I Baudin, J-M Laîné. Appl Catal B Environmental 24:139–155, 2000.
34. RB Clark. Halogenated Hydrocarbons in Marine Pollution. Oxford, UK: Oxford Science, 1989.
35. DE Dorn. Arch Microbiol 99:61–70, 1988.
36. M Koo, WK Lee, CH Lee. Chem Eng Sci 52:1201–1214, 1997.
37. YI Matatov-Meytal, M Sheintuch. Ind Eng Chem Res 37:309–326, 1998.
38. MD LaGrega, PH Buckingham, JC Evans. Hazardous Waste Management. New York: McGraw-Hill, 1994.
39. SD Feast, OM Aly. Adsorption Processes for Water Treatment. Stoneham, MA: Butterworths, 1987.
40. RD Vidic, MJ Suidan, GA Sarialand, RC Brenner. J Hazard Mat 39:373–388, 1994.
41. AAM Daifullah, BS Girgis. Water Res 32:1169–1177, 1998.
42. LS Colella, PM Armenante, O Kafkewitz, SJ Allen, V Balasundaram. J Chem Eng Data 43:573–579, 1998.
43. C Brasquet, J Roussy, E Subrenat, P leCloirec. Env Technol 17:1245–1252, 1996.
44. B Okolo, C Park, MA Keane. J Colloid Interf Sci 226:308–317, 2000.
45. MA Keane. J Catal 166:347–355, 1997.
46. E-J Shin, MA Keane. J Catal 173:450–459, 1998.
47. MS Mubarak, DG Peters. J Electroanal Chem 435:47–53, 1997.
48. SP Zhang, JF Rusling. Environ Sci Technol 29:1195–1199, 1995.
49. AI Tsyganok, K Otsuka. Appl Catal B Environment 22:15–26, 1999.
50. N Couté, JT Richardson. Appl Catal B Environmental 26:265–273, 2000.
51. Kirk-Othmer Encyclopedia of Chemical Technology. 4th ed. Vol 6. New York: Wiley, 1993.
52. DG Ackerman. Destruction and Disposal of PCBs by Thermal and Non-Thermal Methods. Park Ridge, NJ: Noyes Data Corp., 1983.
53. TN Kalnes, RB James. Environ Prog 7:185–191, 1988.
54. A Stanislaus, BH Cooper. Catal Rev Sci Eng 36:75–123, 1994.
55. AR Pinder. Synthesis 425–452, 1980.
56. VV Lunin, ES Lokteva. Russ Chem Bull 45:1519–1534, 1996.
57. JA Manion, R Louw. J Phys Chem 94:4127–4134, 1990.
58. ER Ritter, JW Bozzelli, AM Dean. J Phys Chem 94:2493–2504, 1990.
59. EJ Shin, MA Keane. Chem Eng Sci 54:1109–1120, 1999.
60. EJ Shin, MA Keane. J Chem Technol Biotechnol 75:159–167, 2000.
61. CM King, RB King, NK Bhattacharyya, MG Newton. J Organomet Chem 600: 63–70, 2000.
62. SMH Tabaei, CU Pitman Jnr. Tetrahedron Lett 34:3263–3266, 1993.
63. M Kraus, V Bazant. In: JW Hightower, ed. Proceedings of the 5th International Congress on Catalysis. New York: North-Holland, 1973, pp 1073–1083.
64. P Dini, JC Bart, N Giordano. J Chem Soc Perkins Trans 21:1479–1482, 1975.
65. EJ Shin, MA Keane. J Hazard Mater B 66:265–278, 1999.
66. C Menini, C Park, EJ Shin, G Tavoularis, MA Keane. Catal Today 62:355–366, 2000.

67. BF Hagh, DT Allen. Chem Eng Sci 45:2695–2701, 1990.
68. CW Chan, AJ Gellman. Catal Lett 53:139–143, 1998.
69. MT Buelow, G Zhou, AJ Gellman, B Immaraporn. Catal Lett 59:9–13, 1999.
70. PN Rylander. Catalytic Hydrogenation over Platinum Metals. New York: Academic Press, 1967.
71. RB LaPierre, WL Kranich, AH Weiss. J Catal 52:59–71, 1978.
72. AY Stakheev, LM Kustov. Appl Catal A General 188:3–35, 1999.
73. B Coq, F Figueras, S Hub, D Tournigant. J Phys Chem 99:11159–11166, 1995.
74. G Tavoularis, MA Keane. J Mol Catal A Chemical 142:187–199, 1999.
75. G Tavoularis, MA Keane. J Chem Technol Biotechnol 74:60–70, 1999.
76. G Tavoularis, MA Keane. Appl Catal A General 182:309–316, 1999.
77. J Estellé, J Ruz, Y Cesteros, R Fernandez, P Salagre, F Medina, JE Sueiras. J Chem Soc Faraday Trans 92:2811–2816, 1996.
78. AR Suzdorf, SV Morozov, NN Anshits, SI Tsiganova, AG Anshits. Catal Lett 29: 49–55, 1994.
79. BF Hagh, DT Allen. AIChE J 36:773–778, 1990.
80. J Frimmel, M Zdražil. J Catal 167:286–295, 1997.
81. B Coq, G Ferrat, F Figueras. J Catal 101:434–445, 1986.
82. P Bodnariuk, B Coq, G Ferrat, F Figueras. J Catal 116:459–466, 1989.
83. DI Kim, DT Allen. Ind Eng Chem Res 36:3019–3026, 1997.
84. EJ Creyghton, MHW Burgers, JC Jensen, H van Bekkum. Appl Catal A General 128:275–288, 1995.
85. N Lingaiah, MA Uddin, A Muto, T Iwamoto, Y Sakata, Y Kasano. J Mol Catal A Chemical 161:157–162, 2000.
86. MA Aramendia, R Burch, IM Garcia, A Marinas, JM Marinas, BWL Southward, FJ Urbano. Appl Catal B Environmental 31:163–171, 2001.
87. Y Ukisu, T Miyadera. J Mol Catal A Chemical 125:135–142, 1997.
88. VI Simiagina, VM Mastikhin, VA Yakovlev, IV Stoyanova, VA Likholobov. J Mol Catal A Chemical 101:237–241, 1995.
89. JL Benitez, G Del Angel. React Kinet Catal Lett 66:13–18, 1999.
90. C Schüth, M Reinhard. Appl Catal B Environmental 18:215–221, 1998.
91. JB Hoke, GA Gramiccioni, EN Balko. Appl Catal B: Environmental 1:258–296, 1992.
92. Yu Shindler, Yu Matatov-Meytal, M Sheuntuch. Ind Eng Chem Res 40:3301–3308, 2001.
93. V Felis, C De Bellefon, P Fouilloux, D Schweich. Appl Catal B Environmental 20:91–100, 1999.
94. C Chon, DT Allen. AIChE J 37:1730–1732, 1991.
95. E-J Shin, MA Keane. Appl Catal B Environmental 18:241–250, 1998.
96. EJ Shin, MA Keane. Catal Lett 58:141–145, 1999.
97. EJ Shin, MA Keane. React Kinet Catal Lett 69:3–8, 2000.
98. F Gioia, V Famiglietti, F Murena. J Hazard Mater 33:63–73, 1993.
99. Y Cesteros, P Salagre, F Medina, JE Sueiras. Appl Catal B Environmental 25:213–227, 2000.
100. Z Yu, S Liao, Y Xu. React Funct Polym 29:151–157, 1996.

101. MA Armendia, V Boráu, IM Garcia, JM Jiménez, JM Marinas, FJ Urbano. Appl Catal B Environmental 20:101–110, 1999.
102. L Prati, M Rossi. Appl Catal B Environment 23:135–142, 1999.
103. S Zhuang, J Wu, X Liu, J Tu, M Ji, K Wandelt. Surf Sci 331:42–46, 1995.
104. JW Bozzelli, YM Chen, SSC Chuang. Chem Eng Commun 115:1–11, 1992.
105. M Ocal, M Maciejewski, A Baiker. Appl Catal B Environmental 21:279–289, 1999.
106. A Gampine, DP Eyman. J Catal 170:315–325, 1998.
107. Y Ohtsuka. J Mol Catal 54:225–235, 1989.
108. DJ Moon, MJ Chung, KY Park, SI Hong. Appl Catal A Chemical 168:159–170, 1998.
109. EJ Shin, A Spiller, G Tavoularis, MA Keane. Phys Chem Chem Phys 1:3173–3181, 1999.
110. MA Keane, D Yu Murzin. Chem Eng Sci 56:3185–3195, 2001.
111. S Kovenklioglu, Z Cao, D Shah, RJ Farrauto, EN Balko. AIChE J 38:1003–1012, 1992.
112. MA Keane. Can J Chem 72:372–381, 1994.
113. P Burattin, M Che, C Louis. J Phys Chem B 101:7060–7074, 1997.
114. P Burattin, M Che, C Louis. J Phys Chem B 102:2722–2732, 1998.
115. MA Keane. Microporous Mater 3:93–105, 1995.
116. C Park, MA Keane. J Mol Catal A Chemical 166:303–322, 2001.
117. B Coughlan, MA Keane. Zeolites 11:2–11, 1991.
118. C Hoang-Van, Y Kachaya, SJ Teichner. J Phys Chem B 101:7060–7074, 1997.
119. A Stevenson, JA Dumesic, RTK Baker, E Ruckenstein. Metal Support Interactions in Catalysis, Sintering and Redispersion. New York: Van Nostrand, 1987.
120. MI Zaki. Stud Surf Sci Catal 100:569–577, 1996.
121. M Che, CO Bennett. Adv Catal 36:55–172, 1989.
122. Z Karpinski, K Early, JL d'Itri. J Catal. 164:378–386, 1996.
123. W Juszczyk, A Mallinowski, Z Karpinski. Appl Catal A General 166:311–219, 1998.
124. MA Aramendia, V Borau, IM Garcia, C Jimenez, A Marinas, JM Marinas, FJ Urbano. J Catal 187:392–399, 1999.
125. I Efremenko. J Mol Catal A Chemical 173:19–59, 2001.
126. S Smeds, T Salmi, LP Lindfors, O Krause. Appl Catal A General 144:177–194, 1996.
127. BL Mojet, MJ Kappers, JT Miller, DC Koningsberger. Stud Surf Sci Catal 101: 1165–1174, 1996.
128. ST Homeyer, Z Karpinski, WMH Sachtler. J Catal 123:60–73, 1990.
129. SD Jackson, GJ Kelly, G Webb. J Catal 176:225–234, 1998.
130. C Park, MA Keane. Chem Phys Chem 2:101–109, 2001.
131. Y Ohtsuka. J Mol Catal 54:225–235, 1989.
132. C Menini, C Park, R Brydson, MA Keane. J Phys Chem B 104:4281–4284, 2000.
133. RJ Verderone, CL Pieck, MR Sad, JM Parera. Appl Catal 21:329–250, 1986.
134. A Chambers, RTK Baker. J Phys Chem B 101:1621–1630, 1997.
135. MA Keane. J Mol Catal A Chemical 118:261–269, 1997.
136. MA Keane. J Mol Catal A Chemical 138:197–209, 1999.

137. WH Brown. Organic Chemistry. New York: Saunders College, 1995.
138. U Roland, Th Braunschweig, F Rossler. J Mol Catal A Chemical 127:61–84, 1997.
139. RT Sanderson. Chemical Bonds in Organic Compounds. Scottsdale, AZ: Sun and Sand, 1976.
140. M Blazso. Rapid Commun Mass Spectrum 12:1–4, 1998.
141. Y Shiraga, MA Uddin, A Muto, M Narazaki, Y Sakata, K Murata. Energy Fuel 13:428–432, 1999.

14

TiO$_2$ Nanoparticles for Photocatalysis

HEATHER A. BULLEN and SIMON J. GARRETT Michigan State University, East Lansing, Michigan, U.S.A.

I. INTRODUCTION

Transition metal oxides exhibit a wide range of physical, chemical, and structural properties. One of the most widely studied metal oxides in the semiconducting oxide titanium dioxide (TiO$_2$). Titanium dioxide first attracted significant attention when in 1972 Fujishima and Honda discovered that TiO$_2$ can act as a catalyst for the photocleavage of water, producing H$_2$ and O$_2$ [1]. In the presence of a TiO$_2$ electrode, they observed that water was dissociated using photons with $\lambda \leq 410$ nm, whereas direct photodissociation of water requires photons with $\lambda \leq 185$ nm. This discovery sparked interest in the photocatalytic activity of TiO$_2$ and other metal oxide semiconductors as a possible approach to inexpensively convert solar radiation to chemical energy [2,3]. Subsequent research efforts have focused on understanding the fundamental processes that drive these photoelectrochemical cells and in their application to energy storage applications [4,5].

The ability to oxidatively decompose organic molecules present as pollutants in the environment has recently refocused research attention toward utilizing semiconducting oxides for remediation applications. For example, the TiO$_2$ surface can participate in a wide range of redox chemistries for many types of adsorbed organic molecules, including aromatic, halogenated organic, and commercial dye molecules [6]. The central mechanism involves the photoinduced generation of charge carriers at the surface of a semiconductor, followed by interfacial charge transfer reactions with absorbed molecules. The mechanistic aspects of these reactions have been reviewed [6,7], but in many cases the identities of intermediates have not been firmly established.

The photocatalytic activity of TiO$_2$ in many redox reactions is limited by the relatively large bandgap ($E_g = 3.0 - 3.2$ eV) of the material, which limits absorption to the UV region of the solar spectrum, below about 350 nm. This limitation has led to the development of chemically modified TiO_2 surfaces with better spectral absorption accomplished by dye-sensitization [8,9]. Alternative semiconducting oxide material with smaller bandgaps, such as MoS_2 [10], and composite semiconductors, such as ZnO/ZnS [11] and CdS/PbS [12], have also been investigated.

Recent scientific interest in TiO_2 photocatalysts has been motivated by observations that aqueous solutions of colloidal TiO_2 nanoparticles exhibit significantly enhanced chemical and photochemical reactivity due to so-called quantum size effects (QSE) [13,14]. The chemical and electronic properties of semiconductor nanoparticles are distinct from either extended solids or single molecules and thus represent an exciting new class of materials [15]. It is known that the properties of such nanoparticles vary strongly as a function of particle size (as well as shape) and consequently have "tunable" optical, electronic, and chemical properties [15–17]. In most cases, the TiO_2 nanoparticle surfaces and their role in chemical and photochemical reactivity are still poorly understood.

In this chapter we will outline the basic mechanism of TiO_2 photocatalysis and describe how particle size can influence the photoreactivity of TiO_2. We will also examine current methods being utilized to synthesize TiO_2 nanoparticles and introduce a novel synthetic methodology to grow supported crystalline nanoparticles of TiO_2.

II. BACKGROUND

A. Fundamental Mechanisms of Semiconductor Photocatalysis

1. Bandgap Excitation

The central mechanism of photocatalytic activity in semiconductors relies on absorption of a photon of energy greater than or equal to the semiconductor bandgap energy, E_g. Since solar radiation is a natural and abundant energy source, most photocatalytic strategies have been directed towards exploiting this energy by choosing materials with bandgaps within the range of terrestrial sunlight (approximately 4.1 to <0.5 eV). Many semiconductors have bandgap energies within this desired range and so are potential materials for promoting or photocatalyzing a wide variety of chemical reactions [18].

When a semiconductor absorbs a photon of energy $\geq E_g$, excitation creates an electron (e^-) in the conduction band (CB) and leaves a hole (h^+) in the valence band (VB) [19]. In TiO_2 the CB is composed of empty Ti $3d$ states and the VB is composed of filled O $2p$ states. The e^-/h^+ pair may spontaneously recombine, with thermal or luminescent energy release, or may migrate toward the surface and react with adsorbed acceptor or donor species in reduction or oxidation reac-

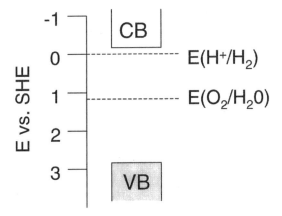

FIG. 1 Approximate band edge positions for rutile TiO$_2$ at pH $= 1$. (From Ref. 4.)

tions, respectively. In order for redox reactions to occur, the energy of the adsorbate orbitals acting as electron acceptors or those acting as electron donors
must lie within the bandgap region of the photocatalyst, as shown in Figure 1.
Hence, the position of these adsorbate energy levels relative to those of the semiconductor surface is crucial. In the absence of redox active surface species, spontaneous e^-/h^+ recombination occurs within a few nanoseconds [20].

2. Electron/Hole Recombination

The reactivity of a photocatalyst is dependent on the rate of e^-/h^+ recombination
in the bulk or at the surface. In order to have an efficient photocatalyst, the photogenerated holes and electrons must have a long lifetime, since recombination is
in direct competition with surface charge transfer to adsorbed species. Therefore,
the recombination of the photoexcited e^-/h^+ pair must be minimized. Surface
and bulk defects can generate electronic states that serve as charge carrier traps.
The presence of these charge carrier trapping sites, such as Ti^{3+} or surface TiOH
sites in TiO$_2$, extend the effective lifetime of the photoexcited e^-/h^+ pair, increasing the probability of an electron transfer process to an adsorbed molecule.

It is generally believed that non-negligible recombination rates limit the overall quantum yield of current photocatalytic systems based on TiO$_2$. Various strategies, such as doping [23,24] and creating Schottky barrier traps [25–27], have
been attempted to extend the lifetime of surface charge carriers and thus improve
overall efficiency. For example, the photocatalytic degradation of rhodamine B
by TiO$_2$ was significantly enhanced when doped with lanthanide metals: Eu^{3+},
La^{3+}, Nd^{3+}, and Pr^{3+} [24]. These dopants create a potential gradient at the surface,
separating the photogenerated e^-/h^+ pairs.

3. Band Bending and the Schottky Barrier

When a semiconductor is in contact with another phase, such as a liquid or gas, there is a redistribution of charge within the semiconductor. As mobile charge carriers are transferred between the semiconductor and contact phase or carriers are trapped at intrinsic or adsorbate-induced surface states, a space charge layer develops and there is no longer a uniform distribution of charge within the semiconductor. The electronic band potentials of the semiconductor are distorted, depending upon whether there is an accumulation or a depletion of charge in the near-surface region. As a consequence, bands may bend upward (n-type semiconductors) or downward (p-type semiconductors) close to the surface. For example, naturally occurring oxygen surface vacancies on TiO_2 create five-coordinate Ti^{3+} sites. The Ti^{3+} sites serve as strong electron traps, causing the surface region to become negatively charged with respect to the bulk of the semiconductor. To compensate for this effect, a positive space charge layer develops within the semiconductor, causing a shift in the electrostatic potential and the upward bending of bands. TiO_2 is therefore considered an n-type semiconductor.

Following bandgap excitation, photogenerated electrons move away from the surface while the holes move toward the surface, due to the potential gradient that has formed from band bending (Fig. 2). This band-bending phenomenon assists in separating the e^-/h^+ pairs and in reducing recombination rates. For TiO_2, the surface holes oxidize adsorbed molecules by electron transfer from the adsorbate into a hole. However, the photocatalytic oxidation of many organic molecules is believed to be mediated by electron transfer from coadsorbed species

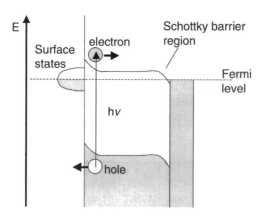

FIG. 2 Diagram showing the surface band bending and Schottky barrier that serve to separate h^+ and e^- following bandgap excitation in an n-type semiconductor.

on the surface [6], such as surface hydroxyl groups Ti^{4+}–OH [21,22], which form surface radicals that can directly oxidize the adsorbed molecule.

By placing a noble metal on the TiO$_2$ surface, the separation rate of the e^-/h^+ pair can be further increased if the metal creates a favorable potential gradient (Schottky barrier) to act as a sink for photogenerated electrons. The metal surface then becomes the site of reduction reactions. Based on this phenomenon, discrete electrochemical cells incorporting small metal islands deposited onto TiO$_2$ nanoparticles have been prepared [28]. For example, Dawson and Kamat determined that the photocatalytic oxidation of thiocyanate ions was increased by 40% using gold-capped TiO$_2$ nanoparticles [25]. The amount of noble metal required to produce an effective Schottky barrier can correspond to less than a few percent of the surface covered.

4. Quantum Size Effects (QSE)

Photocatalytic activity is also affected by particle size. When the physical dimensions of a semiconductor particle fall within the range of 5–20 nm, the diameter of the particle becomes comparable to the wavelength of the charge carriers (e^-/h^+) and quantum size effects (QSE) occur [17,29]. The electronic structure of the semiconductor can no longer be described as an extended solid, with overlapping wavefunctions from each atom giving rise to continuous and delocalized electronic valence and conduction bands. Instead, the charge carriers become localized in the effective potential well of the nanoparticle, and discrete quantized energy states are produced (Fig. 3) that give rise to the strongly size-dependent optical and electronic phenomena. Absorption intensities are perturbed, and the effective bandgap of a semiconductor particle is thought to increase as the particle

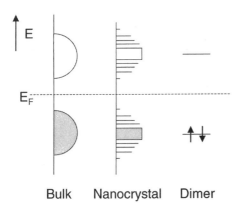

FIG. 3 Density of states for a semiconductor as a function of particular size.

size decreases, corresponding to a blue-shift of the absorption band [15,16]. These phenomena can influence the photocatalytic properties of small semiconductor particles. For example, in the decomposition of 1-butene by SnO_2, 5-nm particles were photoactive, whereas 22-nm particles were not [30]. Similarly, Gao and Zhang discovered that 7.2-nm rutile TiO_2 particles has a much higher photocatalytic activity in the oxidation of phenol compared to 18.5- and 40.8-nm particles [31].

In addition to changes in the electronic structure of the material, other phenomena can occur as particle dimensions are reduced. Smaller particles present more surface adsorption/reaction sites per unit volume and are therefore expected to show increased catalytic activity. Additionally, the formation of unique electronic surface states or reactive defects may become favored. The high curvature of the particle surface creates a large number of low-coordination surface atoms of unique local geometry and bonding, which may also lead to substantial surface relaxation, reconstruction, or faceting. In TiO_2 single crystals, these low-coordination sites have been shown to markedly influence the adsorption and reactivity of small molecules [32,33]. As the volume of the semiconductor becomes very small, the band-bending phenomenon that spatially separates the e^- and h^+ is reduced. Band bending typically operates on the 0.5 to 5-nm distance scale and becomes weak as particle diameters approach these dimensions [4]. A small particle is almost completely depleted of charge carriers, so its Fermi potential is located approximately in the middle of the bandgap. This implies that there is an optimum-size semiconductor nanoparticle for surface photoreactivity, which is dependent upon the material.

B. TiO_2 Photocatalysis

Despite a wide range of materials with suitable bandgaps, titanium dioxide remains a primary candidate as a photocatalyst for environmental remediation applications due to its thermodynamic stability, high abundance, low cost, and nontoxicity. Titanium dioxide exists as three natural crystalline forms (rutile, anatase, and brookite), with rutile being thermodynamically the most stable [19,34]. Most photocatalytic studies have focused on the rutile ($E_g = 3.0$ eV) and anatase ($E_g = 3.2$ eV) forms of TiO_2. Bulk-powder, single-crystal, and thin-film studies of anatase and rutile have helped to elucidate the photocatalytic mechanisms of TiO_2 as well as the application of this semiconductor to technologies of interest [6,7].

Anatase appears to be slightly more photoactive than rutile TiO_2, which is thought to be due to its larger charge carrier diffusion rates [5] and lower recombination rates compared to rutile [35,36]. The photoreactivity for anatase and rutile is highly variable, depending on the exact surface preparation methods. In many cases, the rutile $TiO_2(110)$ surface is seen as "the model system" for surface studies of TiO$_2$. Such single-crystal studies in ultrahigh vacuum (UHV) have complemented ambient powder and film studies and have contributed to the development of a fundamental understanding of the role of the surface in the overall photocatalytic activity of TiO$_2$. These studies have determined the influence of parameters such as surface geometric structure, defect nature, and concentration and the identity of reactive intermediates [7,37,38]. Unfortunately, single-crystal studies of anatase are rare [39,40] due to the difficulty of preparation, but bulk measurements of dispersed particles have been performed [41].

1. Surface Chemistry of TiO$_2$: The Role of Defect Sites

The surface chemistry of TiO$_2$ is significantly influenced by the concentration of oxygen defects. A stoichiometric rutile TiO$_2$(110) surface is quite unreactive, since a fully oxidized surface contains no occupied surface states in the bandgap [42]. However, defects increase the reactivity of the surface, particularly oxygen defects that produce low-coordination Ti^{3+} sites. As mentioned earlier, these Ti^{3+} atoms create surface trap states in the bandgap of TiO$_2$ and ultimately lead to enhanced chemical reactivity [43–47].

Several different types of O atom vacancies have been directly observed by scanning probe microscopies on single crystals [38,48–50], some of which are shown schematically in Figure 4. Chemisorption studies on TiO$_2$ surfaces using various probe molecules (such as H$_2$, CO, O$_2$ and SO$_2$) indicate that adsorption is dependent on oxygen defect sites on the surface [33,46,51,52]. This dependence can influence the nature of the photocatalytic reactions that can take place on the surface. For example, Yates and co-workers have determined that molecularly adsorbed oxygen is essential for photoxidation of methyl chloride [53]. In their study they discovered that substrate-mediated excitation of adsorbed oxygen

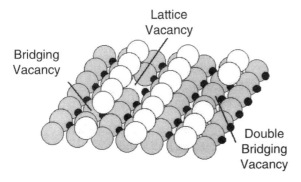

FIG. 4   Oxygen atom vacancies (defect sites) on rutile TiO$_2$(110). Ti = ●, O = ○.

generates an ionic species, probably O_2^{2-}, which directly oxidizes coadsorbed CH_3Cl. They speculate that at the gas–solid interface, adsorbed oxygen may play a more important role in the oxidation of certain organic molecules, such as CH_3Cl, than photocatalytically generated · OH radicals. Defect concentration and adsorbed oxygen has been shown to play a similar role in the photocatalytic dehydrogenation of 2-propanol [54].

2. Nanoparticles: Current Synthetic Methods

Almost all of the studies of particle size–reactivity relationships for TiO_2 have been performed using solutions of colloidal nanoparticles. Anatase and rutile TiO_2 colloid nanoparticles are commonly synthesized by hydrolysis of titanium compounds such as titanium tetrachloride, $TiCl_4$ [31,55–57], and titanium alkoxides, $Ti(OR)_4$ [14,23,58–63], followed by a calcination process. Hydrolysis of $TiCl_4$ in HCl produces both anatase and rutile phases of TiO_2, and the anatase/rutile ratio can be varied [64] by controlling the pH and temperature of calcination. It has also recently been reported by Pottier and co-workers that brookite nanoparticles can be synthesized by carefully controlling the molar ratio of Cl:Ti in solution [57]. In very acidic solutions of $TiCl_4$, Cl^- ions stabilize the brookite nanoparticles. These particles vary in size, with a mean diameter of 5.2 nm.

Using titanium alkoxide precursors such as tetraisopropoxide, $Ti(-OCH(CH_3)_2)_4$ [14,23,59–63], or tetrabutyl titanate, $Ti(OC_4H_9)_4$ [58], TiO_2 colloidal nanoparticles can be generated. By varying the temperature of hydrolysis, the size of the colloid particles can be controlled. For example, Martini has shown that hydrolysis at 1°C produces ~2-nm-size anatase particles and that at 20°C ~20 to 30-nm-size particles are synthesized [63].

During calcination, both the degree of crystallinity and the size of the particles increase, and the colloid composition generally transforms from anatase to rutile. This limits the size of anatase nanoparticles that can be made by such hydrothermal methods. Alternative approaches, such as solvothermal synthesis in organic media instead of water, may be more effective in producing smaller nanoparticles with high crystallinity and large surface areas [65,66].

Despite the ability to synthesize TiO_2 nanoparticles of various size and composition, little is known of the detailed composition or morphology of TiO_2 nanoparticulate surfaces. The particles produced by solution methodologies tend to agglomerate, are nonuniform in size and shape and are generally not amenable to surface characterization by experimental techniques such as electron spectroscopy and scanning probe microscopy. However, a complete understanding of the photocatalytic activity of TiO_2 nanoparticles on a fundamental atomic level is clearly desirable. In the next section, we present a novel approach to growing monodisperse, controlled-size nanoparticles of TiO_2 supported on a substrate. This will allow for a fundamental spectroscopic investigation of the chemical and photoreactivity of TiO$_2$ nanoparticles on both the microscopic and macroscopic scales.

III. NANOSPHERE LITHOGRAPHY APPROACH TO CREATING QUANTUM SIZE TiO$_2$ PARTICLES

A. Background

Our approach to producing ordered arrays of TiO$_2$ nanoparticles is based on the formation of a physical mask using close-packed polystyrene microspheres. The technique, termed *nanosphere lithography*, has been developed and successfully applied by Van Duyne and co-workers [67–70] to create various supported noble metal nanoparticle arrays, but to our knowledge it has not been applied to TiO$_2$ or other oxides. The methodology is simple, intrinsically parallel, relatively inexpensive, and highly precise, producing particles with uniform size and shape. In contrast, standard lithographic methods used to create nanoparticles with controlled size and spacing, such as photolithography [71] and electron beam lithography [72], are limited to the minimum size of features they can produce and are very complex and expensive. The precision of nanostructure fabrication by the nanosphere lithography technique is, in principle, comparable to or better than other nanofabrication approaches [70].

In the nanosphere lithography method, an aqueous, solution of polystrene microspheres ($\sim$50–500 nm in diameter) is drop-coated onto a suitable substrate, and the solvent is allowed to evaporate under controlled conditions. The nanospheres spontaneously order into a hexagonal close-packed array on the surface. Depending upon the initial sphere concentration, different periodic particle array (PPA) masks can be produced: a close-packed single-layer periodic particle array (SL-PPA) or a double-layer periodic particle array (DL-PPA) from a monolayer or bilayer of spheres, respectively [68]. The desired material of interest is then evaporated through the nanosphere mask, coating the exposed surface between the spheres. The microsphere mask is subsequently removed by an organic solvent such as ethanol or methylene chloride. These solvents dissolve the polystyrene spheres, leaving the material deposited through the mask on the substrate (Fig. 5). Complete mask removal becomes difficult if the height of the islands exceeds the radius of the microspheres used to make the mask.

The diameter of the nanoparticle islands is approximately 15% of the diameter of the polystyrene microspheres. Therefore, by changing the size of the microspheres, various-diameter nanoparticles can be made using this technique. Polystyrene spheres with diameters down to 50 nm are commercially available, allowing nanoparticles on the order of 7-nm minimum diameter (i.e., quantum size particles) to be generated. Changing the incidence angle of the evaporative

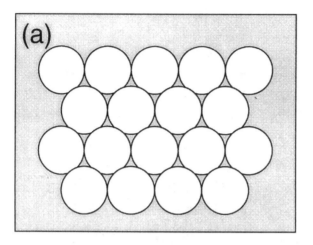

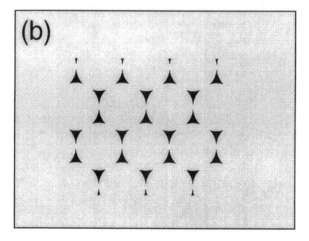

FIG. 5 (a) A close-packed array of polystyrene microspheres and (b) an array of nano-particles produced by evaporation through mask and subsequent mask removal.

source with respect to the substrate normal can alter the shape of the supported nanoparticles somewhat [70], but to date, nanosphere lithography has been limited to triangular and circular particle shapes. Unfortunately, the particle arrays contain up to 1% point and line defects, due to polydispersity in the spheres used to form the mask. These disadvantags may be overcome with continued development.

B. Experimental Results: Preparation of Supported TiO₂ Nanoparticle Arrays

Our research group is currently using nanosphere lithography to grow ordered arrays of supported nanoparticles of TiO_2 on various substrates. These arrays should mimic some of the properties of colloidal nanoparticles but offer specific experimental advantages and the potential for simple control of crystallography and Schottky barrier creation. In this section we discuss methods for producing the polystyrene masks, the deposition procedure, characterization of the nanoparticles produced, and the outlook for this research.

1. Mask Preparation

In order to create TiO_2 nanoparticles of uniform size and shape, the packing of the polystyrene spheres on the support surface must be controlled to maximize the size of defect-free domains. For the deposition of the polystyrene microsphere mask, it is necessary that the spheres be able to freely diffuse in solution across the substrate, seeking their lowest-energy (close packed) configuration during solvent evaporation. A common approach to facilitate surface diffusion is to use surface-modified polystyrene spheres, derivatized with a negatively charged functional group such as carboxylate [70]. These modified spheres are electrostatically repelled by a negatively charged substrate, such as mica or glass, promoting lateral migration of the particles along the surface. At the levels of derivatization of commercially available polystyrene spheres, surface charge does not appear to impede the process of close-packing.

A second important factor in controlling the mask deposition is the hydrophilicity of the support surface. In order to achieve wetting of the aqueous suspension of polystyrene spheres on the substrate, the substrate must be treated to produce a hydrophilic surface. In many previous nanosphere lithography studies, glass substrates have been used, since these are the simplest to prepare. Cleaning glass substrates with piranha solution (3:1 concentrated H_2SO_4:30% H_2O_2) followed by soaking in 5:1:1 H_2O:NH_4OH:30% H_2O_2 functionalizes the glass surface with hydroxyl groups [69]. We have also produced suitable supports from TiO_2 single crystals. It is known that UV irradiation of TiO_2 converts the native hydrophobic surface into a hydrophilic surface [73,74]. This is believed to be caused by photon-stimulated desorption of surface O atoms, creating oxygen defects that act as sites for dissociative water adsorption [75]. Heating the single crystal in air or N_2 can also create these defects. In both cases, there is evidence for surface Ti^{4+}–OH groups that decrease the hydrophobicity of the material.

A third important factor controlling the quality of the polystyrene sphere mask is the rate of solvent evaporation compared with the rate of surface diffusion and packing. It is known that the relative humidity and temperature of the substrate

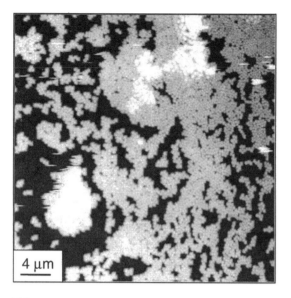

FIG. 6   AFM image (40 × 40 µm) showing poor packing of 420-nm polystyrene spheres on glass, when the evaporation rate is uncontrolled. Images are similar for untreated TiO_2.

can influence the packing order [76–78]. If the evaporation rate is too high, disordered layers do not have a chance to anneal sufficiently, and masks containing agglomerates, poor packing order, uncovered regions, and multiple layers are produced. An example is shown in Figure 6. In our experiments, the 2-D crystallization of the polystyrene spheres was improved by designing a chamber to control the rate of evaporation of the microsphere solution. In this sealed chamber the relative humidity is controlled at approximately 97% using a saturated K_2SO_4 salt solution. A Peltier cell is used to cool the sample to 10–15°C and to minimize spatial temperature gradients on the substrate. Slight tilting of the substrate from the horizontal creates a well-defined liquid front that moves from the higher to the lower portions of the substrate during evaporation of the solvent, further improving the quality of the masks produced.

Once the masks have been prepared, they are examined by atomic force microscopy (AFM) to verify that they contain large ordered domains of spheres, as shown in Figure 7. Typical domain sizes range from 1 to 10 µm using the methodology described here.

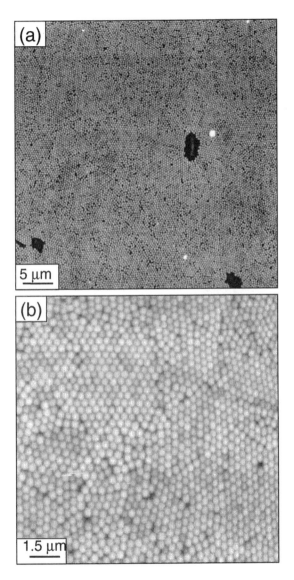

FIG. 7 AFM image of 420-nm polystyrene spheres in a close-packed array on glass under controlled conditions. (a) 50 × 50 μm, (b) 15 × 15 μm.

2. TiO$_2$ Nanoparticle Array Preparation

The mask-covered glass or TiO$_2$(110) substrates are placed into a high-vacuum chamber and evacuated to $<10^{-7}$ torr. The chamber is fitted with a custom-built evaporation source for up to two different metallic samples. For TiO$_2$ deposition, a Ti ribbon filament is resistively heated and allowed to deposit onto a substrate held at room temperature. No attempt is made to collimate the effusive Ti atom beam, although this would presumably affect the reproducibility of the particle shape over widely spaced areas of the substrate. Titanium dioxide can be deposited either by evaporation of Ti in a partial pressure of O$_2$ (10^{-6}–10^{-5} torr) or by deposition of Ti, removal of the mask, and subsequent annealing in O$_2$ at up to 500°C. In this way we have successfully produced ordered arrays of TiO$_2$ particles from 420-nm-diameter sphere masks. Typical AFM images of the TiO$_2$ nanoparticle arrays are shown in Figure 8.

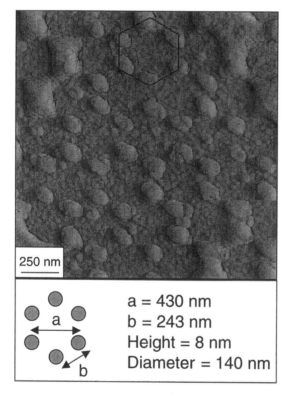

a = 430 nm
b = 243 nm
Height = 8 nm
Diameter = 140 nm

FIG. 8 AFM image (2 $\times$ 2 μm) of TiO$_2$ nanoparticle array on glass formed from a SL-PPA.

Arrays can be made from monolayer or bilayer masks, with the resulting arrays being distinguisable by different-size particles and different interparticular spacing. In Figure 8, each island is approximately 8 nm high and about 140 nm in diameter, consistent with an SL-PPA mask. The diameter is close to that predicted by simple geometric projection arguments considering the packing of the mask. Each island appears to be approximately circular in profile, not the expected triangular features observed in work by Van Duyne et al [67–70]. This suggests that the surface of the TiO_2 nanoparticle undergoes substantial reconstruction or faceting during or following deposition. Arrays grown on glass and on $TiO_2(110)$ single-crystal substrates are similar.

3. Characterization

In addition to the atomic force microscopy images presented earlier, the composition of the particle array was determined by X-ray photoelectron spectroscopy (XPS). Figure 9 shows the Ti $2p$ spectrum of a nanoparticle array grown on glass. Previous XPS studies [79–82] have demonstrated that the various Ti oxides can be distinguished based on their Ti $2p$ XPS spectrum. For example, the Ti $2p_{3/2}$ peak in TiO_2 is at approximately 459.0 eV binding energy (BE), in Ti_2O_3 is at approximately 457.6 eV BE, and in TiO is at approximately 455.3 eV BE. Metallic Ti exhibits a $2p_{3/2}$ peak at 453.8 eV. The shape of the Ti $2p$ photoemission envelope is characteristic of the oxide. Examination of Figure 9 confirms that the dominant species present on the nanoparticle surface is TiO_2. There is also some evidence for Ti^{3+} species, the low-coordination atoms implicated in catalytic and photocatalytic processes for this material. It should be noted that the sampling depth of the XPS technique in the present context is about 30–60 Å, meaning that a substantial fraction of the interior of each nanoparticle is not measured.

4. Outlook

At present we have no information on the crystallinity of the TiO_2 nanoparticles produced, but we are actively engaged in studies to determine this property. It is hoped that control of the crystallinity of the particles can be achieved by choosing a support material that is crystallographically compatible with TiO_2, allowing epitaxial control. The preparation of crystalline nanoparticles is desirable, to ensure reproducible surface properties from particle to particle. There is also evidence that amorphous or mixed TiO_2 phases exhibit reduced photoactivity [31].

The use of single-crystal metal surfaces to impart epitaxial control on growing binary metal oxide films is well documented [83,84], but to our knowledge this has not been successfully applied to generate crystalline TiO_2 films. For successful epitaxy, the lattice parameters and symmetry of the substrate and films should be matched [85]. Unfortunately, no elemental metal adopts a crystallography similar to any known titanium oxide. Several metal oxides adopt a rutile or slightly

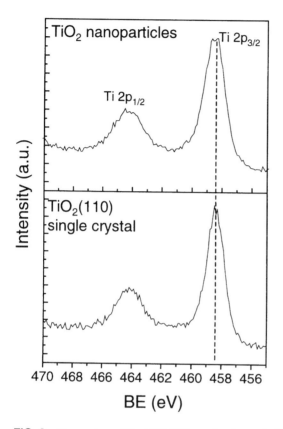

FIG. 9 Comparison of the XPS Ti 2p region for rutile TiO$_2$(110) single crystal and TiO$_2$ nanoparticles on glass.

distorted rutile structure [86], but of these CrO$_2$ (a = 4.41 Å, c = 2.91 Å) and RuO$_2$ (a = 4.51 Å, c = 3.11 Å) most closely match the lattice parameters of rutile (a = 4.59 Å, c = 2.96 Å), as shown in Figure 10. The formation of RuO$_2$(110) surfaces as supports can be achieved by oxidation of a Ru(0001) single crystal at 700 K [87] or molecular beam epitaxy techniques [88]. We have, however, chosen to begin our studies on CrO$_2$(110) surfaces, related to our interest in the ferromagnetic and electronic properties of this material. Although large single crystals of CrO$_2$ are not commercially available, we have succesfully produced highly (110)-textured metallic CrO$_2$ films via a chemical vapor deposition (CVD) process using a rutile TiO$_2$(110) single crystal as a support [89]. The epitaxial control for growing CrO$_2$ on TiO$_2$ is strong enough to stabilize this normally metastable oxide phase for many weeks under ambient laboratory conditions. This suggests that these CrO$_2$ films should be ideal supports for TiO$_2$

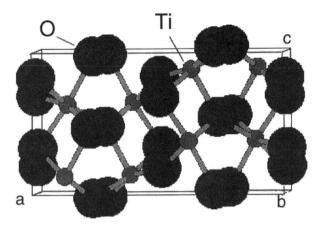

FIG. 10   Idealized rutile crystal structure.

nanoparticle deposition. It is uncertain whether rutile or anatase phases will be formed, but it is believed that the crystallography of the substrate and high surface areas of the particle will thermodynamically favor rutile formation.

An additional property of both RuO$_2$ and CrO$_2$ as supports for TiO$_2$ nanoparticle deposition is their high electronic conductivity. The potential gradients that are generated at the semiconducting nanoparticle/metallic substrate interface may also serve to enhance the photoactivity of the TiO$_2$ nanoparticles by creating a Schottky barrier type of region. The high conductivity will allow the thin TiO$_2$ nanoparticles, which have intrinsically low conductivity, to efficiently transport charge to and from the surface and support. Moreover, the presence of a metallic substrate will enable a detailed surface characterization by surface science techniques such as scanning tunneling spectroscopy (STM).

Unlike solution-phase synthesis of nanoparticles, the TiO$_2$ island arrays produced by the nanosphere lithography approach are nearly identical in size and shape and are expected to have well-defined crystallography and surface structure. In addition, the nanoparticle array films have a large surface area, allowing for macroscopic-scale experiments to relate surface structure to chemical and photochemical reactivity. In this way, it is believed that a fundamental and microscopic view of the role of surface structure in nanoparticle photocatalysis can be achieved.

IV.   SUMMARY

This chapter provides a general background into photocatalytic properties of TiO$_2$ and highlights some interesting examples of its application to environmental remediation. The fundamental aspects of photocatalysis have been described, including: bandgap excitation, electron/hole recombination, band bending, and the formation of a Schottky barrier. Quantum size effects and the role of oxygen defects on the photocatalytic activity of TiO_2 have also been reviewed. It has been determined that there is a need for understanding of surface of TiO_2 nanoparticles on an atomic level to facilitate our understanding of their photocatalytic properties and size-dependent reactivity. Current methods to synthesize TiO_2 nanoparticles often produce agglomerates, making size-dependent experiments difficult. A new experimental method to synthesize monodisperse supported crystalline TiO_2 nanoparticle arrays has been presented.

REFERENCES

1. A Fujishima, K Honda. Nature 238:37, 1972.
2. JS Connolly. Photochemical Conversion and Storage of Solar Energy. New York: Academic Press, 1981.
3. R Memming, ed. Photoelectrochemical Solar Energy Conversion. Berlin: Springer-Verlag, 1988.
4. N Serpone, E Pelizzetti, eds. Photocatalysis: Fundamentals and Applications. New York: Wiley, 1989.
5. M Schiavello, ed. Heterogeneous Photocatalysis. New York: Wiley, 1997.
6. MR Hoffmann, St. Martin, W. Choi, DW Bahnemann. Chem. Rev. 95:69, 1995.
7. AL Linsebigler, G Lu, JT Yates. Chem. Rev. 95:735, 1995.
8. G Benkö, M Hilgendorff, AP Yartsev, V Sundström. J. Phys. Chem. B 105:967, 2001.
9. DW Thompson, CA Kelly, F Farzad, GJ Meyer. Langmuir 15:650, 1999.
10. TR Thurston, JP Wilcoxon. J. Phys. Chem. B 103:11, 1998.
11. J Rabani. J. Phys. Chem. 93:7707, 1989.
12. HS Zhou, I Honma, H Komiyama. J. Phys. Chem. 97:895, 1993.
13. JA Byrne, BR Eggins. J. Electroanal. Chem. 457:61, 1998.
14. Z Zhang, C-C Wang, R Zakaria, JY Ying. J. Phys. Chem. B 102:10871, 1998.
15. AP Alvisatos. J. Phys. Chem. 100:13226, 1996.
16. S Link, MA El-Sayed. J. Phys. Chem. B 103:8410, 1999.
17. A Henglein. Chem. Rev. 89:1861, 1989.
18. P Kamat. Chem. Rev. 93:267, 1993.
19. PA Cox. The Electronic Structure and Chemistry of Solids. Oxford, UK: Oxford University Press, 1987.
20. G Rothenberger, J Moser, M Grätzel, N Serpone, DK Sharma. J. Am. Chem. Soc. 107:8054, 1985.
21. S Goldstein, G Czapski, J Rabani. J. Phys. Chem. 98:6586, 1994.
22. U Stafford, KA Gray, PV Kamat. J. Phs. Chem. 98:6343, 1994.
23. W Choi, A Termin, MR Hoffman. J. Phys. Chem. 98:13669, 1994.
24. Y Wang, H Cheng, L Zhang, Y Hao, J Ma, B Xu, W Li. J. Molec. Catal. A. 151:205, 2000.

25. A Dawson, PV Kamat. J. Phys. Chem. B 105:960, 2001.
26. N Chandrasekharan, PV Kamat. J. Phys. Chem. B 104:10851, 2000.
27. E Stathatos, P Lianos, P Falaras, A Siokou. Langmuir 16:2398, 2000.
28. D Duonghong, E Borgarello, M Grätzel. J. Am. Chem. Soc., 1981.
29. A Henglein, ed. Mechanisms of Reactions on Colloidal Microelectrodes and Size Quantization Effects. Berlin: Springer-Verlag, 1988.
30. L Cao, F-J Spiess, A Huang, SL Suib, TN Obee, SO Hay, JD Freihaut. J. Phys. Chem. B 103:2912, 1999.
31. L Gao, Q Zhang. Scripta Mater. 44:1195, 2001.
32. WJ Lo, YW Chung, GA Somorjai. Surf. Sci. 71:199, 1978.
33. JM Pan, BL Maschhoff, U Diebold, TE Madey. J. Vac. Sci. Technol. A 10:2470, 1992.
34. JK Burdett, T. Hughbanks, GJ Miller, JW Richardson, JV Smith. J. Am. Chem. Soc. 109:3539, 1987.
35. H Kawaguchi. Environ. Technol. Lett. 5:471, 1984.
36. KM Schindler, M Kunst. J. Phys. Chem. 94:8222, 1990.
37. GE Brown, Jr., VE Henrich, WH Casey, DL Clark, C Eggleston, A Felmy, DW Goodman, M. Grätzel, G Maciel, MI McCarthy, KH Nealson, DA Sverjensky, MF Toney, JM Zachara. Chem. Rev. 99: 77, 1999.
38. DA Bonnell. Prog. Surf. Sci. 57:187, 1998.
39. GS Herman, Y Gao, TT Tran, J Osterwalder. Surf. Sci. 447:201, 2000.
40. R Hengerer, B Bolliger, M Erbudak, M Grätzel. Surf. Sci. 460:162, 2000.
41. KI Hadjiivanov, DG Klissurski. Chem. Soc. Rev. 25:61, 1996.
42. AK See, RA Bartynski. J. Vac. Sci. Technol. A 10:2591, 1992.
43. VE Henrich, G Dresselhaus, HJ Zeiger. Phys. Rev. Lett. 36:1335, 1976.
44. VE Henrich, RL Kurtz. Phys. Rev. B 23:6280, 1981.
45. S Eriksen, RG Egdell. Surf. Sci. 180:263, 1987.
46. W Göpel, G Rocker, R Feirerbend. Phys. Rev. B 28:3427, 1983.
47. G Lu, A Linsebigler, JT Yates. J. Chem. Phys. 102:3305, 1995.
48. H Onishi, Y Iwasawa. Surf. Sci. 313:L783, 1994.
49. GS Rohrer, VE Henrich, DA Bonnell. Surf. Sci. 278:146, 1992.
50. H Maeda, K Ikeda, K Hashimoto, K Ajito, M Morieta, A Fujishima. J. Phys. Chem. B 103:3213, 1999.
51. H Kobayashi, M Yamaguchi. Surf. Sci. 214:466, 1989.
52. KE Smith, VE Henrich. Phys. Rev. B 32:5384, 1985.
53. G Lu, A Linsebigler, JT Yates. J. Phys. Chem. 99:7626, 1995.
54. D Brinkley, T. Engel. J. Phys. Chem. B 102:7596, 1998.
55. E Santacesaria, M Tonell, G Storti, RC Pace, SJ Carra. J. Colloid Interface Sci. 111: 44, 1986.
56. A Safrany, R Gao, J Rabani. J. Phys. Chem. B 104:5848, 2000.
57. A Pottier, C Chanéac, E Tron, L Mazerolles, J-P Jolivet. J. Mater. Chem. 11:1116, 2001.
58. WF Zhang, MS Zhang, Z Yin. Phys. Stat. Sol. 179:319, 2000.
59. HN Ghosh, S Adhikari. Langmuir 17:4129, 2000.
60. X-M Wu, L Wang, Z-C Tan, G-H Li, S-S Qu. J. Solid State Chem. 156:220, 2001.
61. S-Y Kwak, SH Kim, SS Kim. Environ. Sci. Technol. 35:2388, 2001.

62. ST Martin, CL Morrison, MR Hoffman. J. Phys. Chem. 98:13695, 1994.

63. I Martini, JH Hodak, GV Harland. J. Phys. Chem. B 103:9104, 1999.

64. R Cavani, E Foresti, F Parrinello, F Trifiro. Appl. Catal. 38:311, 1988.

65. B Ohtani, K Iwai, H Kominami, T Matsurra, Y Kera, S Nishimoto. Chem. Phys. Lett. 242:315, 1995.

66. H Kominami, Y Takada, H Yamagiwa, Y Kera, M Inoue, T Inui. J. Mater. Sci. Lett. 15:197, 1996.

67. JC Hulteen, RP Van Duyne. J. Vac. Sci. Technol. A 13:1553, 1995.

68. JC Hulteen, DA Treichel, MT Smith, ML Duval, TR Jensen, RP Van Duyne, J. Phys. Chem. B 103:3854, 1999.

69. TR Jensen, MD Malinsky, CL Haynes, RP Van Duyne. J. Phys. Chem. B 104:10549, 2000.

70. CL Haynes, RP Van Duyne. J. Phys. Chem. B 105:5599, 2001.

71. GM Wallraff, WD Hinsberg. Chem. Rev. 99:1801, 1999.

72. T Ito, S Okazaki. Nature 406:1027, 2000.

73. R Wang, N Sakai, A Fujishima, T Watanabe, K Hasimoto. J. Phys. Chem. 103: 2188, 1999.

74. A Nakajima, S Koizumi, T Watanabe, K Hashimoto, Langmuir 16:7048, 2000.

75. MA Henderson. Surf. Sci. 355:151, 1996.

76. S Rakers, LF Chi, H Fuchs. Langmuir 13:7121, 1997.

77. F Burmeister, W Badowsky, T Braun, S Wieprich, J Boneberg, P Leiderer. Appl. Surf. Sci. 144–145:461, 1999.

78. R Micheletto, H Fukuda, M Ohtsu. Langmuir 11:3333, 1995.

79. N Beatham, AF Orchard, G Thronton. J. Phys. Chem. Solids 42:1051, 1981.

80. RL Kurtz, VE Henrich. Surf. Sci. Spectra 5:179, 1998.

81. JM McKay, VE Henrich. Surf. Sci. 137:463, 1984.

82. R Zimmermann, P Steiner, R Claessen, F Reinert, S Hüfner. J. Electron. Spec. Rel. Phenom. 96:179, 1998.

83. SC Street, C Xu, DW Goodman. Ann. Rev. Phys. Chem. 48:43, 1997.

84. DW Goodman. J. Vac. Sci. Technol. A 14:17, 1996.

85. DC Sayle, SA Maicaneanu, B Slater, CRA Catlow. J. Mater. Chem. 9:2779, 1999.

86. CNR Rao, B Raveau. Transition Metal Oxides. New York: Wiley-VCH, 1998.

87. J Wang, CY Fan, K Jacobi, G Ertl. Surf. Sci. 481:113, 2001.

88. YJ Kim, Y Gao, SA Chambers. App. Surf. Sci. 120:250, 1999.

89. HA Bullen, SJ Garrett. Chem. Mater. 14:243, 2002.

15

Use of a Pt and Rh Aerosol Catalyst for Improved Combustion and Reduced Emissions

TREVOR R. GRIFFITHS The University of Leeds, Leeds, United Kingdom

I. INTRODUCTION

Today we are much more conscious about preserving our planet than some 20 years ago. At that time, global warming, the greenhouse effect, the ozone hole, and worrying about pollution and emissions were beginning to be the concerns of scientists; now such phrases worry the general public, even though the topics are not often properly understood and the media usually exaggerate scientific concerns and hence public worry can be out of proportion. That being said, scientists are not always right and, being human, can miss an advance or development, or not recognize its worth until years later. How that can be is more for the historians of science, and the psychologists. This is the story of an invention for reducing the emissions of automotive and diesel engines that was initially neglected, because it was born before its time, and now may well be the simple but significant and economic means for reducing emissions to levels that governments are wanting to achieve.

Searles and Bertelsen [1] reviewed the existing and emerging technologies that will be available to meet the exhaust emission regulations for diesel-powered trucks and other commercial vehicles, or "heavy-duty vehicles," adopted or being considered by the European Union (EU) and the United States for implementation during the 21st century. These technologies include diesel oxidation catalysts, DeNOx [2] catalysts and nitrogen oxide (NOx) adsorbers, selective catalytic reduction (SCR), and diesel particulate filters (DPFs), as well as filter technology for particulate matter crankcase emission control.

They noted that exhaust emission control technology, in the form of the catalytic converter, or auto-catalyst, was first introduced in the United States in 1974 in passenger cars and appeared on European roads in 1985. Currently, more than 275 million of the world's 500 million cars and nearly 90% of all new cars produced worldwide are equipped with auto-catalysts. However, exhaust emission control technology for diesel-powered heavy-duty engines has not yet experienced a similar, widespread application. Recent regulatory developments in Europe and in the United States, however, will result in the widespread use of a variety of exhaust emission technologies on heavy-duty engines to control diesel particulate matter, NOx, carbon monoxide, and hydrocarbons, including those hydrocarbon species regarded by health experts as toxic. They concluded that a technological solution for meeting these very stringent emission standards will require an engineered systems approach combining advances in engine designs, advanced exhaust control equipment, and low-sulfur diesel fuel. They did not, however, discuss or consider chemical advances for improving combustion.

The legislation for the European Union provided new test cycles and tougher emission standards for heavy-duty diesel vehicles for 2000 and 2005. The 2005 (Euro IV) emission standards set carbon monoxide, hydrocarbon, and NOx (3.5 grams per kilowatt-hour (g/kWh)) limit values that can, they believe, probably be achieved by engine improvements, but the particulate limit value set is intended to force the use of DPFs. The particulate matter limits are 0.02 g/kWh on the steady-state cycle and 0.03 g/kWh on the transient cycle. The new limit values are a 30% reduction in carbon monoxide, hydrocarbons, and NOx and an 80% reduction in particulate matter from Euro III limit values.

In 2008 (Euro V) a 2.0-g/kWh NOx limit reflects the need for deNOx or SCR catalysts, but the limit is subject to a European Commission study, reporting by the end of 2002 on the progress of the technologies necessary for the achievement of a 2.0-g/kWh NOx limit in 2008 [2].

In July 2000, the U.S. Environmental Protection Agency (EPA) reconfirmed standards originally adopted in 1997 that require 2004 and later model-year on-road heavy-duty vehicles to meet a 2.5-gram per brake-horsepower-hour (g/bhp-hr) NOx plus nonmethane hydrocarbon standard. (The relationship between these two units is that grams per brake-horsepower-hour = 3/4 of grams per kilowatt-hour.) The EPA also added a new steady-state test cycle not to exceed standards to accompany the existing transient certification test cycle; the added certification test requirements will take effect with the 2007 model year [3].

In December 2000, the EPA finalized emission standards for 2007 and later model-year on-road heavy-duty vehicles and placed limits on the allowable levels of sulfur in diesel fuel. The regulations establish a 0.2-g/bhp-hr NOx standard, a 0.01-g/bhp-hr particulate matter standard, and a 15 parts per million (ppm) sulfur maximum for diesel fuel used by on-road vehicles. The emission standards represent a 90% reduction in particulate matter in NOx from the emission standards applicable in 2004. The particulate matter standard is applicable to all 2007 model-year heavy-duty vehicles, and the NOx standard will be phased in between

2007 and 2010. The sulfur limit would represent a 97% reduction in allowable levels of sulfur, compared with the current limit of 500 ppm.

The Californian Air Resources Board (CARB) in 1998 had declared particulate matter emissions from diesel-fueled engines and vehicles to be a toxic air contaminant. In response, in September 2000 the CARB adopted a comprehensive plan to reduce particulate matter emissions from diesel-fueled engines by 75% in 2010 and by 85% in 2020. The next step will be for the CARB to adopt a series of individual rules requiring the reduction of particulate matter emissions from new and existing on-road, off-road, and stationary diesel engines, as well as setting a sulfur limit of 15 ppm for all diesel fuel sold for use in California beginning in 2006 [4].

Another area of concern with regard to diesel fuel is its sulfur content. Sulfur in diesel fuel has a major negative impact on catalyst performance in several ways. It inhibits catalyst performance by strong adsorption on the catalyst surface and by competing for space on the surface with pollutants, and it limits the amount of nitrogen dioxide formed on an oxidizing catalyst—a problem for some DPFs and NOx adsorbers that rely on nitrogen dioxide for their regeneration. Further, it reacts with chemical NOx traps more strongly than NOx, thereby decreasing NOx storage capacity and requiring more vigorous and frequent regeneration and hence increasing fuel consumption. Finally, it can create sulfate particles. These, measurable by current sampling and measurement techniques, arise with any emission control system that includes a precious metal catalyst with an oxidizing function. These sulfate species participate in coating the catalyst surface, which reduces catalyst performance.

It should be noted that a diesel oxidation catalyst converts carbon monoxide and hydrocarbons to carbon dioxide and water, decreases the mass of particulate matter emissions, but has little effect on NOx emissions. An oxidation catalyst will reduce the soluble organic fraction of diesel particulate by up to 90% [5].

The destruction of the soluble organic fraction is important because this portion of the particulate contains numerous chemicals of concern to health experts. Control of the soluble organic fraction enables the oxidation catalyst to reduce total particulate emissions by 25%–50%, depending on the constituents that make up the total particulate. This technology also reduces diesel smoke and eliminates the pungent diesel exhaust odor as well as making significant reductions in carbon monoxide and hydrocarbons of up to 90%. However, the number of particles is unchanged, and issues associated with the effects of ultrafine particulate control are still being investigated. Diesel oxidation catalyst technology has been successfully used on all diesel cars sold in Europe since 1996, but not many heavy-duty vehicles are equipped with diesel oxidation catalysts.

Diesel oxidation catalysts may also be used in conjunction with NOx adsorbers, deNOx catalysts, DPFs, or SCR to increase nitrogen dioxide levels or to "clean up" any bypass of injected hydrocarbons or ammonia. Searles and Bertelsen [1] conclude that emission control requirements in both the EU and the United States that will be implemented during the next 10 years will pose significant challenges for manufacturers of, in particular, diesel-powered heavy-duty engines. The ultimate solution will require a systems approach combining the best in engine design and exhaust emission-control technology with low-sulfur fuel, and, they note, existing and emerging technologies are being developed and optimized to help achieve these rigorous emission-reduction standards. However, although these approaches are interfacial applications in environmental engineering, applications involving the chemistry of catalytic combustion in the vapor phase rather than on surfaces have received little attention.

Fuel engineers consider that we are very near the limits of technological developments for emissions reduction and fuel efficiency and that novel chemistry approaches are likely to be the way to future progress. One development with potential would be to have the chemistry that takes place in the catalytic converter take place in the combustion chamber. That way, levels of carbon monoxide, CO, and oxides of nitrogen, NOx, would be reduced, and unburned (and thus wasted) hydrocarbons, UHCs, can be largely eliminated, having been converted to useful energy in the engine. To achieve this would be a significant contribution to environmental engineering, and the way to do so involves the application of interfaces at various stages of a novel overall system. The system will be described, and the involvement of interfaces will be highlighted.

The conventional automotive catalytic converter consists essentially of a ceramic honeycomb through which the exhaust gases have to pass and whose insides are coated with a fine layer of platinum (nowadays palladium), rhodium, and cerium catalyst. The back-pressure engendered reduces engine efficiency, but the platinum component of the catalyst oxidizes CO to CO_2 and UHCs, and the rhodium reduces the levels of NOx formed. The role of platinum can be readily recognized by those who remember that many years ago the domestic gas lighter used a platinum wire heated momentarily to red heat by a battery. Placing this in the fuel–air mixture would ignite the gas.

II. CATALYSIS USING AN AEROSOL

The problems to be solved are thus to have or introduce these catalyst elements into the engine in the right form, in the right amounts, in the right place, and at the right time in the combustion cycle. These are considered in turn.

A. The Right Form

This is dependent upon either the construction of the engine or the fuel management system. It would be possible to line or plate parts of the components of the combustion chamber with platinum metal. The cylinder walls could be covered with Pt, but the action of the cylinder rings would gradually erode away the Pt long before engine life was exhausted. Platinum on the cylinder head would be more effective. Earlier, Mobil [6] had introduced platinum into a reactor and improved CO combustion from 66 to 99.6%. To reduce or remove NOx, rhodium would have to be incorporated with the platinum on the cylinder head.

One possible alternative is thus to introduce platinum with the fuel. Attempts have been made to add platinum compounds to the fuel. Not many standard platinum compounds are soluble in hydrocarbons, and they have to be transformed into usable species in the short time between entry into the combustion chamber, compression, and spark ignition. So far this approach has not been successful. One variation has been the addition of metal balls into the fuel tank, and success has been claimed. But it is generally found that its effects are minimal and are slow to become apparent.

The only other route we are left with therefore is the introduction of a catalyst into the air stream that mixes with the fuel. This requires the production of an aerosol containing appropriate chemicals. An aerosol involves interfacial chemistry.

B. The Right Amount

Aerosols contain minute amounts of salts, in the ppm and ppb range. Fortunately, only small levels are required to act as catalysts. Aerosols are generated by bubbling a gas, usually air, through a solution of, in this case, the catalyst. The loading of the air depends on a variety of parameters, including bubble rate and bubble size.

C. The Right Place

Aerosols have limited stability. Further, a small amount of the salt contained therein can be deposited on surfaces with which the gas stream comes into contact. It is therefore necessary that the distance between the source of the aerosol and the combustion chamber be as short as possible; hence the delivery tube is also short. The aerosol must therefore remain intact when it is mixed with the fuel and until it enters the combustion chamber. For S.I. (spark ignition) engines, the aerosol is made to enter the air stream just before it enters the carburetor so that it is mixed with the vaporized fuel, going thence into the engine. For diesel engines, the aerosol is introduced into the air stream after it has passed through the air filter.

D. The Right Time

For the catalyst to become effective it must undergo the necessary chemical changes, largely while heating up, during the time between entering the hot combustion chamber and the firing of the spark or commencement of compression ignition. Accumulated catalyst within the combustion chamber is also a source of interface reactions.

III. PUTTING IT ALL TOGETHER: THE TECHNOLOGY

The foregoing are somewhat critical conditions, but they have all been accomplished, even though not all of them are fully understood and research is in progress. A detailed account is now given, together with examples of the experiences of the drivers of a wide range of vehicles in applying this technology.

A. The Catalyst Solution

Patents [7] protect the exact composition of the catalyst solution, but it is stated that it includes salts of platinum, rhodium, and rhenium, the last said to improve flame propagation after ignition.

The catalyst solution, when first developed, contained enough material for the average car to travel 6000 miles (10,000 km), or for 200 h running of a stationary engine, after which a new dose of catalyst was added to the liquid in the aerosol dispenser. At that time the recommended service interval was 6000 miles. It is not currently possible to use a very concentrated solution that would be equivalent to the mileage associated with a car's catalyst converter. Vehicle service intervals are commonly 10,000 miles (16,000 km) and it is now possible to tailor the system to extended intervals: This has been researched and established by Emissions Technology, LLC, in the United States and is available to its UK partner, Emissions Technology Europe.

B. The Aerosol Dispenser

Simply stated, this consists of a plastic container for the catalyst solution, designed so that air bubbles pass upward from the bottom of the solution, picking up the catalyst at the appropriate concentration. The bubbles are of close-to-uniform size, and all pass through the same depth of liquid (Fig. 1). The bubbles also constantly stir the solution. At the interface between the bubble and the catalyst solution, some solvated ions of the catalyst move across from the liquid to become essentially solvated ions in the gas phase and then on to the engine combustion chamber. Behind this description is some interesting chemistry. The procedure is not that of gas-stripping, which removes species completely and generally quickly; the catalyst is removed slowly and steadily at trace levels, using the vacuum on an S.I. engine to create the bubbles and an additional small pump for diesel engines, which do not use a vacuum system.

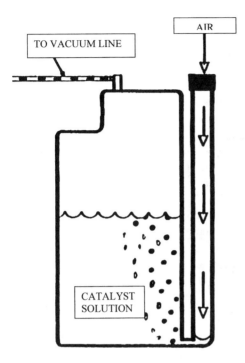

FIG. 1 Example of a dispenser designed for use with S.I. engines. Note that the air enters the bottom of the container, and hence the air bubbles effectively pass through the same depth of catalyst solution and, on leaving, contain effectively the same levels of catalyst in aerosol form until no more catalyst remains in solution.

C. Nonfoaming Bubble Fractionation

The process that provides the aerosol catalyst having essentially constant catalyst levels is based on the phenomenon known as *nonfoaming bubble fractionation*. As the name suggests, this involves the selective separation of solute species by means of the passage of gas bubbles up through the solution that additionally contains a surfactant at concentration levels that do not produce foam.

The technique was pioneered by Lemlich [8], who, with his group, developed the theoretical and mathematical understanding of the process. The basic requirements are that bubbles enter at the bottom of a narrow column, to avoid solution recirculation, and a concentration gradient is then slowly set up along the column. The bubbles convey the solute up through the solvent, the solute species being carried largely at the gas–liquid interface. The bubbles exiting at the top of the column, traversing an interface, thus carry a small amount of solute in aerosol form. The effect can be dramatically observed if a dye is used as the solute (Lemlich [9] used Gentian violet). After about 30 min, the bottom of a sufficiently long column, 1–2 m, is colorless, the purple color increasing toward the top, and traces of Gentian violet can be trapped on filter paper above the liquid surface [10].

The EmTech system (Fig. 1) uses a liquid depth of 10 cm and a bubble rate of around 2–4 s^{-1}, and the dimensions of the dispenser, unlike in nonfoaming bubble fractionation, result in recirculation of the solution; in any case, the vibrations of the engine to which it is fitted will ensure thorough mixing at all times. However, this vibration effect is thought to have a necessary beneficial effect, in that it disrupts the bubbles as they break at the surface so that more of the solute attached to the wall of the bubble is released in the aerosol form. Tests in the laboratory and on vehicles have shown that more of the catalyst is released from dispensers subjected to slight but continuous vibration.

D. Controlling the Bubble Rate

In use, the bubble rate in the EmTech and earlier systems is controlled, for S.I. vehicles, by fitting a T-piece in the vacuum line near the carburetor; the T-piece contains a ceramic insert with a hole of diameter 8 thou and connected to the dispenser. This provides the required bubble rate at engine tick-over and a slightly increased rate at speed and under acceleration. For diesel engines, air drawn through the dispenser by a small electrical pump is delivered into the air stream, which has passed through the air filter, close to a point where it will be evenly mixed with the air and also almost immediately enters the engine.

IV. THE AEROSOL CATALYST

A detailed understanding of the nature of the aerosol catalyst awaits further research. But from its effects on reducing emissions and improving combustion, certain conclusions and probabilities can be reached. From the time taken to transfer all the catalyst into the engine, the concentration of the component ions in the aerosol is around 50–70 ppb. The ions containing platinum and rhodium, for example, have first to be reduced to the elemental state before they can act in the required way. They first enter the combustion chamber when the piston is traveling downwards to suck the fuel-air (plus aerosol) mix into the cylinder; the piston then rises to compress it before firing. Under these conditions it is considered that the ions containing platinum and rhodium are reduced by hydrogen atoms to the metallic state. The temperature of the mixture is also steadily increasing. For best performance, the ignition timing now needs to be advanced. This results in improved engine performance and indicates that when compression is almost complete there are platinum atoms present that have reacted with the mix to form oxygen atoms. The high-temperature spark initiates even more oxygen atoms, continuing and setting off the chain reactions that effect the oxidation of the fuel to, ultimately, carbon dioxide and steam. More details, including the role of rhodium and rhenium, are now discussed.

A. The Role of Platinum

The role of a platinum surface to improve and accelerate combustion processes has been much studied. In the gas phase, for a hydrocarbon to be oxidized, among the species it has to encounter are oxygen radicals (oxygen atoms) and OH radicals. Simply stated, the platinum present is believed to increase the concentration of the O atoms and OH radicals, thus facilitating essentially complete combustion in the combustion chamber, effectively ensuring that unburned hydrocarbons (UHCs) and CO levels are as low as possible. The mechanism for combustion involving an aerosol catalyst has yet to be investigated, but the preceding seems the most plausible at this time.

The presence of platinum is also thought to cause the flame in diesel engines to burn at a lower temperature; thus, it does not go out before the piston has reached the end of its travel. At present, there have been no in-cylinder experiments or confirmatory observations, but the reasoning is obtained from diesel engine results. A feature of diesel engines is their "rattle," better termed *harmonics*. This arises when the energy of the expanding gases has decreased beyond a certain point, partly due to the drop in temperature arising from adiabatic expansion and, it is thought, when the flame goes out. When this occurs, the piston still has approximately one-quarter of its travel to go. The piston movement at this point changes from a push to a pull action, creating rattle. When the EmTech system is installed, the harmonics are much reduced and the engine is much quieter and smoother. Table 1 shows recent results obtained on Caterpillar diesel engines before and after the EmTech system was in use. Under normal circumstances, when combustion is initiated the resulting expansion pushes the cylinder down. The expansion will cause more or less the same temperature drop from whatever combustion temperature, so the flame is vulnerable to quenching. It is therefore currently supposed that the platinum aerosol present is still involved in producing oxygen atoms and thus keeps the flame and expansion going longer. Alternatively, and probably in addition, the platinum makes for a faster burn, thereby sustaining the flame and contributing to the increased power levels experienced. Experiments could resolve this and should be performed.

B. The Role of Rhodium

The optimum conditions for operating an S.I. engine for minimum CO and NOx emissions are somewhat opposed. The higher the temperature in the combustion chamber, the lower the amount of CO formed. Therefore, operating the engine

TABLE 1 DC-XHD Dynamometer Tests on Vibration and Torsion Harmonics of a
D3408 Caterpillar Engine at 85% Load Before and After Installation of EmTech
Aerosol Catalyst, 26–28 November 2001

Harmonics	Vibration			Torsion		
	Before	After	% change	Before	After	% change
0.50	0.0652	0.0774	19	0.0146	0.0107	−27
1.00	0.318	0.484	52	0.0149	0.0248	66
1.50	0.0445	0.0659	48	0.0248	0.0195	−21
2.00	0.0418	0.0214	−49	0.007	0.0054	−23
2.50	0.0841	0.0766	−9	0.042	0.0397	−5
3.00	0.140	0.122	−13	0.0116	0.0085	−27
3.50	0.153	0.119	−22	0.0214	0.0163	−24
4.00	0.0859	0.0791	−8	0.079	0.0602	−24
4.50	0.0862	0.084	−3	0.0051	0.0048	−6
5.00	0.0924	0.102	10	0.0079	0.0075	−5
5.50	0.105	0.0611	−42	0.0055	0.0059	7

Source: Ref. 15.

at as high a temperature as possible will minimize CO emissions. However, the
reverse is true for the formation of NOx, and for minimum emissions the temperature in the combustion chamber should be as low as possible. Currently, engines
are tuned to be close to the crossover in the plots of CO and NOx formation as
a function of temperature and in conjunction with the efficiency of the catalytic
converter.

The unique advantage of the EmTech system is that combustion temperatures
no longer have to be programmed to range around the crossover temperature for
minimizing CO and NOx levels in the exhaust gases from the combustion chamber. Because the platinum assists in the spark-initiated oxidation process, the
oxidation is essentially complete, with no unburned (unoxidized) carbon monoxide; this process takes place at a lower temperature. The lower-temperature re-
gime consequently means that the amount of NOx formed is less, and this quantity
is further reduced by the presence of rhodium in the combustion chamber. Another consequence is that the workload on the catalytic converter is now much
reduced, so its efficiency and lifetime are extended. In the future, if converters
can be redesigned so that less back-pressure is set up, then miles per gallon will
be improved even further in many vehicles.

C. The Role of Rhenium

The role of rhenium is less well understood at this time. Its inclusion in the
catalyst concentrate appears to result in smoother and quieter engine running;

hence its role is considered to be that of making for a smoother flame burn and a smooth front to the flame as it travels and pushes the piston downward. Further research, using engines with optical ports, could usefully confirm this and readily enable the action and effectiveness of other or alternative additives.

V. SAFETY AND OTHER CONSIDERATIONS

Normal care is all that is needed with the chemicals involved. The catalyst solution is made up by adding a vial of catalyst concentrate to a base liquid. This latter consists largely of aqueous ethylene glycol, and thus, since it is essentially the chemical used as antifreeze in vehicle radiators, it does not freeze in winter or boil away in summer. At the end of 6000 miles, the level in the dispenser has dropped by the volume of the vial or vials previously added (more than one is required for large diesel engines and is recommended when initially fitted to an S.I. engine). The vials contain about 20 ml of basically water-soluble salts of platinum, rhodium, and rhenium, and a fresh vial can be readily poured into the dispenser. The end product of using the catalyst solution is essentially retained in the engine, the exhaust pipes, and the catalytic converter. One test did not detect any in the exhaust gas of a stationary test diesel engine [11].

Catalytic converters degrade with time and use; platinum and rhodium have been detected alongside autobahns in Germany [12,13], but only after 1984, when the converter was required on new cars in that country. With the EmTech system, either the catalyst is thus retained in the combustion chamber, assisting combustion, or that which gets through to the catalytic converter is trapped there within its honeycomb structure. One can therefore expect this to be beneficial to the converter, and there are reports by drivers that, upon using the catalyst aerosol system, inefficient converters have been revived and emissions tests passed without the expensive need to replace a catalytic converter. A case was reported [10] some years ago that a motorist, who had somehow put leaded fuel into his car, fitted the catalyst aerosol system and regained the required emission levels. It is thus supposed either that the lead that had poisoned his converter had been removed by the catalyst-containing exhaust gases or, more likely, that the lead had been covered by essentially a platinum–rhodium mixture.

VI. RESULTS

The foregoing basic description of the application of the interface between a gas, air, and a liquid to remove trace amounts of catalyst and deliver them into a combustion chamber where they improve combustion and simultaneously reduce NOx formation is now examined for its efficacy. Applied in the early 1980s to inefficient American large "gas-guzzlers," 350-cubic-inch (5.7-liter) engines, improvements in miles per gallon of 20% were common. A controlled test on 20 school buses in Concord, Conn., showed this level of improvement on all but one of the buses [14]. Fitted to diesel vehicles, a smaller mpg improvement was obtained, and expected, since this engine type is intrinsically more efficient. But there was generally a marked reduction in oil consumption, a problem of diesels with time, and less impurities detected on subsequent oil analysis. A reduction in oil consumption has been observed of up to 500% and generally of more than 100%. Another associated feature is that hard carbon buildup is removed and injectors remain clean and hence last longer (Fig. 2). Improved combustion has also resulted in a marked reduction in the emission levels of smoke, CO, and NOx (Table 2). A reduction or removal of the odor and eye-watering effects of the exhausts of city diesel buses has also been noted.

FIG. 2 Photograph of the head of an Albuquerque, NM, bus, number T650. This bus had traveled 175,000 miles, followed by 20,000 miles with the aerosol catalyst system fitted. The maintenance personnel of the bus company, Sun Tran, initially noted a total lack of carbon and ash. An injector (shown) was removed with very little effort, and all the injector tips were entirely devoid of any buildup or deposit. The cylinder on the right illustrates an interesting point: The rings on the piston in this cylinder had "frozen," and this piston had started to experience a noticeable bypass of lubricating oil. The aerosol catalyst was thus clearing the residue in this cylinder, despite this mechanical problem. The "etching" of the spray pattern can be clearly seen. The head was removed from another bus engine, number T625, after operating for 195,000 miles, and the two heads compared. The latter head showed the normal buildup of hard carbon and ash deposit expected by the Sun Tran maintenance personnel. The injector tips were also surrounded by carbon that was reported as extremely hard and would require considerable effort to clean. Report by RM Montano, National Director Quality Assurance.

TABLE 2 Results for Los Angeles County Buses Obtained in 1989

Bus no.	NOx/ppm			CO/ppm		
	Baseline	With catalyst	% reduction	Baseline	With catalyst	% reduction
43361	600	160	73	500	100	80
43362	2000	180	91	350	100	71
55900	5000	200	96	1200	100	92
N6619	1600	325	80	300	80	73
47544	800	250	69	500	400	20
43547	800	280	65	500	110	78
	Average reduction		78		Average reduction		69

Source: Ref. 16.

In the current emissions test procedures for diesels around the world, the smoke emissions, when measured, are recorded at various engine speeds but not under load, or "snap," conditions. We have all seen large diesels smoking on start-up and belching smoke when climbing hills and when the engine is suddenly revved, or "blipped"; the aerosol catalyst system is effective in reducing smoke under these engine conditions. Table 3 contains some examples for Los Angeles County buses [7]. Tests have also been conducted using a dynamometer on a

TABLE 3 Opacity Results after a Six-Month Demonstration Project on Six Los Angeles County Buses in 1989

Bus no.	Baseline mileage with catalyst	Start	Idle	Snap acceleration	Full throttle
43361	Baseline	41	5	28	30
	After 14,013 mi	29	2	22	8
43362	Baseline	41	1	74	50
	After 12,427 mi	22	1	31	6
55900	Baseline	13	0	69	57
	After 22,564 mi	1	0	88	15
N6619	Baseline	67	2	57	73
	After 6,207 mi	60	1	47	10
47544	Baseline	63	0	31	84
	After 19,811 mi	64	3	65	37
43547	Baseline	51	0	32	72
	After 14,304 mi	64	1	39	20

Source: Ref. 16.

newly overhauled diesel 8V71 bus engine owned and operated by the Washington Metropolitan Area Transit Authority (WMATA) in Washington, DC. Table 3 shows the effect of the aerosol catalyst system on NOx emissions and the levels experienced for CO reduction.

A more detailed study was reported [11] involving a Deutz MWM-916-6 diesel engine belonging to the Kerr-Magee Mining Corporation. Such engines are employed underground, and it was therefore of interest to minimize emissions from such engines. Measurements were made before and after using the aerosol catalyst of CO and NOx levels with the engine cold, and after reaching normal operating temperature. Table 4 lists the results for the engine running at 2300 RPM at normal load (42 kW) and maximum load (69 kW). The numbers in parentheses indicate the number of determinations. The levels with the aerosol catalyst on start-up and when hot are very similar, indicating that this system is effective within moments of start-up, at both normal and maximum load. Further, the reductions with the aerosol catalyst are generally a third or less of baseline values. This engine was operated for 150 h using the catalyst system and then for another 153 h with it disconnected. At the end of this time the CO levels had increased slightly and the NOx levels had decreased by more than a further 50%. This shows that the platinum and rhodium deposited in the combustion chamber were now continuing to effect emissions reductions. In confirmation of this, Table 5 reports the recommended exposure limits and maximum concentrations of platinum, rhodium, and rhenium found in the exhaust gases, using a fine paper filter;

TABLE 4 Effect of Aerosol Catalyst on Exhaust Emissions Using a Deutz MWM-916-6 Diesel Engine at 2300 RPM

Aerosol catalyst usage	Engine temperature/°C	Engine load/kW	CO volume/ppm	NOx volume/ppm
Before	Cold	42	900 (1)	200 (1)
During	Cold	42	33 (3)	67 (3)
Before	Cold	69	—	100 (1)
During	Cold	69	29 (3)	72 (3)
Before	82	42	90 (3)	175 (1)
During	82	42	35 (7)	45 (3)
After[a]	82	42	50 (1)	20 (1)
Before	82	69	500 (1)	300 (1)
During	82	69	33 (4)	57 (3)
After[a]	82	69	49 (4)	25 (2)

[a] Up to 153 h operation after stopping using aerosol catalyst; note no apparent increase in exhaust emissions.

Source: Ref. 11.

TABLE 5 Effect of PVI on Exhaust Particulates Using a
Deutz MWM-916-6 Diesel Engine at 2300 RPM

Aerosol catalyst usage	Engine load/kw (HP)	Average particulates rate/g/h ± SD	Number of runs
Before	42 (56)	66 ± 59	9
During and after	42 (56)	12 ± 5	10
Before	69 (93)	88 ± 72	9
During and after	69 (93)	43 ± 38	9

Source: Ref. 11.

the amounts were below detectable levels and several orders of magnitude below
the recommended exposure limits.

Further evidence of the rapid effect of using an aerosol catalyst on emission
levels is illustrated in Figure 3. Here, a stationary diesel engine was operated
with cycling of the system, and NOx levels were monitored. After running for
40 min to establish a baseline level of 2500 ppm, the catalyst was switched on
for 20 min, after which time the NOx level was 1000 ppm. It was switched off
for 5 min, and the NOx concentration increased to 1750 ppm, but further opera-

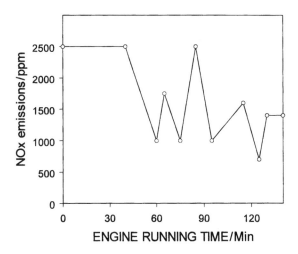

FIG. 3 Graph showing the immediate effect of the aerosol catalyst on a stationary gasoline engine. NOx levels were measured after the engine was fully warmed up and at various
times after the aerosol supply had been continued and discontinued (see text).

tion with the aerosol catalyst for 10 min returned the level to 1000 ppm, after which it was discontinued for 10 min and the NOx level returned to its baseline value. Cycling the system on again for only 10 min was sufficient to regain the 1000-ppm level, but then when switched off for 20 min the level had returned to only 1600 ppm, indicating that the rhodium content of the catalyst was now being retained in the combustion chamber. Another 10-min "on" cycle then lowered the NOx level further to 700 ppm. After another 5 min "off," it had risen to only 1400 ppm. Motorists have reported similar findings: The odor of their exhaust had disappeared after half an hour's running with the EmTech system, and their S.I. cars now passed emissions tests when they had failed earlier in the day. A comparable datum concerns a large City of Santa Fe refuse truck with a Caterpillar diesel engine. Its baseline mileage was 1.996 mpg (U.S.). With the aerosol catalyst installed, this increased by just under 15%. But when the system was removed, the fuel mileage returned to 2.012 mpg. Reinstalling the system increased the fuel mileage, this time by just over 15% [10].

The very latest results, achieved in 2001 on a stationary Caterpillar diesel engine, were obtained by Martin Marietta [15]. They found that the exhaust temperature was lower and said that the engine must have been using this heat on the power stroke. They also found approximately 10% less fuel was used at 50% load and 6.7% less at 85% load. They further concluded that there must now be an increase in available static torque or horsepower; on a water brake dynamometer test they recorded an increase in maximum horsepower (Hp) from 447 Hp to 520 Hp, or 16.3%. They also expected a reduction in emissions due to improved combustion and noted opacity down by 79%, carbon monoxide by 60%, and hydrocarbons by 66%.

They also argued that with improved combustion, the vibration and torsional measurements from the engine should go down and the ultrasonic energy from combustion should go up. They therefore performed the first study on these engine characteristics upon using an aerosol catalyst. They obtained a marked reduction in torsion at harmonics between 0.5 and 5.0, often in excess of 20%, and concluded that reductions in the torsion, or the "dynamic torque," of an engine means that the engine is running smoother, and that is less strain on the rod bearings and main bearings, resulting in longer engine life.

VII. LOOKING AHEAD

Table 6 lists the main findings and benefits reported for S.I. and diesel engines. There are many other examples that could be adduced, including official tests in Poland, Austria, the Czech Republic, and Mexico. In the few instances where an improvement in performance or emission levels has been negligible or negative compared with the rest of the group, subsequent investigation has generally been associated with needful engine maintenance or leaks in the vacuum section such

TABLE 6 Experiences and Benefits Claimed Using Aerosol Catalyst

---

Diesel and S.I. engines

Improved starting (and cold-starting) and warm-up: extended engine life
Improved engine performance from more complete combustion
Reduction of vibration and torsion harmonics in diesels: quieter running
Lower exhaust temperature: improved efficiency
Removal of hard carbon and prevention of its buildup, particularly in diesels:
  On injectors gives longer life
  In the piston head region: allows reduced cylinder liner wear and less oil use
  Gives improved ring sealing: preventing blowby
  Leads to less pitting of valve and seats: maintaining performance longer
  Keeps engine cooling efficiency: reduces overheating problems
Can extend maintenance intervals after establishing by regular oil analysis

Fuel and oil usage

Reduced fuel consumption: up to 10% with older S.I. engines, up to 5% with diesels
Can use lower octane (S.I. engines) for same performance: cost savings
Can blend in waste oil (diesel where allowed) for up to additional 5% savings
Diesel oil consumption reduced on average by 75%, but wide variation here
Generally marked reduction in oil contaminants upon analysis: extended oil life

Environmental impact

Fuel burned more efficiently: less CO_2 generated for same mileage
Less lubricating oil used: less waste oil dumped or burned
Less smoke (opacity) from diesels: cleaner air
Polluting NOx emissions reduced by 30–90%: less smog and cleaner air
Poisonous CO emissions reduced by 25–90%: cleaner air
Exhaust smell and eye irritation reduced: cleaner city air

---

that the bubble rate had become too slow (recent design changes now effectively eliminate this problem). The modern diesel and S.I. engines are much more efficient than the "gas-guzzlers" of yesteryear, but obviously they could benefit from the EmTech system, a much cheaper remedy than the expensive devices that engineers are considering and developing in the hope of achieving the future mileage and emission levels set or indicated for the future, as mentioned earlier. The trend has been to encourage vehicle owners to purchase new vehicles, but not everyone can afford to do so, particularly in Third World countries. These countries also have high pollution levels, and the EmTech system is a rapid and simple system that would reduce CO and NOx emissions and assist them in attaining CO_2 emission levels laid down in the Kyoto Agreement. The attack on the World Trade Center towers on September 11, 2001, has resulted in a marked downturn in the economy of the West; hence one can anticipate a reluctance to buy new vehicles. The EmTech aerosol catalyst system can thus be seen at this time, and probably for some time ahead, as a remedy to enable owners to continue to run their existing vehicles and, importantly, pass emissions tests and see some savings in their fuel bills. The owners of classic and similar cars can also use the EmTech system to run their old cars, having passed current emissions tests. Another advantage of the EmTech system is that in improving combustion it also raises the octane number of fuel, a feature that would be appreciated by classic car owners, particularly if they now have to use lead replacement petrol (LRP).

The aerosol catalyst system is still undergoing refinement. The current method, which applies interfacial interactions to supply catalyst into the combustion chamber, and the catalyst formulation, can be improved and investigated. The management of the dispenser and the associated equipment can be made more robust. Research on the effects of bubble size, bubble rate, and depth of liquid in relation to the pickup of catalyst could prove useful. At present, the explanation concerning how the aerosol catalyst works in the combustion chamber is based upon our knowledge of combustion and the improvement observed. The mechanism involved is not yet understood, because studies in catalytic combustion have not included the role of aerosols containing catalytic materials. The role of catalytic surfaces during combustion has been widely investigated, and these are doubtless formed when the aerosol catalyst system is operating. The conversion of the aerosol into an active catalyst upon entry into the combustion chamber and the identity of the catalytic particles and their role have yet to be established. Experiments have been commenced in Leeds, UK.

The formulation of the catalyst concentrate has been changed over the years. However, once the mechanism of action of the aerosol to improve combustion and reduce NOx is better understood, then the various concentrations of the catalyst components can be optimized, beneficial additives can be identified, and possibly cheaper and equally or more effective chemicals can be employed.

VIII. BENEFITS

The current and future modifications to S.I. and diesel engines to improve performance and reduce emissions for coming and anticipated legislation are having to be more elaborate and expensive. The EmTech system does not entail elaborate and expensive hardware and is quickly fitted to existing vehicles. The cost of it and of the vials of catalyst required for regular maintenance is normally less than half the running costs, defined as including fuel and oil saved, extended maintenance intervals, cleaner engine and injectors (no hard carbon buildup), longer engine life, easier starting, less smoking (particularly diesels), and so on. With regular use, and hence platinum and rhodium buildup, some owners may find that the mileage intervals between concentrate addition may be increased. When the catalyst solution has been depleted, it is usually readily apparent, and once renewed the vehicle performance improves within a few minutes to half an hour. Various other benefits have been detailed or mentioned in the preceding sections and tables.

It is encouraging to report that a government based in the Middle East is currently testing the EmTech system with a view to installing these units at the manufacturing stage within their car industry and on all previously built vehicles as an aftermarket install. Also, diesel refuse vehicles belonging to the City of York Local Council, UK, are shortly to be utilized in testing the EmTech system, with the expectation of wider Council adoption.

IX. CONCLUSIONS

A very promising approach for reducing vehicular emissions arising from combustion is now available and has the potential for further development and expansion from S.I. and diesel engines to all plants where fossil fuel is burned. These would include fossil fuel–fired power plants and stationary turbines and probably, at some future date, aero engines. With the EmTech aerosol catalyst system, carbon dioxide emission would be reduced, thereby making it easier for countries to achieve the levels planned in the Kyoto Agreement. Although carbon dioxide is the main product of the combustion of carbon-containing fuels, less will be produced because, with the system, the combustion process is more efficient and thus more energy is generated and less fuel is used.

With the present downturn in the world economy, entrenchment, and need for burning less fuel and reducing emissions, now is the right time for using and developing the aerosol catalyst system. The chemical basis of this system is a novel application of the interfacial interaction between air bubbles and a catalyst solution. The design of the catalyst mixture has required some sophisticated chemistry, and the details of the chemical mechanism of action within the combustion chamber require further elucidation.

ACKNOWLEDGMENTS

The author is grateful to the developers and users of the aerosol catalyst system for much information and the opportunity to test the system. The many discussions over recent years with Matt Bair, Emissions Technology Inc., Phoenix, Arizona, have been invaluable, particularly on the engineering aspects. Richard Tarry, Emissions Technology Europe Ltd., has also freely supplied examples of the results of tests involving the EmTech system installed on various vehicles. Helpful discussions with my namesake colleague, Professor John F. Griffiths, are also gladly acknowledged.

REFERENCES

1. RA Searles, BI Bertelsen. In: Business Briefing: Global Truck and Commercial Vehicle Technology. London: World Markets Research Centre, 2000, pp 97–102.
2. EU Directive 1999/96/EC.
3. Emission Control, Air Pollution from 2004 and Later Model Year Heavy-Duty Highway Engines and Vehicles. 40 CFR Parts 85 and 86. 65 FR 59896, October 6, 2000.
4. California's Air Resources Board. Risk Reduction Plan to Reduce Particulate Matter Emissions from Diesel-Fueled Engines and Vehicles, September 28, 2000.
5. M Horiuchi, S Makoto, K Saito, S Ichihara. The Effects of Flow-Through Type Oxidation Catalysts on the Particulate Reduction of 1990's Diesel Engines. Society of Automotive Engineers Paper 900600, February 1990.
6. AW Chester, AB Schwartz, WA Stover, JP Williams. CO Oxidation Promoter in Catalytic Cracking. Mobil Research and Development Report, Paulsberg, NJ, 1981.
7. For example: JW Haskew, U.S. Patents applied for.
8. R Lemlich. Am Inst Chem Eng 12:802–804, 1966.
9. DO Harper, R Lemlich. Am Inst Chem Eng 12:1220–1221, 1966.
10. M Bair. Emissions Technology. Phoenix, AZ, personal communication, 1999.
11. EP Doane, JAL Campbell, FH Lee, JW Haskew. Minimizing diesel engine emissions by catalysis. Proceedings of the 7th Mine Ventilation Symposium, Lexington, KY, 1995, pp 95–98.
12. JE Helmers, N Mergel, R Barchet. USWF—Z Umweltchem Ökotox. 6:130–134, 1994.
13. JE Helmers. Fresnius' J Anal Chem 362:522–524, 1998.
14. J Robinson. Homogeneous catalysis of gasoline combustion by platinum and rhenium. ACS Annual Meeting, St. Louis, MO, 1984.
15. R Tarry. Emissions Technology Europe Ltd., personal communication, 2001.
16. RM Montano, J Robertson, JW Haskew, CD Schivley. Simultaneous reduction of soot and NOx in diesel engines by homogeneous catalysis of group platinum metals. Future Transportation Conference, Vancouver, BC, 1989, SAE paper 891634.

16

Polymer Waste Recycling over "Used" Catalysts

SALMIATON ALI and ARTHUR GARFORTH University of Manchester Institute of Science and Technology, Manchester, United Kingdom

DAVID H. HARRIS Engelhard Corporation, Iselin, New Jersey, U.S.A.

RON A. SHIGEISHI Carleton University, Ottawa, Ontario, Canada

Polymer waste can be regarded as a potentially cheap source of chemicals and energy, although its recycling varies widely across Europe [1,2]. Most polymer waste is difficult to decompose naturally, but most polymers are still discarded by open dumping. The destruction of wastes by incineration is widespread, but it is expensive and often aggravates atmospheric pollution. Open dumping brings rat infestation and related diseases, but sanitary landfill remains one of the most commonly used methods to control the waste. Unfortunately, disposing of the waste to landfill is becoming undesirable due to rising disposal costs, the generation of explosive greenhouse gases (such as methane, which is formed from the decomposition of organic material in landfill), and the poor biodegradability of plastic. Legislation effective on 16 July 2001 states that all wastes sent to the landfill must be reduced by 35% over the period from 1995 to 2020 [3].

I. REVIEW ON POLYMER RECYCLING OVER CATALYSTS

A. Background

Plastic materials are among the best modern products of the chemical process industry and offer a unique range of benefits that make an improved standard of living more accessible for everyone. Involving continuous innovation related to new products, systems, manufacturing technologies, and markets, the plastics industry itself is one of the great industrial successes of the 20th century. The consumption of plastics is predicted to show an annual growth rate of approximately 4% (from 25.9 million tons in 1996 to 36.9 million tons in 2006) [1]. However, consumption in 1999 increased by 5.4% over 1998 figures [2], showing that the consumption of plastics might exceed this prediction for 2006. This may be due to the fact that plastics are resource efficient and affordable and are increasingly replacing traditional materials.

However, with growing consumption, the main challenge is to keep pace in waste recovery [1,2,4–6]. Every year, the UK produces 29 million tons of municipal waste, which is equivalent to half a ton per person [7]; the total amount is between 170 million and 210 million tons when waste from households, commerce, and industry is included. From this total waste figure, about 60% is disposed of to landfill sites [8], in 1999, about 6 million tons of plastic wastes were landfilled [9]. As well as landfill, there are two main alternatives in treating municipal and industrial plastic wastes: energy recycling, in which plastics are incinerated, with some energy recovery, and mechanical recycling, in which plastics are regranulated and reused. However, these methods are less desirable due to the need for the collection and sorting of waste, which is one of the most serious problems for the plastic recycling industry [10]. Table 1 shows the total plastics consumption and total plastics waste recovery in Western Europe [8].

In the UK itself, total plastic waste collectable was 3.6 million tons in 1999. However, of that amount only 12.1%, or about 438 thousand tons, of wastes were recovered by mechanical recycling (6.2%) and energy recovery (5.9%) [2].

Even though these common methods are practical in handling wastes, they have their own drawbacks. Besides poor biodegradability, landfill treatment is also less desirable because of the European Union Landfill Directive to the United Kingdom on 16 July 2001, which states that all waste sent to the landfill must be reduced by 35% over the period from 1995 to 202 [3]. Incineration of plastic waste meets strong societal opposition due to possible atmospheric contamination [11–13]. In addition, under the legally binding Kyoto Protocol, the UK must reduce its gas emissions by 12.5% by 2008–2010 and move toward the domestic

TABLE 1 Total Plastics Consumption and Waste Recovery, Western Europe ($\times$ 1000 tons)

	1991	1993	1995	1997	1999
Total plastics consumption	24,600	24,600	26,100	29,000	33,600
Total plastics waste	13,594	15,651	17,505	16,975	19,166
Total plastics waste recovered	3,218	3,340	4,019	4,364	6,113
Mechanical Recycling	1,080	915	1,222	1,455	1,800
Feedstock Recycling	0	0	99	344	364
Energy Recovery	2,138	2,425	2,698	2,575	3,949
% total plastics waste recovered	22	21	26	26	32

goal of reducing carbon dioxide emissions by 20% by 2010 [3]. Mechanical recycling (the conversion of scrap polymer into new products) is a popular recovery path and most preferred by manufacturers, but the recycled plastic products often cost more than virgin plastic [14]. Also, this method can be performed only on single-polymer plastics waste, since a market for recycled products can be found only if the quality is close to that of the original [15].

B. Feedstock Recycling: Current State of the Art

The latest Association of Plastics Manufacturers in Europe (APME) report claims that feedstock recycling, sometimes known as chemical recycling or tertiary recycling, has great potential to enhance plastics waste recovery levels. In addition, this method does not have the negative public impact of incineration, and the recovered materials may have broader applications than mechanically recovered plastics [16–17]. In 1999, however, only 364,000 tons of wastes were treated by this method. This has not changed significantly since 1997, as shown in Table 1 [2]. Mixed plastics waste can also be recovered by this new approach, as long as its halogen organic compound content does not exceed 2–6 wt % [18].

In the past few years, feedstock recycling, which has appeared as a reliable option and alternative strategy, has attracted the attention of many scientists [19–62] whose aim is to convert waste polymer materials into original monomers or into other valuable chemicals. These products are useful as feedstock for a variety of downstream industrial processes or as transportation fuels. Two main chemical recycling routes are the thermal and catalytic degradation of waste plastics. In thermal degradation, the process produces a broad product range and requires high operating temperatures, typically more than 500°C and even up to 900°C [40–44]. On the other hand, catalytic degradation might provide a solution to these problems by controlling the product distribution and reducing the reaction temperature [40,41,45–48].

Plastics are divided into two groups: (1) condensation polymers and (2) addition polymers. Condensation polymers, which include materials such as polyamides, polyesters, nylon, and polyethylene terephthalate (PET), can be depolymerized via reversible synthesis reactions to initial diacids and diols or diamines. Typical depolymerization reactions such as alcoholysis, glycolysis, and hydrolysis yield high conversion to their raw monomers [63]. In contrast, the second group of materials, addition polymers, which include materials such as polyolefins, are not generally reversible, and therefore they cannot easily be depolymerized into the original monomers. However, they can be transformed into hydrocarbon mixtures via thermal and catalytic cracking processes [22,26,29–33,46,47,56,63,64].

Plastic wastes treated by catalytic degradation processes are limited mainly to waste polyolefins and polystyrene (PS). Waste polyvinyl chloride (PVC), which is probably the most heat sensitive, with initial thermal degradation temperatures between 100 and 150°C, has been excluded because of the emission of hazardous gases such as hydrogen chloride, which is the main volatile product [4,65,66]. A few researchers, however, have tried to investigate the effect of PVC waste on the recycling of PET and PE [47,49,50]. One study on the dechlorination and chloro-organic compounds from PVC-containing mixed plastic-derived oil has given an encouraging result using an iron–carbon composite catalyst in the presence of helium [51]. A number of authors have reported promising results on the cracking of PS at operating temperatures from 350 to 500°C over acid catalysts such as HMCM-41, HZSM-5, amorphous SiO_2-Al_2O_3, BaO powder, mordenite (HMOR), zeolite-Y, and a sulfur-promoted zirconia [25,52–54]. The results have been compared with those from thermal degradation processes, which require higher operating temperatures.

The largest plastic constituent in municipal waste stream consists of polyolefins, mostly derived from high-density polyethylene (HDPE), low-density polyethylene (LDPE), linear low-density polyethylene (LLDPE), and polypropylene (PP). These are among the most abundant polymeric waste materials, typically making up 60–70% of municipal solid waste plastics [2,13,15,55]. Reports on the degradation of polyolefin derivatives under various operating conditions and cracking methods also give promising results. An example is given in a report on HDPE being pyrolyzed in the temperature range 290–430°C in a fluidized-bed reactor using HZSM-5, HMOR, Silicalite, HUS-Y, and SAHA. The yield of volatile hydrocarbon products was in the order HZSM-5 > HUS-Y ≈ HMOR > SAHA ≫ Silicalite [40]. Another report [56] describes a two-stage catalytic degradation process consisting of amorphous silica alumina and HZSM-5 in series to convert PE into high-quality gasoline-range fuels. The author found that a silica alumina : HZSM-5 weight ratio of 9 : 1 gave improved gasoline yield with a high octane number in spite of low aromatic content.

Aguado et al. obtained a high conversion of 40–60% and good selectivity to C_5–C_{12} hydrocarbons of 60–70% [57] when PP and PE were catalyzed by zeolite beta at 400°C and atmospheric pressure in a batch reactor under N_2 flow. Tertiary recycling of HDPE and PP over catalysts amorphous silica-alumina and F9 (a silica-alumina catalyst with sodium oxide) using a powder-particle fluidized-bed reactor gave liquid fuels, gas products, and solid residue [58]. In a further example, PE and PS were degraded in a two-stage process, first over the catalyst and then hydrogenated over platinum catalysts. This gave more than a 90% yield of gas and liquid fractions with boiling point less than 360°C [59]. Volatile-product distributions of PE and PP were also investigated with respect to the effect of catalyst activity and pore size of HZSM-5, HY, and MCM-41 using a fixed-bed microreactor. This concluded that HZSM-5 and MCM-41 gave higher-olefin products in the range C_3–C_5 and HY gave higher-paraffin products in the range C_3–C_8 [60,61].

Most of the polymer degradation studies using solid catalysts involve pure zeolites, a crystalline, porous aluminosilicate made up of a linked framework of $[SiO_4]^{4-}$ and $[AlO_4]^{5-}$ edge-sharing tetrahedra, or amorphous silica alumina [67–69]. Figure 1 shows typical zeolites of different structure used in polymer cracking [70]. Different zeolites have different channels and pore sizes, which control product distribution. For example, zeolite ZSM-5 has smaller channels 5.3×5.5 Å, with unique three-distributional pore structures that consist of straight channels and interconnecting sinusoidal channels that increase shape selectivity in petrochemical reactions. Zeolite Y has larger pore openings of 7.4-Å diameter with three-dimensional connecting cavities of about 13-Å diameter, thus permitting diffusion of hydrocarbon molecules into the interior of the crystals and accounting for the high effective surface area of these material [71–82].

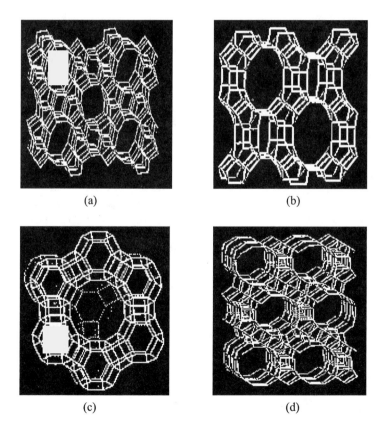

(a)   (b)

(c)   (d)

FIG. 1 Typical zeolites used in polymer cracking: (a) H-ZSM-5, (b) H-MOR (Mordenite), (c) H-Y or HUS-Y, and (d) H-Beta. (From Ref. 70.)

In previous studies, catalytic cracking of polymers was carried out over fresh catalysts. There is only one report where one spent fluid catalytic cracking (FCC) catalyst was used, and the results compared with those on pure zeolites and silica alumina [62]. Polypropylene was catalytically degraded in a semibatch stirred reactor at 380°C. The authors found that the spent catalyst they used generated very low product yield and that the yield and selectivity were similar to those of silica alumina; however, the amount used could be increased due to its very low cost.

In this work, we evaluate fresh and also "used" FCC catalysts with different rare earth oxides and heavy metal loadings [3–91], for the recycling of polymer wastes, and compare with pure catalysts. This study will help in refining our initial economic model of the polymer recycling process, which as been published elsewhere [92], and will enable a comparison with current process technology [42]. Previously, almost all studies dealt with pure catalysts as cracking catalysts in the degradation of polymers. Even though pure catalysts generate good product distributions and selectivity, their costs are much higher. This makes polymer waste recycling unrealistic [18]. However, as presented in this chapter, using zero-cost "used" catalysts, which give sufficient cracking products and reliable selectivity, will make catalytic polymer waste recycling more economically viable.

II. THERMAL VISUAL ANALYSIS TO STUDY THE MOLTEN POLYMER/ZEOLITE CATALYST INTERFACE

A. Introduction

Before design predictions can be made for a pyrolysis process on an industrial scale, an understanding of the interface between the polymer and the catalyst might be developed in order to study the reaction of the mixture. The mechanism of interaction is highly complex, with three phases (liquid polymer, solid catalyst, and gaseous products), mass transfer by diffusion, convection, and bulk flow, as well as cracking-type reactions with a large number of products.

In this section, the degradation of high-density polyethylene (HDPE) over zeolite ZSM-5 has been investigated using a heated stage microscope and scanning electron microscopy to examine partially and fully reacted polymer and zeolite mixtures. The aim of the study was to elucidate the physical behavior of the system: how the polymer melts at the interface and how the molten polymer wets the catalyst.

B. Experiments

The polymer used was HDPE in a powder form with an average molecular weight of 75,000 (Grade HMLJ200MJ8, BASF), while the zeolite was ZSM-5 with a

Si/Al ratio of 17.5 and average particle size of 1 μm (BP, Sunbury-on-Thames, London). Equal amounts of polymer and zeolite (pelletized and sieved) were blended together by grinding, and samples of this blend were mounted between glass plates. Each mount was then placed on the heated stage (Linkam model THMS 600 with controller TMS 91) of a microscope (Olympus model BH2). The stage was then programmed to heat at 300°C at 20°C min^{-1}, followed by a hold at 300°C for various times to a maximum of 120 min.

During the heating process, mounts were removed from the heating stage and allowed to cool to room temperature. After cooling, the upper glass plate was removed gently so that the bulk of the heated mixture remained on the lower glass. The lower glass was then attached to an aluminum sample stub of a standard scanning electron microscope (SEM) and coated with gold (Polaron Equipment Ltd., Watford, England, model E 5000). In addition, samples of fresh ZSM-5 and HDPE were prepared for the SEM (Hitachi Ltd., Tokyo, 9, model S-520 SEM) and micrographs were taken at 200× magnification.

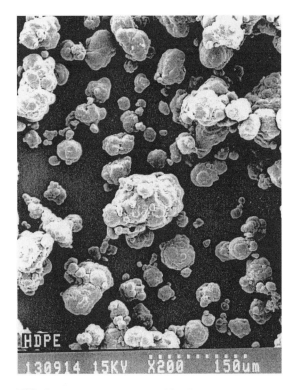

FIG. 2 HDPE at 200× magnification.

C. Results and Discussion

In viewing the micrographs, it is important to remember that all the samples were cooled to room temperature prior to examination under SEM. Since the sample could not be reheated under the microscope after SEM examination, Figures 4–7 show different sample mixtures heated to different temperatures or held at 300°C for various times.

Figure 2 shows fresh HDPE particles with an average particle size estimated to be in the range of 25–125 μm. Figure 3 shows fresh zeolite ZSM-5 after being pelleted and sieved at 125–180 μm.

Figures 4–7 show SEM micrographs of a series of samples taken at four different stages in the pyrolysis of the polymer over ZSM-5. Figure 4 portrays the mixture of HDPE/ZSM-5 after heating from room temperature to 205°C The polymer particles could be seen to have melted partially, with a tendency to stick together. The individual particles were still noticeable, but the polymer surface

FIG. 3   Fresh zeolite ZSM-5 at 200× magnification.

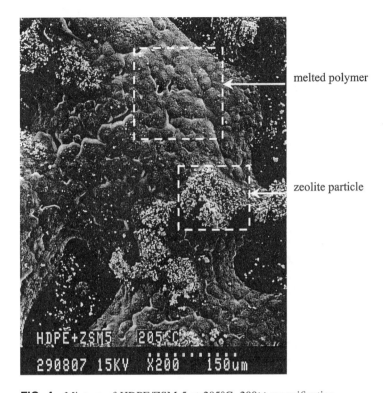

melted polymer

zeolite particle

FIG. 4 Mixture of HDPE/ZSM-5 at 205°C, 200× magnification.

was smooth compared to fresh HDPE (Fig. 2). Figure 5 shows that the 300°C, as the polymer melted, the zeolite particles moved under the microscope slide, eventually, all the zeolite particles were well "wetted" with liquid polymer.

It is known that the reaction of HDPE begins at less than 300°C [93]. As the heating time at 300°C increased to 40 minutes (Fig. 6), it appeared that the polymer was being drawn into the zeolite so that more zeolite became visible. As the reaction progressed and more polymer diffused into the zeolite, holes or bubbles appeared on the polymer surface. Gaseous products could escape from the interior of the sample through these holes or bubbles. After 120 minutes at 300°C, the amount of unreacted polymer had decreased significantly and was hardly visible on the surface, as shown in Figure 7.

Therefore, in the degradation of HDPE's over ZSM-5, the molten polymer wetted the surface of the zeolite particles and was then pulled into the interior of the catalyst particles, where the reaction took place. Two simple models have been suggested [94] for fluid–solid reactions.

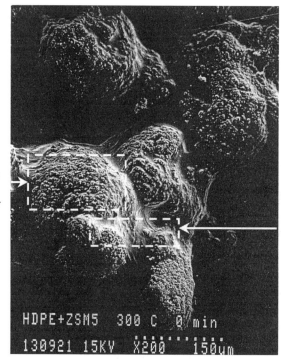

FIG. 5 Mixture of HDPE/ZSM-5 at 300°C, 200× magnification.

1. *Progressive-conversion model*: The reactants are imagined to enter and react with the particle at all positions at once.
2. *Unreacted-core model*: The reactant is imagined to enter the outer layer of the particle and then move toward the middle, leaving behind converted material and inert solid (ash), Therefore, at any time, there exists an unreacted core of material that shrinks in size during the reaction.

Though the two models were not developed for catalytic reactions, the experimental results suggest that the unreacted-core model applies to this case. As the polymer diffused into the aggregate of zeolite particles, a reaction took place and products were produced. Pyrolysis of the polymer gave both solid products (coke) and gaseous products. Gaseous products escaped and coke was deposited on the zeolite, at which point the zeolite began to deactivate. Figures 6 and 7 show that the polymer was being pulled into the aggregate of zeolite particles while the particle size remained unchanged. This process is consistent with the unreacted-core model. The deactivation of zeolite as the reaction proceeded could be related

Holes or bubbles
appear

Small amount of
unreacted
polymer

Zeolite coated
with melted
polymer

FIG. 6 Mixture of HDPE/ZSM-5 at 300°C, isothermal for 40 minutes, at 200× magnification.

to the amount of converted material that was left behind, as the front of the polymer core was moving toward the middle of the zeolite particles.

III. HDPE DEGRADATION OVER CATALYSTS

A. Introduction

Polymers can be pyrolyzed to lower-molecular weight hydrocarbons either catalytically, using a variety of reactor types, including batch [22,26–28, 32] and fixed bed [20,21,23,29,34–36,47,56,58], or noncatalytically, using thermal degradation in a fluidized-bed reactor or kiln [42,66,95–98]. However, using batch reactors leads to predominantly secondary reactions, which have a broad range of products, including heavy aromatics and coke as well as saturated hydrocarbons. Fixed-bed reactors are prone to blocking due to the viscous nature of melted polymer. These create problems in scaling up to industrial dimensions [99]. Non-

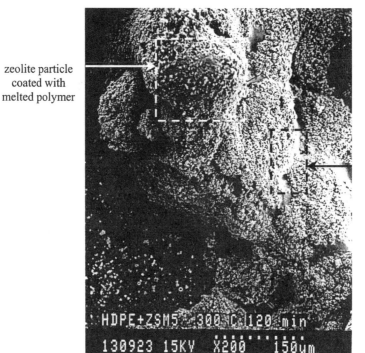

zeolite particle
coated with
melted polymer

unreacted
polymer

FIG. 7 Mixture of HDPE/ZSM-5 at 300°C, isothermal for 120 minutes, 200× magnification.

catalytic thermal cracking using a fluidized-bed reactor, with sand as a fluidizing agent, or kiln requires a higher operating temperature and produces products in a very broad range.

On the other hand, the use of a fluidized-bed reactor has advantages in terms of heat and mass transfer, as well as constant temperatures throughout the reactor [40,41,55,92,99,100]. Therefore, to minimize the operating temperature and to maximize the product selectivity and yield, compared to thermal cracking, HDPE degradation has been investigated in this work using a laboratory-scale catalytic fluidized-bed reactor [99]. A few pure catalysts were used initially, and then the work was extended to real commercial fluid catalytic cracking catalysts. Results will be compared to simple thermal cracking experiments.

B. Experiments

1. Materials and Reaction Preparation

Pure HDPE (ex. BASF) with an average weight of 75,000, as determined by gel permeation chromatography, and a density of 960 kg m^{-3} was used in this study.

TABLE 2 Catalysts Used

Catalyst	Commercial name	Supplier
ZSM-5	ZSM-5 zeolite	BP Chemicals, Sunbury-on-Thames, UK
US-Y	Ultrastabilized Y zeolite	Crosfield Chemicals, Warrington, UK
SAHA	Amorphous silica alumina (high alumina)	Crosfield Chemicals, Warrington, UK
CAT-A, B, C, D	Fresh commercial FCC catalyst	Engelhard Corp., NJ
E-Cat 1, 2, 3	Equilibrium catalysts	Engelhard Corp., NJ

The catalysts employed in this study are described in Table 2, with relevant characterization details shown in Table 3. Prior to use, all the catalysts were pelleted using a press (compression pressure = 160 MPa), crushed, and sieved to give particles ranging from 125 to 190 μm. The catalysts (0.25–0.30 g) were then activated by heating in the reactor in flowing nitrogen (50 mL min^{-1}) to 120°C at 60°C h^{-1}. After 2 h, the temperature was increased to 500°C at a rate of 120°C h^{-1}. After 5 h at 500 °C, the reactor was cooled to the desired reaction temperature. The particle size of both catalysts (125–180 μm) and polymer (75–250 μm) were chosen to be large enough to avoid entrainment and small enough to be accurately fluidized in a flow of high-purity nitrogen at 450–500 mL min^{-1}.

Two types of polymer cracking experiments were carried out:

1. 360°C with a catalyst/polymer (C/P) ratio of 2:1 (typical laboratory conditions)
2. 450°C with a C/P ratio of 6:1 (more closely resembling FCC conditions).

2. Product Analysis

A schematic diagram of the fluidized bed reactor is shown in Figure 8. A detailed operation procedure has been reported previously [40,55]. Volatile products leaving the reactor were passed through a glass-fibre filter to capture catalyst fines. A three-way valve was used to route products either into a sample gas bag or to an automated sample valve system with 16 loops. Tedlar bags, 15 liters in capacity, were used to collect time-averaged gaseous samples. The bags were replaced at intervals of 5 and 10 minutes throughout the course of the reaction. The multiport sampling valve allowed frequent, rapid sampling of the product stream at 0.5- and 1-minute intervals. Gaseous products were analyzed using a gas chromatograph (VARIAN 3400) equipped with (1) a thermal conductivity detector (TD) fitted with a molecular sieve 13× packed column (1.5-m × 0.2-mm i.d.) and (2) a flame ionization detector (FID) fitted with a PLOT Al$_2$O$_3$/KCl capillary column (50-m × 0.32-mm i.d.). A calibration cylinder containing 1% C$_1$–C$_5$

TABLE 3 Summary of Catalyst Details

Catalyst	wt%			m²/g			ppm		μmol/g catalyst		Particle size (microns)
	Si/Al	Al₂O₃	REO	TSA	MSA	ZSA	Ni	V	Bronsted	Lewis	
ZSM-5	17.5	—	—	391	—	391	—	—	182.5	65.0	Hexagonal
US-Y	6.2	—	—	603	—	603	—	—	254.9	185.8	Tetragonal
SAHA	2.6	25	—	274	274	—	—	—	12.3	134.9	15–160
Catalyst A	0.6	32.5	1.2	384	118	266	nd	nd	216.1	146.7	30–100
Catalyst B	0.6	30.7	4.8	437	112	325	nd	nd	168.5	114.3	6–60
Catalyst C	0.2	53.1	1.9	335	172	163	nd	nd	248.9	109.4	15–60
Catalyst D	0.5	35.5	3.1	198	54	144	nd	nd	136.3	40.1	15–70
E-Cat 1	nd	42.5	1.3	175	76	99	171	217	10.1	18.1	30–100
E-Cat 2	nd	29.4	1.6	127	32	95	5400	6580	8.1	8.4	25–210
E-Cat 3	nd	37.4	2.8	155	56	99	1520	3920	9.9	16.3	25–90

nd—not detected; REO—rare earth oxide; TSA—total surface area; MSA—matrix surface area; ZSA—zeolite surface area; Ni—nickel; V—vanadium; Bronsted and Lewis acidity—FTIR measurement.

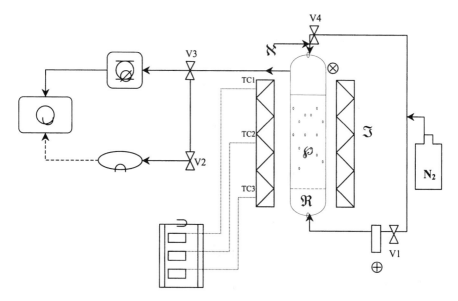

FIG. 8 Schematic diagram of the fluidized-bed reactor. ℵ polymer feeder; ℑ 3-zone furnace; ℜ sintered distributor; ℘ fluidizing catalyst; ⊗ fluidized-bed reactor; ⊕ flowmeter; ∅ 16-loop automated sample box; ∩ gas bag; ∪ gas chromatograph; ⊃ 3-zone digital controller.

hydrocarbons (Linde Gas, UK) was employed to help identify and quantify the gaseous products.

The remaining solids deposited on the catalyst after the catalytic degradation of the polymer were considered "residues" that contained involatile products and coke. The amount and nature of the residues were determined by thermogravimetric analysis (TA Instruments, DST 2960 Simultaneous DTA-TGA), as described elsewhere [55,99].

C. Results and Discussions

1. Overall Analysis

Catalytic pyrolysis products (P) are grouped together as hydrocarbon gases ($<C_5$), gasoline (C_5–C_9), coke, and residues to simplify the description of the overall pyrolysis processes. The term *yield* as used here is defined by this relation:

$$\text{yield (wt\%)} = \frac{P(g)}{\text{polymer feed (g)}} \times 100$$

At the reaction temperatures used in this study, the products from polymer cracking were mostly gases in the region C_1–C_9. Product characterization by gas chromatography included amount, carbon chain length, and degree of unsaturation. This, combined with TGA analysis of coke and unreacted polymer in the catalysts after the cracking reaction, led to mass balances of 90 ± 5% for most of the experiments.

2. Thermal Cracking in Polymer Degradation

Thermal cracking of polymer waste as carried out at the BP pilot plant in Grangemouth [42,66] was investigated in comparison with catalytic cracking at 450°C with a C/P ratio of 6:1. However, instead of using sand, as in the BP process, silica (SiO_2) with no activity was used as a fluidizing agent in the process. The SiO_2 was derived from a commercial 40% aqueous solution of Ludox by evaporation at 70–100°C. Ludox AS40 is commonly used as a source of silica in the synthesis of zeolites. In this section, the thermal cracking process using SiO_2 was compared with catalytic cracking over catalysts SAHA, E-Cat 1, and E-Cat 2. The overall results are tabulated in Table 4. Figure 9a shows the yields of gas, liquid, wax, and coke; Figure 9b shows selected olefin product distributions for both catalytic and thermal cracking.

TABLE 4 Wt% of Product Distributions for Catalytic Compared to Thermal Cracking at $T = 450°C$; C/P = 6:1

	Catalytic			Thermal
	SAHA	E-Cat 1	E-Cat 2	SiO_2
Gaseous	79.2	83.3	83.8	15.1
Liquid	0.0	1.8	1.7	5.9
Coke	5.0	1.9	1.3	0.0
Wax	0.0	0.0	0.0	50.8
Involatile	15.8	13.0	13.2	28.3
Total	100.0	100.0	100.0	100.0
Gaseous product distribution				
H_2	0.0010	0.0008	0.0012	0.0000
C_1–C_4	41.5	35.2	34.4	44.2
C_5–C_9	58.0	63.4	64.3	55.7
BTX[a]	0.5	1.4	1.3	0.0
Total	100.0	100.0	100.0	100.0
Total gaseous product				
Paraffins	15.8	23.6	23.7	6.7
Olefins	83.7	74.9	75.0	93.3

[a] BTX—benzene, toluene, and xylene.

(a)

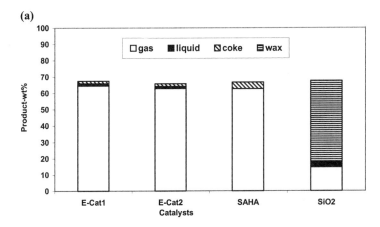

(b)

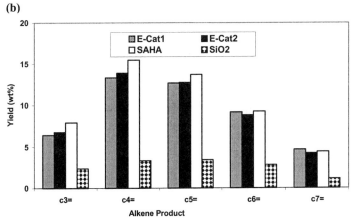

FIG. 9 Catalytic versus thermal cracking at $T = 450°C$; C/P $= 6:1$ (based on 0.1 g polymer feed): (a) total product and (b) alkene product distribution.

For all reactions, the 30 wt% of unreacted polymer was taken into account for mass balances. At 450°C, thermal cracking gave yields of 50 wt% wax product with a hydrocarbon range of $C_{14}–C_{30}$, at 360°C, no products were detected. It will be shown later that catalytic cracking of polymer was better than thermal cracking even when using less reactive catalysts, such as E-Cats.

3. Catalytic Cracking: Two Operating Conditions

Two operating conditions were investigated in this study: (1) temperature 360°C and catalyst-to-polymer (C/P) ratio 2:1, and (2) temperature 450°C and C/P ratio 6:1. The overall results are presented in Tables 5 and 6, respectively.

TABLE 5 Wt% of Product Distributions at $T = 360°C$; C/P = 2:1

	ZSM-5	US-Y	SAHA	Cat-A	Cat-B	Cat-C	Cat-D	E-Cat1	E-Cat2	E-Cat3
Gaseous	81.1	70.4	86.9	79.8	76.4	77.3	78.3	53.6	42.4	48.9
Liquid	0.0	0.0	0.0	0.0	0.0	0.0	0.0	3.1	4.1	3.7
Coke	1.5	7.3	2.2	9.1	9.8	8.4	7.9	0.6	0.4	0.8
Involatile	11.4	22.2	10.8	11.1	13.9	14.3	13.8	42.7	53.0	46.7
Total	100.0	100.0	100.0	100.0	100.0	100.0	100.0	100.0	100.0	100.0
Gaseous product distribution										
C_1–C_4	72.6	38.1	45.7	42.6	40.2	43.4	41.8	28.67	21.4	26.4
C_5–C_8	24.6	59.4	54.3	54.6	57.0	54.7	56.1	71.0	74.2	72.8
BTX[a]	2.7	2.5	0.0	2.8	2.8	1.8	2.2	0.3	4.5	0.8
Total	100.0	100.0	100.0	100.0	100.0	100.0	100.0	100.0	100.0	100.0
Total gaseous product										
Paraffins	16.1	48.7	6.8	48.7	49.1	38.5	42.7	19.5	20.2	26.4
Olefins	81.2	48.8	93.2	48.4	48.1	59.7	55.2	80.2	75.4	72.8

[a] BTX—benzene, toluene, and xylene.

TABLE 6 Wt% of Product Distributions at $T = 450°C$; $C/P = 6:1$

	ZSM-5	US-Y	SAHA	Cat-A	Cat-B	Cat-C	Cat-D	E-Cat1	E-Cat2	E-Cat3
Gaseous	83.7	69.6	79.2	75.8	66.0	71.4	75.0	83.3	83.8	82.6
Liquid	2.0	0.6	0.0	0.7	0.3	0.7	0.8	1.8	1.7	1.8
Coke	2.4	5.6	5.0	13.3	13.3	11.7	11.0	1.9	1.3	1.5
Involatile	11.9	24.2	15.8	10.2	20.4	16.2	13.1	13.0	13.2	14.2
Total	100.0	100.0	100.0	100.0	100.0	100.0	100.0	100.0	100.0	100.0
Gaseous product distribution										
H_2	0.0070	0.000	0.0010	0.0012	0.0015	0.0008	0.0007	0.0008	0.0012	0.0019
C_1–C_4	68.6	36.6	41.5	55.4	53.2	51.4	50.8	35.2	34.4	32.7
C_5–C_9	23.1	60.2	58.0	40.5	42.7	45.0	45.6	63.4	64.3	66.4
BTX^a	8.3	3.2	0.5	4.1	4.1	3.7	3.6	1.4	1.3	0.9
Total	100.0	100.0	100.0	100.0	100.0	100.0	100.0	100.0	100.0	100.0
Total gaseous product										
Paraffins	27.0	48.8	15.8	63.8	63.4	53.7	54.7	23.6	23.7	25.8
Olefins	64.7	47.8	83.7	31.5	32.5	42.7	41.6	74.9	75.0	73.3

^{a} BTX—benzene, toluene, and xylene.

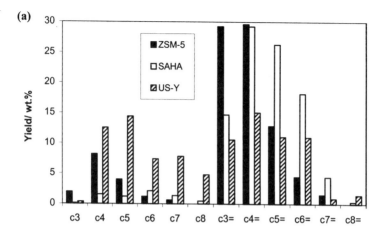

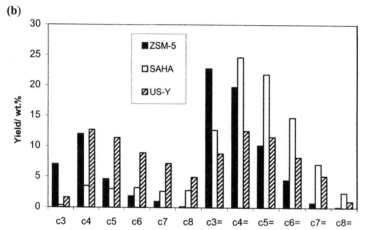

FIG. 10 Product distribution for model catalysts at (a) $T = 360°C$ (C/P = 2 : 1) and (b) $T = 450°C$ (C/P = 6 : 1).

(a) Model Catalysts. The products from polymer cracking shown in Figures 10a and 10b reflect the different natures of both the zeolitic pore structure and the predominant type of acidity found in the catalysts. At temperature 360°C, ZSM-5 yielded predominantly olefinic materials (~83 wt%) in the very narrow range of carbon number C_3–C_5 due to its restricted channels and smaller pore openings. The less reactive amorphous silica-alumina, high alumina (SAHA), which contains largely Lewis acid sites, yielded almost 95 wt% unsaturated olefin products over the C_3–C_7 carbon range. US-Y, with its large pore openings and internal supercages, allowed bulky bimolecular reactions to occur and, conse-

quently, gave a C_3–C_8 carbon number distribution with a 50:50 balance of paraffins and olefins. These results were comparable to those of previous studies [40,41,99].

At 450°C and higher catalyst-to-polymer ratio, the results were similar but with increased paraffin yields. Higher temperature increases the reactivity of the catalysts, and hence more primary products (olefins) were converted to secondary products (paraffins). However, zeolite US-Y at higher temperature showed rapid deactivation and high coke yield at very short reaction times, consequently, the product yield of US-Y looked similar at both temperatures.

(b)   US-Y and FCC Catalysts (Cat-A, -B, -C, -D).   FCC commercial catalysts (Figs. 11a and 12a) produced a balanced paraffinic and olefinic product stream (mainly C_3–C_8) similar to that of US-Y. This is expected, since the major catalytic component in FCC catalysts is zeolite Y, with approximately 10 wt% ZSM-5 [101]. All catalysts had high levels of coke, due to catalyst deactivation. However, when laboratory conditions more closely resembled FCC process conditions, which are at temperature 450°C and catalyst-to-polymer ratio of 6:1, all FCC catalysts produced significantly larger amounts of paraffins (54–64 wt%, Figs. 11b and 12b). This increased activity in FCC catalysts compared to US-Y may be explained by the presence of rare earth oxides and matrix material, which are present in FCC catalysts. Their role is to moderate the activity loss, to maximize the cracking activity, and to reduce the effect of containment metals and coke formation, respectively [83,84,90,101]. Also, US-Y deactivated very rapidly at the high temperature.

(c)   Equilibrium Catalysts (E-Cats).   Equilibrium catalysts are "used" FCC catalysts with different levels of metal poisoning, which came from previous FCC processes, as presented in Table 3. The metal poisoning, namely, nickel (Ni) and vanadium (V), originate from porphyrins present in the heavy oils (crude oil) converted to lighter products in the FCC process. Nickel causes unwanted dehydrogenation reactions, which produce byproducts, namely, coke and hydrogen; and vanadium destroys the zeolite component, which decreases the zeolite surface area and activity [88,89]. Typically, all E-Cats yielded a predominantly olefin product ($\geq 80\%$) in the range of C_3–C_8 (Tables 5 and 6, Figs. 13a and 13b) and compared very favorably with SAHA (similar to a previous batch reactor study [62]). Since a used catalyst is deactivated, predominantly primary products, namely, olefins, are produced. More hydrogen transfer was observed when E-Cats were used, with approximately double the amount of paraffins being generated at 360°C than with SAHA. While increased temperature generated more paraffins with SAHA, an even greater yield was observed when HDPE was degraded over E-Cats. Most importantly, the level of nickel and vanadium contamination trapped in the "used" catalysts did not greatly affect the product stream generated. The results are very encouraging and have significant impact on the economics

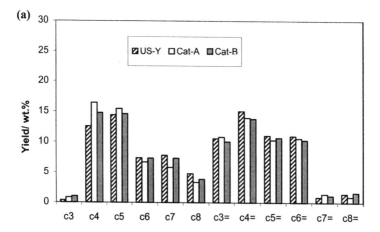

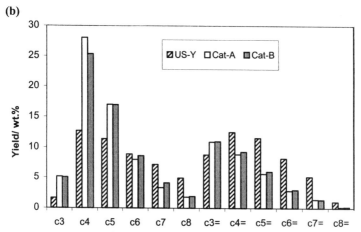

FIG. 11 Product distribution for US-Y and FCC catalysts (Cat-A and -B) at (a) $T = 360°C$ (C/P $= 2:1$) and (b) $T = 450°C$ (C/P $= 6:1$).

of a catalytic polymer degradation process employing catalysts of zero market value.

Figures 14a and 14b present total product yields at 360 and 450°C for zeolite US-Y compared with one of the FCC catalysts, Cat-A, and SAHA compared with one of the E-Cats, E-Cat 2, respectively. It is evident from these figures that the cracking of HDPE over the various catalysts at a set temperature is very similar. At 360°C, the cracking time required is of the order of 10 minutes, whereas at the higher temperatures, cracking is complete within 2.5 minutes.

(a)

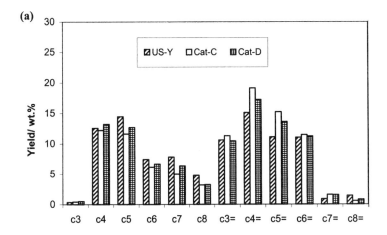

(b)

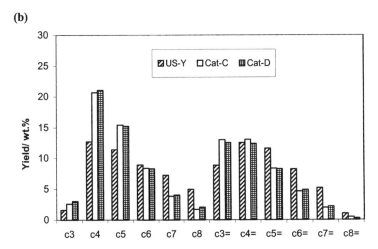

FIG. 12 Product distribution for US-Y and FCC catalysts (Cat-C and -D) at (a) $T = 360°C$ (C/P $= 2:1$) and (b) $T = 450°C$ (C/P $= 6:1$).

4. A Study on Catalyst Regeneration

In this section, some of the catalysts from the previous experiment were reused to compare to the industrial FCC process. Clearly, polymer waste recycling would be more economical if catalysts can be reused.

After the previous cracking process, catalyst residues were regenerated in air, to remove all the coke, and reactivated in nitrogen gas by a similar activation process. Four regenerated catalyst residues (zeolite US-Y, FCC Cat-A, E-Cat 1,

(a)

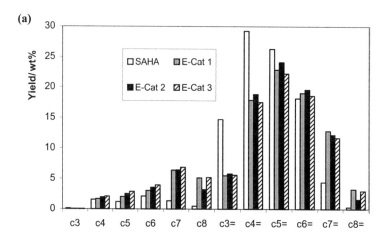

(b)

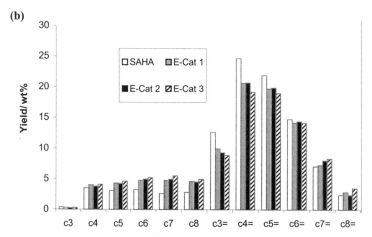

FIG. 13 Product distribution for SAHA and E-Cats at (a) $T = 360°C$ (C/P $= 2:1$) and (b) $T = 450°C$ (C/P $= 6:1$).

and E-Cat 2) were studied to crack HDPE at 450°C and C/P ratio of 6:1. The overall results for the four regenerated catalysts were compared with those using fresh catalysts and are presented in Table 7. These show that the product distributions for the fresh and used catalysts are similar. This is very encouraging.

Figure 15 illustrates a few selected product distributions on fresh and regenerated catalysts CAT-A, E-Cat 1, and E-Cat 2. Regenerated E-Cat 1 catalyst gave slightly higher olefin products, possibly indicating a slight lowering of activity when reused. As expected, a more significant difference between the fresh and

(a)

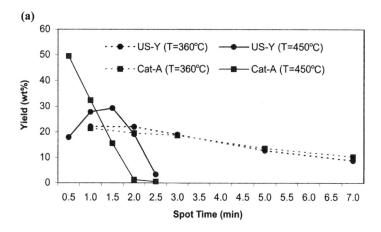

(b)

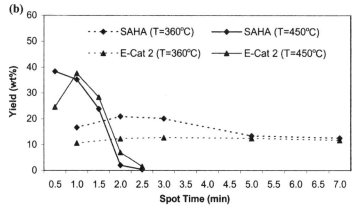

FIG. 14 Total product yields at both temperatures for (a) US-Y and FCC Cat-A and (b) SAHA and E-Cat 2.

regenerated catalysts (Cat-A) was observed with a marked drop in *iso*-butane and a rise in olefin products. This irreversible drop in activity from fresh catalyst to reused catalyst would result in less conversion of primary olefin products to secondary paraffin products via hydrogen transfer.

5. A Study on Catalyst Deactivation in Multiple Addition Series

In order to achieve better and more economic polymer waste recycling, multiple-step additions of polymer were made throughout the cracking process. It was noted that even after five additions, the catalyst was still active, although less

TABLE 7 Wt% of Product Distributions for Fresh (1) and Regenerated (2) Catalysts at $T = 450°C$; C/P = 6:1

	US-Y		Cat-A		E-Cat 1		E-Cat 2	
	(1)	(2)	(1)	(2)	(1)	(2)	(1)	(2)
Gaseous	69.6	66.7	75.8	75.4	83.3	83.7	83.8	60.5
Liquid	0.6	2.0	0.7	0.0	1.8	0.4	1.7	17.1
Coke	5.6	6.5	13.3	12.6	1.9	1.6	1.3	1.1
Involatile	24.2	24.8	10.2	12.0	13.0	14.2	13.2	21.2
Total	100.0	100.0	100.0	100.0	100.0	100.0	100.0	100.0
Gaseous product distribution								
H_2	0.0012	0.014	0.0000	0.0000	0.0008	0.0000	0.0012	0.0020
C_1-C_4	36.6	38.2	55.4	51.2	35.2	37.6	34.4	34.2
C_5-C_9	60.2	58.7	40.5	45.1	63.4	60.9	64.3	64.5
BTX[a]	3.2	3.1	4.1	3.7	1.4	1.5	1.3	1.3
Total	100.0	100.0	100.0	100.0	100.0	100.0	100.0	100.0
Total gaseous product								
Paraffins	49.0	49.6	63.8	56.2	23.6	23.3	23.7	21.2
Olefins	47.9	47.1	31.5	39.6	74.9	75.2	75.0	77.5

[a] BTX—benzene, toluene, and xylene.

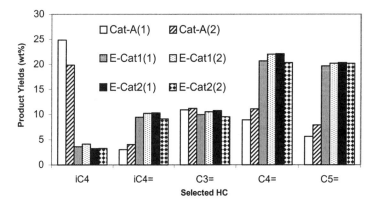

FIG. 15 Comparison of fresh (1) and regenerated (2) catalysts at $T = 450°C$; C/P = 6:1—selected product distributions.

than at the beginning. This implies that the catalyst-to-polymer ratio can be reduced from 6:1, thus making the process more economical.

Effects due to a series of polymer additions were investigated on FCC Cat-A and FCC E-Cat 2 at 450°C and C/P ratio of 6:1. As discussed earlier, Cat-A is the most active FCC catalyst and E-Cat 2, with the highest metal contamination, is the least active E-Cat. This analysis was carried out to study the deactivation of the catalysts during the multiple addition of polymer in the cracking process. One, three, and five additions were made with the same amount of polymer feed for every addition at five-minutes interval during the cracking process.

Figure 16a shows that for Cat-A, the olefin products are directly proportional to the additions of polymer. Similarly, the paraffins are decreasing. Both results indicate that, as the addition of polymer increased, the catalyst deactivated and was slightly blocked with coke. As the deactivation of the catalyst rose, the reactivity of the catalyst decreased and limited the conversion of primary product (olefins) to paraffins. Also, as the coke increased during the addition of polymer, it decreased the pore space and produced a more narrow carbon number distribution. Figure 16b gives similar results for E-Cat 2. It shows that *iso*-butane and total butane remained unchanged during the addition, while *iso*-butane and total butane decreased slightly. Propane yield was essentially zero with the addition of polymer.

The percentage of coke increased with polymer addition for Cat-A but was essentially constant for E-Cat 2, as shown in Figure 17a. For Cat-A, the yield of *iso*-butane decreased while that for butylene increased with coke content (Fig. 17b). As discussed earlier, the increasing deactivation of Cat-A leads to less hy-

(a)

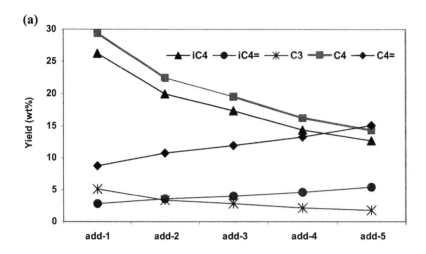

(b)

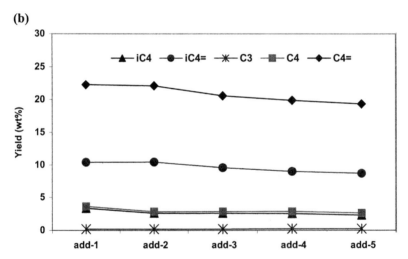

FIG. 16 Multiple addition series at $T = 450°C$, C/P = 6:1 for (a) Cat-A and (b) E-Cat 2.

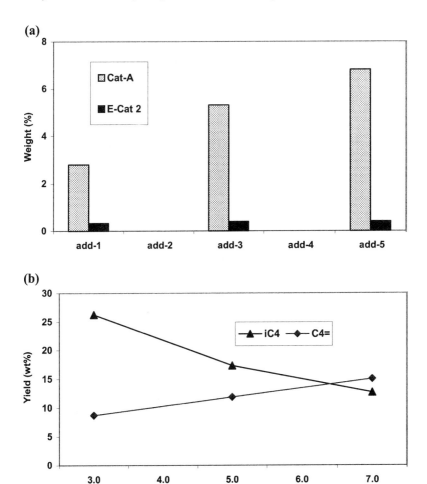

FIG. 17 Multiple addition series at $T = 450°C$: (a) % coke versus addition (for Cat-A and E-Cat 2) and (b) % yield versus coke content (Cat-A).

drogen transfer and a decreased conversion of olefins to paraffins. There is no equivalent plot for E-Cat 2, since its coke content was constant.

IV. CONCLUSION

The initial results obtained with zero-cost used catalysts are promising and suggest improvements to current technology that uses fluidized sand to thermally crack polymer waste. The use of a catalytic recycling process based on used catalysts would lower the operating temperature, improve the yield and selectivity of volatile products, and produce potentially more valuable olefinic feedstocks in the range of C_3–C_8. A simple economic study based on catalyst, process cost, and value of the products (presented in a previous publication [91]), suggest that a gate fee of £135–170 per ton of plastic waste compared competitively with current thermal cracking technology [42].

Thus, catalytic recycling of plastic wastes from postconsumers as well as from industrial production has the potential of supplying valuable chemical feedstock to refineries and the petrochemical industries.

ACKNOWLEDGMENTS

The financial support of the Malaysian Government (Universiti Putra Malaysia) is acknowledged. In addition, thanks are also due to the Center for Microporous Materials for the use of facilities and to Miss S. Maegaard for her contribution on thermal analysis. The authors would also like to thank Mr. M. Hiam, Mr. R. J. Plaisted, Dr. C. S. Cundy, and Dr. D. Cresswell, for helpful discussions during the preparation of this chapter.

REFERENCES

1. APME. Association of Plastics Manufacturers in Europe 1998, pp 1–14.
2. APME. Association of Plastics Manufacturers in Europe 2001, pp 1–16.
3. Department for Environment, Food and Rural Affairs: Waste Strategy 2000 for England and Wales, Parts I & II, Chap 5, pp 1–28.
   <http://www.defra.gov.uk/environment/waste/strategy/part2/html/9.htm>
   Date Reviewed: 04 July 2001; Date Searched: 3 September 2001.
4. E Colyer. Waste Management, (May):36–38, 1999.
5. M Lee. Chemistry in Britain 1:5–6, 1994.
6. J Brandrup. Macromolecular Symposia 135:223–235, 1998.
7. R Kingston. Chemistry in Britain 4:30–33, 2000.
8. P Swallow, A Redfern, M. Bunyan, C Ward, eds. The Environment in Your Pocket 2000. West Yorkshire: Department of the Environment, Transport and the Regions, 2001, pp 45–49.
9. K Harper. Great Britain Plc: The Environmental Balance Sheet. United Kingdom: Biffa, 1999, pp 88–93.
10. W Kaminsky, F Hartmann. Angewandte Chemie International Edition 39:331–333, 2000
11. F Rodriguez. Principles of Polymer Systems. 3rd ed. Hemisphere, 1989, pp 241–260.
12. G Scott. Waste Management (May): 38–39, 1999.
13. British Plastic Federation. Plastic Waste Management. BPF Homepage.

<http://www.bpf.co.uk/bpf/wastemanagement>
Date Reviewed: 12 January 2001; Date Searched: 25 May 2001.

14. M Lee. Chemistry in Britain 7:515–516, 1995.

15. J Brandrup. In: J Brandrup, M Bitter, W Michaeli, G Menges, eds. Recycling and Recovery of Plastics. Cincinnati: Hanser/Gardner, 1996, pp 393–412.

16. A Miller. Chemistry and Industry 1:8–9, 1994.

17. S Hardman, DC Wilson. Macromolecular Symposia 135:115–120, 1998.

18. J Brandrup. In: J Brandrup, M Bitter, W Michaeli, G Menges, eds. Recycling and Recovery of Plastics. Cincinnati: Hanser/Gardner, 1996, pp 31–45.

19. T Yoshido, A Ayame, H Kanoh. Bull Chem Soc Japan 17:218–225, 1975.

20. Y Uemichi, A Ayame, Y Kashiwaya, H Kanoh. Journal of Chromatography 259: 69–77, 1983.

21. Y Uemichi, Y Kashiwaya, M Tsukidate, A Ayame, H Kanoh. Bull Chem Soc Japan 56:2768–2773, 1983.

22. G Audisio, A Silvani, P Carniti. Journal of Analytical Pyrolysis 7:83–90, 1984.

23. C Vasile, P Onu, V Barboiu, M Sabliovshi, D Ganju, M. Florea. Acta Polymerica 36:543–550, 1985.

24. H Nambu, Y Ishihara, T Takesue, T Ikemura. Polymer Journal 18:871–975, 1986.

25. H Nambu, Y Sakuma, Y Ishihara, T Takesue, T Ikemura. Polymer Degradation and Stability 19:61–76, 1987.

26. Y Ishihara, H Nambu, C Iwata, T Ikemura, T Takesue. Bull Chem Soc Japan 62: 2981–2988, 1989.

27. PL Beltrame, P Carniti, G Audisio, F Bertini. Polymer Degradation and Stability 26:209–219, 1989.

28. Y Ishihara, H Nambu, K Saido, T Ikemura, T Takesue. Journal of Applied Polymer Science 38: 1491–1501, 1989.

29. G Audisio, F Bertini, PL Beltrame, P Carniti, Polymer Degradation and Stability 29:191–200, 1990.

30. Y Ishihara, H Nambu, T Ikemura, T Takesue. Fuel 69:978–984, 1990.

31. Y Ishihara, H Nambu, K Saido, T Ikemura, T Takesue. Bull Chem Soc Japan 64: 3585–3592, 1991.

32. G Audisio, F Bertini, PL Beltrame, P Carniti. Macrom Chem, Macromolecular Symposia 57:191–209, 1992.

33. Y Ishihara, H Nambu, K Saido, T Ikemura, T Takesue, T Kuroki. Fuel 72(8):1115–1119, 1993.

34. H Ohkita, R Nishiyama, Y Tochihara, T Mizushima, N Kakuta, Y Morioka, Y Namiki, S Tanifuji, H Katoh. Industrial and Engineering Chemistry Research 32: 3112–3116, 1993.

35. AR Songip, T Masuda, H Kuwahara, K Hashimoto. Applied Catalysis B: Environmental 2:153–164, 1993.

36. RC Mordi, R. Fields, J Dwyer. Journal of Analytical and Applied Pyrolysis 29: 45–55, 1994.

37. R Lin, RL White. Journal of Applied Polymer Science 63(10):1287–1298, 1997.

38. VJ Fernandez Jr. AS Araujo, GJT Fernandez. Progress in Zeolite and Microporous Materials. Studies in Surface Science and Catalysis 105:941–947, 1997.

39. N Hovart, FTT Ng. Fuel 78:459–470, 1999.

40. AA Garforth, YH Lin, PN Sharratt, J Dwyer. Applied Catalysis A General 169: 331–342, 1998.
41. AA Garforth, YH Lin, PN Sharratt, J Dwyer. Science and Technology in Catalysis 28:197–202, 1998.
42. I Dent, S Hardman. IChemE Environmental Protection 1996: Issue 44 (September): 1–8.
43. AG Buekens, JG Schoeters. Macromolecular Symposia 135:63–81, 1998.
44. JA Conesa, R Font, A Marcilla, AN Garcia. Energy and Fuels 8:1238–1246, 1994.
45. Y Sakata. Macromolecular Symposia 135:7–18, 1998.
46. Y Sakata, MA Uddin, K Koizumi, K Murata. Chemistry Letters, 245–246, 1996.
47. Y Uemichi, K Takuma, A Ayame. Chemical Community, 1975–1976, 1998.
48. Y Sakata, MA Uddin, K Koizumi, K Murata. Polymer Degradation and Stability 53:111–117, 1996.
49. G Giannotta, R Po, N Cardi, E Tampellini, E Occhiello, FGL Nicolais. Polymer Engineering and Science 34:1219–1226, 1994.
50. M Paci, FP La Mantia. Polymer Degradation and Stability 63:11–14, 1999.
51. N Lingaiah, MA Uddin, A Muto, Y Sakata, T Imai, K Murata. Applied Catalysis A General 207:79–84, 2001.
52. DP Serrano, J Aguado, JM Escola. Applied Catalysis B Environmental 25:181–189, 2000.
53. H Ukei, T Hirose, S Horikawa, Y Takai, M Taka, N Azuma, A Ueno. Catalysis Today 62:67–75, 2000.
54. G de la Puente, U Sedran. Applied Catalysis B. Environmental 19:300–311, 1998.
55. YH Lin, PN Sharratt, AA Garforth, J Dwyer. Thermochimica Acta 294:45–52, 1997.
56. Y Uemichi, J Nakamura, T Itoh, M Sugioka, AA Garforth, J Dwyer. Industrial & Engineering Chemistry Research 38:385–390, 1999.
57. J Aguado, DP Serrano, JM Escola, E Garagorri, JA Fernandez. Polymer Degradation & Stability 69:11–16, 2000.
58. G Luo, T Suto, S Yasu, K Kato. Polymer Degradation & Stability 70:97–103, 2000.
59. J Walendziewski, M Steininger. Catalysis Today 65:323–329, 2001.
60. DL Negelein, R Lin, RL White. Journal of Applied Polymer Science 67:341–348, 1998.
61. ND Hesse, R Lin, E Bonnet, J Cooper III, RL White. Journal of Polymer Science 82:3118–3125, 2001.
62. SC Cardona, A Corma. Applied Catalysis B: Environmental 25:151–156, 2000.
63. DD Cornell. In: CP Rader, SD Baldwin, DD Cornell, GD Sadler, RF Stockel, eds. Plastics, Rubber, and Paper Recycling—A Pragmatic Approach. ACS Symposium series 609. Washington, DC: American Chemical Society, 1995, pp 72–79.
64. J Aguado, DP Serrano, MD Romero, JM Escola. Chemical Community 1: 725–731, 1996.
65. RH Burnett. In: CP Rader, SD Baldwin, DD Cornell, GD Sadler, RF Stockel, eds. Plastics, Rubber, and Paper Recycling—A Pragmatic Approach. ACS Symposium series 609. Washington, DC: American Chemical Society, 1995, pp 97–103.

66. JH Brophy, S Hardman. In: J Brandrup, M Bitter, W Michaeli, G Menges, eds. Recycling and Recovery of Plastics. Cincinnati: Henser/Gardner, 1996, pp 422–433.
67. J Dwyer, A Dyer. Chemistry & Industry 7:237–240, 1984.
68. A Dyer, Chemistry & Industry 7:241–245, 1984.
69. LB McCusker, C Baerlocher. In: H van Bekkum, EM Flanigan, PA Jacobs, JC Jansen, eds. Introduction to Zeolite Science and Practice. Studies in Surface Science and Catalysis. The Netherlands: Elsevier Science, 2001, pp 37–68.
70. International Zeolite Association. Atlas of Zeolite Structure Types. <http://www.iza_structure.org/database> Date Revised: 28 February 2001; Date Searched: 06 June 2001.
71. J Dwyer. Chemistry & Industry 7:258–269, 1984.
72. A Dyer. An introduction to zeolite molecular sieves. New York: Wiley, 1988, pp 1–145.
73. A Humphries, DH Harris, P O'Connor. In: JS Magee, MM Mitchell Jr, eds. Fluid Catalytic Cracking: Science and Technology, Studies in Surface Science and Catalysis. vol 76. The Netherlands: Elsevier Science, 1993, pp 41–82.
74. AA Avidan. In: JS Magee, MM Mitchell Jr., ed. Fluid Catalytic Cracking: Science and Technology, Studies in Surface Science and Catalysis. vol 76. The Netherlands: Elsevier Science, 1993, pp 1–40.
75. FG Dwyer, TF Degman. In: JS Magee, MM Mitchell Jr, eds. Fluid Catalytic Cracking: Science and Technology, Studies in Surface Science and Catalysis. vol 76. The Netherlands: Elsevier Science, 1993, pp 449–530.
76. DW Breck. Zeolite Molecular Sieves. New York: Wiley, 1974, pp 25–300.
77. RM Barrer. Zeolites and Clay Minerals as Sorbents and Molecular Sieves. London: Academic Press, 1978, pp 120–257.
78. DEW Vaughan. In: RP Townsend, ed. The Properties and Applications of Zeolites. London: The Chemical Society, 1980, pp 294–320.
79. JW Ward. In: BE Leach, ed. Applied Industrial Catalysis. vol. 3. New York: Academic Press, 1984, pp 271–278.
80. BW Wojciechowski, A Corma. Catalytic Cracking: Catalysts, Chemistry and Kinetics. New York: Marcel Dekker, 1986, pp 1–121.
81. S Bhatia. Zeolite Catalysis: Principles and Applications. Boca Raton, FL: CRC Press, 1989, pp 1–330.
82. JA Marters, PA Jacobs. In: H van Bekkum, EM Flanigan, PA Jacobs, JC Jansen, eds. Introduction to Zeolite Science and Practice. Studies in Surface Science and Catalysis. The Netherlands: Elsevier Science, 2001, pp 633–672.
83. J Dwyer, DJ Rawlence. Catalysis Today 18:487–496, 1993.
84. JD Rawlence, J Dwyer. In: ML Occelli, ed. ACS Symposium Series 452. Washington, DC: American Chemical Society, 1991, pp 56–61.
85. TF Petti, D Tomczak, CJ Pereira, WC Cheng. Applied Catalysis A: General 169: 95–109, 1998.
86. PB Venuto, ET Habib Jr. Chemical Industries: 1, 1979.
87. PB Venuto, ET Habib Jr. Fluid Catalytic Cracking with Zeolite Catalysis. New York: Marcel Dekker, 1979, pp 35–210.
88. MM Mitchell Jr, JF Hoffman, HJ Moore. In: JS Magee, MM Mitchell Jr, eds. Fluid

Catalytic Cracking: Science and Technology, Studies in Surface Science and Catalysis. vol. 76. The Netherlands: Elsevier Science, 1993, pp 293–338.

89. RH Nielsen, PK Doolin. In: JS Magee, MM Mitchell Jr, eds. Fluid Catalytic Cracking: Science and Technology, Studies in Surface Science and Catalysis. vol. 76. The Netherlands: Elsevier Science, 1993, pp 339–384.

90. HW Kouwenhoven, B de Kroes. In: H van Bekkum, EM Flanigan, PA Jacobs, JC Jansen, eds. Introduction to Zeolite Science and Practice. Studies in Surface Science and Catalysis. The Netherlands: Elsevier Science, 2001, pp 673–706.

91. IE Maxwell, WHJ Stork. In: H van Bekkum, EM Flanigan, PA Jacobs, JC Jansen, eds. Introduction to Zeolite Science and Practice. Studies in Surface Science and Catalysis. The Netherlands: Elsevier Science, 2001, pp 747–820.

92. S Ali, AA Garforth, DH Harris, DJ Rawlence, Y Uemichi. Catalysis Today 75: 247–255.

93. SL Madorsky. Thermal Degradation of Organic Polymers. New York: Wiley, 1964, pp 96–114.

94. O Levenspiel. Chemical Reaction Engineering. 3rd ed. New York: Wiley, 1999, pp 300–343.

95. W Kaminsky, H Rossler. ChemTech. February: 108–113, 1992.

96. W Kaminsky. Advanced Polymer Technology 14:337–344, 1995.

97. W Kaminsky, B Schlesselmann, C Simon. Journal of Analytical and Applied Pyrolysis 32:19–27, 1995.

98. S Hardman, SA Leng, DC Wilson. BP Chemicals Limited, Polymer Cracking. European Patent Application 567292, 1993.

99. YH Lin. Theoretical and Experimental Studies of Catalytic Degradation of Polymers. PhD dissertation, University of Manchester Institute of Science and Technology, Manchester, UK, 1998.

100. W Kaminsky, JS Kim. Journal of Analytical and Applied Pyrolysis 51:127–134, 1999.

101. GM Woltermann, JS Magee, SD Griffith. In: JS Magee, MM Mitchell Jr, eds. Fluid Catalytic Cracking: Science and Technology, Studies in Surface Science and Catalysis. vol 76. The Netherlands: Elsevier Science, 1993, pp 105–144.

17

Catalytic Dehalogenation of Plastic-Derived Oil

AZHAR UDDIN* and YUSAKU SAKATA Okayama University, Tsushima Naka, Japan

I. PLASTIC WASTES AND ASSOCIATED TREATMENT

The widespread use of plastics in modern industry has resulted in a huge volume of waste plastic requiring treatment. In the years between 1975 and 1998 in Japan, the amount of postconsumer plastic waste increased 3.8 times, plastic production 2.7 times, and consumption of plastics 3.2 times (Table 1) [1]. It can be anticipated that this trend will continue unless more intensive recycling of plastic waste is undertaken. In the past, landfilling and incineration (without energy recovery) were widely practiced in many countries, but these options have been criticized in recent years for their adverse effects on the environment. Not only do the waste plastics make the environment unsightly, but also their energy and chemical content are lost. Presently, the alternative options for the treatment of plastic waste are: (1) mechanical recycling, (2) energy recycling, and (3) feedstock or chemical recycling. Mechanical recycling by melting and remolding of the used plastics is limited to particular types of polymers and applications of the recycled materials. If mechanical recycle is not feasible, energy recovery by incineration is another option, since the calorific value of plastics is very high. However, incineration may lead to the formation of pollutants, such as dioxins and other toxins, depending on the composition of the plastic waste and the nature of combustion. Feedstock or chemical recycling of plastic waste by pyrolysis or thermal degradation permits the recovery of valuable hydrocarbons, which can be used as feedstock materials or fuels. The plastic wastes are transformed into useful chemicals by thermal degradation in an inert atmosphere. Generally the gaseous and liquid products obtained are complex mixtures of hydrocarbons and other organic compounds, whose composition depends on the composition of the plas-

---

* *Current affiliation*: The University of Newcastle, Callghan, New South Wales, Australia.

TABLE 1   Production, Consumption, and Disposal Rates of Polymer Resins in Japan

	1975	1998	Increase
Resin production	517	1391	2.7 times
Domestic consumption of plastic goods	315	1020	3.2 times
Disposal of plastic waste	261	984	3.8 times

tic waste. The potentials and prospects of chemical recycling of plastic waste have been addressed in a recent review [2].

II. DEHALOGENATION ISSUES IN THE CHEMICAL RECYCLING OF PLASTIC WASTE

Most of the practical research on plastic degradation to fuel oil is limited to PE, PP, and PS but excludes PVC, since it contains chlorine (56 wt%) and releases toxic and corrosive hydrogen chloride gas during the early stage of degradation at 280–320°C. Upon dehydrochlorination, PVC forms a polyene macromolecular structure and it decomposes at higher temperatures (380–600°C) to produce volatiles containing aliphatic, olefinic, and aromatic hydrocarbons and solid chars [3]. Dehydrochlorination of PVC has been studied extensively, in particular the kinetics and mechanism of PVC degradation [4–9]. Although more than 99% of the chlorine content of PVC is removed as HCl in the early stages, the remaining chlorine in the polyene macromolecular structure may lead to the formation of unwanted Cl-containing organics. Detailed studies on the identification of the products evolved from PVC degradation and their mechanisms of formation have also been reported [10–12]. We have reported that various types of organic chlorine compounds are produced when PVC mixed with PE, PP, and PS is thermally degraded at 430°C, and we have clarified the route of formation of these compounds [13–14]. Hydrogen chloride released from PVC degradation reacts with the hydrocarbons produced from the other polymers to form organic chlorine compounds. Generally, municipal plastic waves (MPW) contain all kinds of plastics, including PVC. In Japan, municipal plastic wastes contain about 5–10 wt% PVC [15]. Therefore, thermal degradation of MPW produces unwanted organic chlorine compounds in the oil. The waste plastic–derived oil that contains organic chlorine compounds cannot be used safely as a fuel oil because there is a high possibility of producing toxic compounds, such as dioxins, dibenzofurans, and biphenyls, during combustion. It is necessary to remove the organic chlorine compounds from the oil before use.

Halogens other than chlorine may also be present in the polymer as additives, for instance, brominated flame-retardant compounds, such as polybrominated benzene compounds, are added to PS and ABS to inhibit or modify polymer combustion when heated in an oxidative atmosphere. Electrical and electronic home appliances are made of high-impact polystyrene (HIPS) containing brominated flame retardants. Some of these brominated plastics end up in municipal plastic wastes [16]. Thus debromination of brominated hydrocarbons in the plastic-derived oil is another important issue. Dehalogenation of the plastic-derived oil is a key technology for the success of chemical recycling of mixed plastics containing chlorinated and brominated polymers into fuel oil or chemicals. Compared to PVC, very little work has been reported on the pyrolysis of brominated flame retardants containing plastic waste. The separation of halogenated flame retardants from polymer matrixes with extraction using supercritical fluids such as supercritical carbon dioxide (SC-CO$_2$) has been studied [17]. Thermal-behavior and degradation-mechanism studies on brominated polystyrenes have been studied mainly by thermogravimetry, thermal volatilization analysis (TVA), and Py-GC/MS technique and show that the degradation occurs mainly to the monomer via radical polymerization [18]. We have reported the formation of brominated hydrocarbons when brominated-flame-retardant-containing HIPS and its mixture with other plastics are thermally degraded to obtain fuel oil [19].

III. OUR APPROACH TO THE DEHALOGENATION OF PLASTIC-DERIVED OIL

Various methods are used to remove the chlorine by coprocessing catalysts or absorbent during the pyrolysis [20–22]. All of these efforts succeed to some extent in removing chlorine content in PVC-containing plastics before or during the pyrolysis. However, even the presence of a small amount of chlorine (<1%) leads to the formation of chloro-organic compounds. We observed that even when the PVC is dechlorinated to >98%, chloro-organic compounds are still produced [23]. There has been no detailed study of the removal of chloro-organic compounds from the plastic-derived oil thus far.

Catalytic dechlorination is a promising method for the removal of organic chlorine compounds, compared to other methods, such as combustion [24]. There are many research papers on the catalytic dechlorination of various organic chlorine compounds [25–26], but most of them deal with the removal of chlorine by noble metal catalysts in the presence of hydrogen. We have been developing iron oxide–carbon composite catalysts for the dehydrochlorination and dehydrobromination of plastic-derived oil. In our proposed catalytic process, waste plastic–derived oil containing chlorinated and brominated hydrocarbons is treated over solid catalysts in a fixed-bed flow-type reactor in a temperature range of 300–400°C at atmospheric pressure in He or N$_2$ flow in order to decompose the halogenated hydrocarbons into hydrocarbons and hydrogen halides. We have demonstrated the effectiveness of our catalysts for the dehydrodechlorination of model compounds, such as chlorocyclohexane and 1-chloroheptane, in a fixed-bed reactor. These iron oxides are reported to be effective in the destruction or reduction of dioxins from municipal solid-waste incineration products [27]. In ABS degradation, iron oxides showed a catalytic effect in decomposition of N-containing heterocyclic compounds from the degradation oil [28–29]. This review summarizes our recent studies on (1) the catalytic dechlorination of PVC-containing mixed plastics–derived oil [30], (2) municipal waste plastic–derived oil [31], and (3) simultaneous dechlorination and debromination from the pyrolysis products of PVC and brominated flame-retardant-containing high-impact polystyrene (HIPS) mixed plastics over iron-base catalysts.

A. Catalytic Dechlorination of Chloro-Organic Compounds from PVC-Containing Mixed Plastic-Derived Oil over Iron Oxide Catalysts

We have reported the catalytic activity of various iron oxide and iron oxide–carbon composite catalysts for the dechlorination of chloro-organic compounds formed during the thermal degradation of PVC-containing mixed plastics. PVC-containing mixed plastic-derived oil was prepared by thermal degradation in a separate facility, and the dechlorination of the derived oil was performed in a fixed-bed flow-type reactor. Emphasis was put on the stability of iron oxide-based catalysts in the presence of HCl gas produced during dechlorination of mixed plastic–derived oil. The PVC-containing waste mixed plastic (MX/PVC)–derived oil was prepared by degrading mixed plastics containing PE (33%), PP (33%), PS (33%), and PVC (1%) as a model sample at 410°C. Details of the experimental procedure for the preparation of mixed plastic (MX/PVC)–derived oil is given elsewhere [32]. Toda Kogyo Corporation, Japan, supplied the catalysts α-Fe_2O_3 [PDC-03 (2)], γ-Fe_2O_3 (TR99701), and iron oxide–carbon composite (TR97305 and TR99300) catalysts used in this study. TR97305 and TR99300 were prepared from physical mixtures of iron oxide (Goethite: FeOOH) and phenol resins in a ratio of 9:1 by heat treatment. After the heat treatment at 500°C in N_2 flow, the product catalysts were identified as Fe_3O_4 and carbon composites. The physical characteristics of the catalysts used in this study are presented in Table 2. TR97305 and TR99300 are two iron oxide–carbon composites with similar composition but prepared by different methods. These composite catalysts were prepared in order to increase the physical strength of the catalyst pellets.

The dechlorination of mixed plastic–derived oil was carried out using a fixed-bed reactor at atmospheric pressure with a reaction temperature of 350°C. In a typical experiment, about 1 mL (0.1-mm average size) of the catalyst was loaded in between two quartz wool beds and treated in He atmosphere (60 cc/min) at reaction temperature for 1 h before feeding the mixed plastic–derived oil (10

TABLE 2 Physical Characteristics of Iron Oxide and Iron Oxide–Carbon Composite Catalysts

Catalyst	Surface area $(m^2 \cdot g^{-1})$	Iron oxide content (wt%)	Carbon content (wt%)	XRD
γ-Fe$_2$O$_3$	13	100	0	γ-Fe$_2$O$_3$
α-Fe$_2$O$_3$	4	100	0	α-Fe$_2$O$_3$
TR97305	74	93.3	6.7	Fe$_3$O$_4$
TR99300	60	93.2	6.8	Fe$_3$O$_4$

mL/h) by a microfeeder. The product was collected in a trap downstream of the reactor. A separate coldwater condenser was provided to ensure the condensation of all volatile liquid products. The products were analyzed by gas chromatography equipped with a flame ionization detector (FID; YANACO G6800; column, 100% methyl silicone; 50 m $\times$ 0.25 mm 0.25 μm; temperature program, 40°C (hold 15 min) $\to$ 280°C (rate 5°C/min; hold 37 min). The distribution of chlorine compounds and the quantity of chlorine content (organic) in the liquid products were analyzed by a gas chromatograph equipped with atomic emission detector (AED; HP G2350A; column, HP-1; cross-linked methyl siloxane; 25 m $\times$ 0.32 mm $\times$ 0.17 μm) using 1, 2, 4-trichlorobenzene as an internal standard.

The conversion by the catalysts in the dechlorination of the mixed plastic–derived oil was calculated as follows:

$$\frac{[\text{Cl content in mixed plastic–derived oil–Cl content in product}]}{[\text{Cl content in mixed plastic–derived oil}]} \quad (1)$$

The physicochemical properties of the MX/PVC-derived oil estimated by standard procedures and are tabulated in Table 3. The oil derived from mixed plastic-degradation oil contained 1894 ppm of organic chlorine compounds. The carbon number distribution of all compounds (C-NP gram) in MX/PVC derived oil and the carbon number distribution of chloro-organic compounds (Cl-NP gram) are shown in Figure 1a and 1b, respectively. The C-NP gram was obtained by plotting the weight percent of carbon-containing compounds in the MX/PVC oil against the carbon number of equivalent b.p. of normal paraffin. The Cl-NP gram was also obtained by plotting the content of Cl-containing compounds in the MX/PVC oil against the carbon number of each normal paraffin (equivalent to boiling points) [33]. The chlorine compounds were distributed mainly in the b.p. range of nC$_6$ to nC$_{19}$. The main organic chloro compounds are in the range of nC$_6$–nC$_{11}$. The major organic chlorine compounds were identified as 2-chloro, 2-methyl-propane, 2-chloro,2-methyl pentane, chloroethyl benzene, and 2-chloro, 2-phenyl propane [13].

TABLE 3 Physicochemical Properties of MX/
PVC-Derived Oil

Property	
Density (g/cm^3)	0.8520
Kinematic viscosity (cst @ 50°C)	1500
Flash point (°C)	7
Pour point (°C)	10
Moisture content (wt%)	0.24
Conradson carbon residue (wt%)	0.11
Calorific value (cal/g)	10600
Ash content (wt%)	<0.01
Chlorine content (wt%)	0.2
Nitrogen content (wt%)	<0.03
Sulfur (mg/kg)	<1
Carbonic acid (ppm)	42
Terephthalic acid (ppm)	<5

Firstly, we carried out the dechlorination of MX/PVC-derived oil over TR97305 catalyst, which had shown high dehydrochlorination activity in our earlier studies on model chloro alkanes [34]. Here the experiment was carried out only with the plastic-derived oil feed without any carrier gas. The activity pattern against time is presented in Figure 2. The catalyst showed an initial high removal of chlorine content, but the dechlorination activity stabilized after initial deactivation. It is well known that in catalytic dechlorination the catalyst will be deactivated due to the adsorption of HCl produced during the reaction on the catalyst surface [35]. The used catalyst was characterized by XRD. The XRD patterns of the fresh and used TR97305 catalysts are presented in Figure 3. The Fe_2O_3 phase is converted to $FeCl_2$ by interacting with the absorbed HCl. Blazso and Jakab, in their study of decomposition of PVC, reported that metal oxides with a large enough metal ion radius, such as iron oxide, are able to dehydrochlorinate the PVC by attracting chlorine and weakening the C–Cl bonds [36]. The conversion of iron oxide to chloride suggests that the oxide may act only as an absorbent. In order to elucidate this, a separate experiment was carried out by first treating the catalyst in HCl gas at 400°C for 3 h. During this treatment the iron oxide converted to iron chloride phase, and the dechlorination of MXPVC-derived oil was carried out using this catalyst. The dechlorination results on the HCl-treated catalysts are shown in Figure 4. A similar activity pattern to that of pure TR97305 catalyst is observed. This result suggests that the iron oxide initially acts as catalyst and converts to iron chloride phase by reacting with the

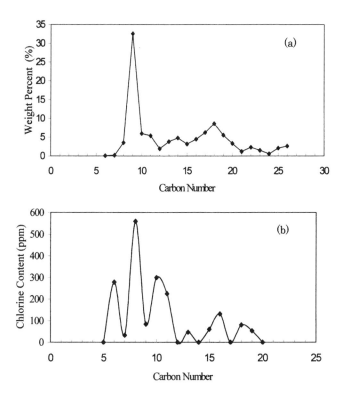

FIG. 1 (a) Carbon number distribution of all compounds (C-NP gram) in MX/PVC derived oil and (b) carbon number distribution of chlorine organic compounds (Cl-NP gram).

produced HCl, and that this iron chloride phase is also active for the dechlorination of chloro-organic compounds.

Based on the foregoing results, it is anticipated that the physically adsorbed HCl might be responsible for the initial decrease in activity. We carried out an experiment where the reaction was stopped after 10 h onstream; later, the catalyst was treated in He for 1 h at reaction temperature. These results are shown in Figure 5. After the first He treatment, the catalyst regained most of its activity. This behavior was found after this procedure was repeated twice. These results suggest that a continuous removal of reversible adsorbed HCl from the catalyst surface will be necessary to maintain a stable dechlorination activity of the catalyst. Further experiments were carried out using He as a carrier gas (5 mL/cm^3) over iron oxides α- and γ-Fe$_2$O$_3$ and iron oxide–carbon composites TR97305 and TR97300 at a liquid hourly space velocity (LHSV) of 40 h^1. Because the true

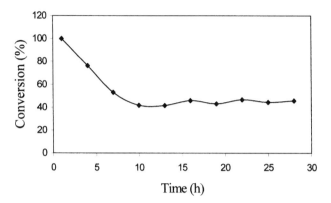

FIG. 2 MX/PVC-derived oil dechlorination activity over TR97305 catalyst with time in the absence of carrier gas.

value of gas hourly space velocity (GHSV) is difficult to determine due to the presence of a large number of hydrocarbons, only the LHSV, without considering He carrier flow rate, was used; the results are presented in Figure 6. The iron oxides and iron oxide–carbon composite showed a high activity in dechlorination. These catalysts showed similar high activity in the dechlorination of model chloroalkane, like chlorocyclohexane and 1-chloroheptane [37]. The effect of LHSV over γ-Fe$_2$O$_3$ catalyst was studied; the results are presented in Figure 7. The removal of chlorine compounds at lower space felocity was very high. None of the catalysts studied in this reaction greatly affected the carbon number distri-bution (C-NP gram) during the dechlorination. It is reasonable to think that the removal of chlorine may result in a change in the carbon number (equivalent to the b.p. of normal paraffin) distribution. However, the change in carbon number distribution was negligible, since the amount of chlorine-containing compounds in the original mixed plastics–derived oil was not high enough to produce any significant change in the C-NP gram.

Figure 8 shows the time-on-stream analysis over TR97305 and TR99300 cata-lysts. It is worth mentioning that these catalysts are very active and that no appre-ciable deactivation was observed in 24 h of time on stream. It has also reported on a Ni/SiO$_2$ catalyst system that a rapid deactivation (60%) occurred within 4 h during the dechlorination of chloro alkanes in the presence of He carrier gas, but the catalyst deactivation was suppressed by hydrogen [38]. It is well known that the suppression of catalyst deactivation in the presence of H$_2$ is due to a displacement of the hydrogen halide by dissociated hydrogen, which acts to clean the surface of the metal catalyst [40]. Surprisingly, in the present iron oxide catalyst system, in the presence of He carrier, high dechlorination activity with higher stability was achieved without using any hydrogen atmosphere.

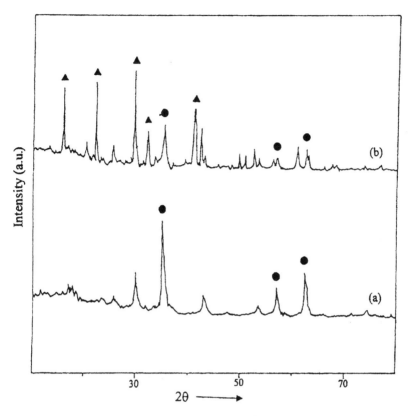

FIG. 3   X-ray diffraction patterns of the (a) fresh and (b) used TR97305 catalyst. (●) Fe_3O_4; (▲) $FeCl_2 \cdot 4H_2O$.

B.   Removal of Organic Chlorine Compounds from Municipal Waste Plastic–Derived Oil by Catalytic Dehydrochlorination over Iron Oxides (Fe_3O_4-Carbon), Zinc Oxide, Magnesium Oxide, and Red Mud

In this study, the dehydrochlorination of chloro-organic compounds from the municipal waste plastic–derived oil was carried out using various metal oxides, such as iron oxides, MgO, ZnO, and Red mud. The dehydrochlorination of model chloro alkanes, like chlorocyclohexane and 1-chloroheptane, was also carried over these catalysts as a test reaction. The municipal waste plastics (MWP) col-

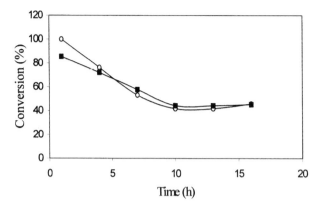

FIG. 4 Activity patterns of TR97305 catalysts before and after HCl treatment in the dechlorination of MX/PVC-derived oil. (○) Fresh catalyst; (■) HCl-treated catalyst.

lected from Kamagaya City of Japan were thermally graded to obtain MWP-derived oil. The composition of the municipal waste plastics from Kamagaya City is shown in Table 4. The thermal degradation was carried out at 410°C in a continuous-flow stirred-tank reactor. The oil obtained from municipal waste plastics contained about 600 ppm organic chlorine compounds. The physico-chemical characteristics of the derived oil were estimated using available standard methods; the results are summarized in Table 5.

The catalysts iron oxide (TR99701) and iron oxide–carbon composite (TR99300) were obtained from Toda Kogyo Corporation, Japan. The Red mud,

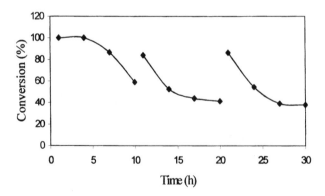

FIG. 5 Activity patterns during the sequential regeneration of the TR97305 catalyst during the dechlorination of MX/PVC-derived oil in the absence of carrier gas.

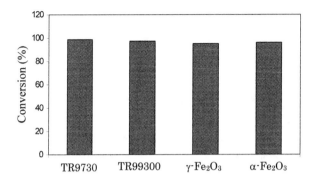

FIG. 6 Conversion data during the dechlorination of MX/PVC-derived oil over various iron oxide catalysts in the presence of He carrier gas.

a waste byproduct of alumina production was provided by Seydisehir Alumina Plant, Turkey. The ZnO and MgO were prepared by precipitation of their corresponding nitrates with ammonia solution. The physical characteristics of the catalysts used in this study are shown in Table 6. The catalyst TR99701 contains mainly the γ-Fe_2O_3 phase. The dehydrochlorination (DHC) reaction of MWP-derived oil and model chloro alkanes were carried out using a microreactor at atmospheric pressure. The details of the experimental procedure have been given in the previous section. In the use of dehydrochlorination of model compounds, about 0.5 mL catalyst was loaded in the reactor and pretreated in He flow (60 mL/min) at 300°C for 1 h. The reactant was fed into the reactor at a flow rate of 1.2 mL/min along with He carrier gas (30 mL/min).

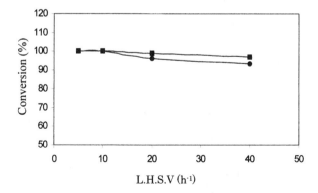

FIG. 7 Effect of space velocity during the dechlorination of MX/PVC-derived oil over TR97305 and TR99701 catalysts. (■) γ-Fe_2O_3; (●) TR97305.

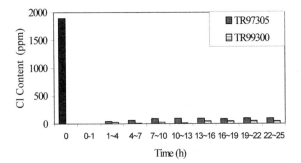

FIG. 8   Time-on-stream analysis: Dechlorination of MX/PVC-derived oil over different iron oxide–carbon composite catalysts.

The distribution of all the carbon compounds (C-NP gram) in MWP oil and the carbon number distribution of chloro-organic compounds (Cl-NP gram) are presented in Figures 9a and 9b, respectively. The C-NP gram shows that the hydrocarbons in the MWP-derived oil were distributed mainly in the b.p. range of nC_7 to nC_{11}. The Cl compounds are also distributed in almost the same range. All the characteristics of the MWP-derived oil indicate that it can be used as a fuel oil or upgraded to gasoline if the chloro-organic compounds are removed. The steady-state activity patterns of various catalysts in the dehydrochlorination of chloro-organic compounds of MWP-derived oil at the reaction temperature of 350°C are presented in Figure 10. All the catalysts except MgO showed high activity in the DHC of chloro-organic compounds. The TR99701 catalyst, which consists of γ-Fe_2O_3, is most active in this reaction. It is able to remove about 95% of chloro-organic compounds from the MWP-derived oil.

To understand more about the DHC activity of these catalysts, a model DHC reaction was carried using chlorocyclohexane (CCH) as a model compound. The results of the DHC reaction of CCH over these catalysts are shown in Figure 11. The TR99701, TR99300, and Red mud showed almost stable activity with reaction time, whereas the ZnO and MgO catalysts' activity decreased drastically. It is reported that in the DHC reaction, the catalytic activity decreased due to the presence of HCl produced during the reaction [38]. The XRD of the used ZnO catalyst show patterns related to $ZnCl_2$ and ZnO. It suggests that part of the ZnO converted to $ZnCl_2$ by reacting with HCl. In the case of MgO catalyst it is difficult to find the $MgCl_2$ phase due to its low stability at reaction temperature. However, we observed the leaching of MgO catalyst from the reactor, suggesting that a part of MgO is converted to its chloride phase. The decreased activity in the case of ZnO and MgO catalysts might be due to the formation of their corresponding chloride phases. These oxides showed high activity at initial

TABLE 4 Composition of Municipal Waste Plastics Collected from Kamagaya City of Japan

Plastics composition (wt%)					Elemental composition (wt%)						
PE	PP	PS	PVC	PET	ABS	C	H	N	S	Cl	O	Ash
30	17	27	8	13	5	77.02	10.39	0.46	0.01	1.10	4.94	6.08

TABLE 5 Physicochemical Properties of MX/
PVC-derived Oil

Property	
Density (g/cm^3)	0.8456
Kinematic viscosity at 50°C (mm^2/s)	1.337
Flash point (°C)	-3
Pour point (°C)	10
Moisture content (wt%)	1.5
Conradson carbon residue (wt%)	0.42
Calorific value (J/g)	43120
Ash content (wt%)	<0.05
Chlorine content (ppm)	600
Nitrogen content (wt%)	<0.12
Sulfur (mg/kg)	<1
Carbonic acid (ppm)	42
Terephthalic acid (ppm)	490

stages, and the decrease in activity with time suggests that these oxides might be acting as adsorbents rather than catalysts. The basic oxides, like MgO and ZnO, easily catch the liberated HCl during the dehydrochlorination. The iron oxide is also converted to $FeCl_2$ during the reaction. The conversion of iron oxide to chloride during the reaction suggests that the iron oxides are acting as adsorbents. However, the iron oxide catalyst activity remains constant with time, even after iron oxide converted to iron chloride. This result suggests that the iron chloride ($FeCl_2$) also acts as a dehydrochlorination catalyst. The Red mud catalyst, which contains mainly iron oxide, also shows a similar activity pattern to that of the iron oxide catalyst.

TABLE 6 Physical Characteristics of Iron Oxide and Iron
Oxide–Carbon Composite Catalysts

Catalyst	Surface area (m^2-g^{-1})	XRD
TR99701	13	γ-Fe_2O_3
TR99300	60	Fe_3O_4
ZnO	45	ZnO
MgO	72	MgO
Red mud	16	Fe_2O_3, SiO_2, Al_2O_3

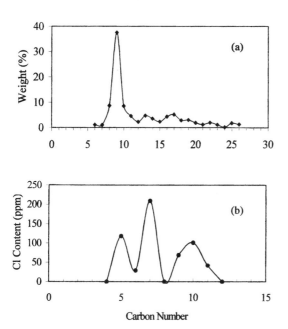

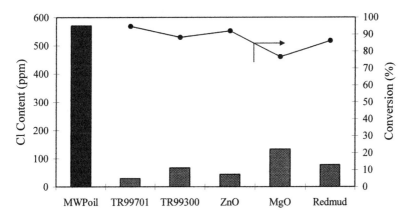

FIG. 9   (a) Carbon number distribution of all compounds (C-NP gram) in MWP-derived oil and (b) carbon number distribution of chloro-organic compounds (Cl-NP gram).

FIG. 10   Steady-state activity of various catalysts in the dehydrochlorination of MWP-derived oil (reaction temperature: 350°C).

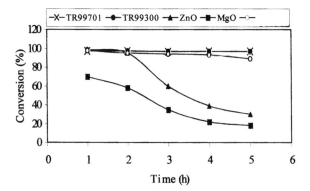

FIG. 11 Activity patterns of various catalysts in the dehydrochlorination of chlorocyclo-hexane with time (reaction temperature: 300°C).

Time-on-stream analyses over TR99701, ZnO, and MgO catalysts in the DHC of MWP-derived oil are shown in Figure 12. The amount chlorine removal decreased with reaction time in the case of the ZnO and MgO catalysts. This result suggests that these catalysts are deactivated with time. The effect of deactivation is more intense in the case of MgO. As reported earlier in the model of DHC reaction, in the dechlorination of MWP-derived oil also, the ZnO and MgO catalysts are deactivated with time. Here also these oxides showed high activity at the initial stages. As observed in the model compound DHC reaction, the decrease in activity with time suggests that these oxides act as adsorbents only. It is noteworthy that the iron oxide catalyst exhibits stable activity over the 16-h period

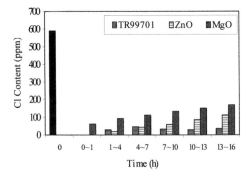

FIG. 12 Time-on-stream analysis: dehydrochlorination of MWP-derived oil over different catalysts (reaction temperature: 350°C).

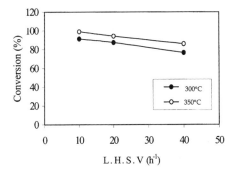

FIG. 13 Effect of space velocity during the dehydrochlorination of MWP-derived oil over TR99701 catalyst at different temperatures.

of study. Further studies were carried out using the more active TR99701 catalyst. The effects of space velocity at two different temperatures are given in Figure 13. In reporting the space velocities, once again only the LHSV is considered. It was found that DHC activity decreased with increase in LHSV from 10 to 40 h^{-1}. The change in activity with different space velocities also suggests that the iron oxides are acting as catalysts rather than adsorbents. The activity of the catalyst was found to be high at higher temperatures. All these results suggest that iron oxide catalysts effectively dehydrochlorinate the organic chlorine compounds from MWP-derived oil.

Figure 14 shows the C-NP gram of MWP-derived oil before and after the

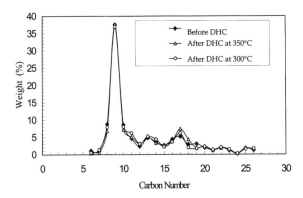

FIG. 14 Carbon number distribution of all compounds (C-NP gram) in MWP-derived oil before and after dehydrochlorination over TR99701 catalyst at different reaction temperatures with LHSV: 40 h^{-1}.

DHC. It is important to note that during the removal of chlorine compounds there is no considerable change in the carbon number distribution of MWP-derived oil. This result suggests that the catalyst has activity only in the DHC and no significant cracking activity. During the removal of chloro-organic compounds at different temperatures and space velocities, the iron oxide catalysts did not affect the carbon number distribution. Even if there is a slight change in the distribution at high temperature, it is a beneficial effect, because the higher hydrocarbons are degraded to lower hydrocarbons.

C. Catalytic Dehalogenation of Brominated Flame-Retardant-Containing High-Impact Polystyrene (HIPS-Br) Mixed with Polyvinyl Chloride–Derived Oil over Fe_3O_4-Carbon Composite Catalyst (TR00301)

Here we report for the first time the catalytic degradation of brominated flame-retardant-containing high-impact polystyrene (HIPS-Br) mixed with polyvinyl chloride (PVC) into halogen-free hydrocarbons. The effect of catalyst weight on the complete removal of halogen content during the degradation and the distribution of hydrocarbons in the waste plastic–derived fuel oil were discussed. The HIPS-Br [brominated high-impact polystyrene (Br-HIPS) contained decabromodiphenylethane as a flame retardant with antimony trioxide as a synergist; SAYTAX 8010] was obtained from Asahi Chemical Co., Ltd., and poly(vinyl chloride) (PVC Mw: 100,000) from Geon Chemical Co., Ltd., Japan. HIPS-Br contains 10.5 wt% bromine, and PVC contains 54 wt% chlorine. TR-00301 is an iron oxide carbon–composite catalyst, as mentioned in the previous two sections. In this case, thermal degradation of plastic was carried out in a glass reactor under atmospheric pressure by batch operation, and the thermal degradation products are brought into contact with the solid catalysts in the same reaction system. A schematic diagram of the experimental setup is shown in Figure 15. Briefly, 5 g of mixed plastics [weight ratio: HIPS-Br/PVC = 4/1)] was loaded into the reactor for catalytic degradation in vapor-phase contact. In a typical run, after setting the reactor, the reactor was purged with nitrogen gas at a flow rate of 30 mL/min and held at 120°C for 60 min to remove the physically adsorbed water from the catalyst and plastic sample. The nitrogen flow was then cut off, and the ractor temperature was increased to the degradation temperature (430°C) at a heating rate of 3°C/min. In a similar way, the thermal degradation of plastics was carried out in the absence of catalyst. The gaseous products were condensed to liquid products and trapped in a measuring jar. The hydrogen bromide and hydrogen chloride evolved from the degradation of the plastic mixture was trapped in a flask containing an aqueous solution of NaOH. The organic compounds containing chlorine and bromine atoms cannot be trapped in the flask,

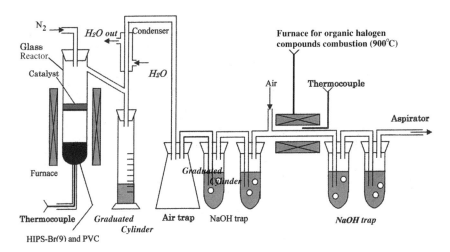

FIG. 15 Schematic experimental setup for waste plastic degradation.

so these gases were passed through the furnace at 900°C with air and converted into inorganic compounds (HBr and HCl) and trapped in NaOH solution. The quantitative analysis of the liquid products was performed using a gas chromatograph equipped with a flame ionization detector for hydrocarbons and a gas chromatograph equipped with atomic emission detector (AED; HP G2350A) for chlorinated and brominated hydrocarbons. A gas chromatograph with a mass selective detector [GC-MSD; HP 5973] was used for identification of the products. The amount of inorganic Br and Cl content was analyzed using an ion chromatograph (DIONEX, DX-120 Ion Chromatograph). The amount of Br and Cl in the residue was analyzed using a combustion flask and the solution further analyzed using the ion chromatograph.

The thermal and catalytic degradation of mixed plastics HIPS-Br and PVC in a weight ratio of 4 : 1 was carried out under atmospheric pressure in a batch process. The products of waste plastic degradation were classified into three groups: gas, liquid, and degradation residue. Table 7 shows the yield of products (gas, liquid, and residues) and the halogen (Cl and Br) content of the liquid products, average carbon number, and density of liquid products obtained during thermal and catalytic degradation of HIPS-Br/PVC at 430°C. The thermal degradation yielded more liquid products (48 wt%) than the catalytic degradation. The liquid products obtained during the catalytic degradation were found to decrease with the increase of catalyst amount from catalyst 1 g (44%) to catalyst 8 g (34%). However, the degradation residue increased with the catalyst amount, and it was

TABLE 7 Products Obtained During Thermal and Catalytic Degradation (Vapor Phase) of HIPS-Br/PVC(4/1) at 430°C; Catalyst: TR-00301

Method	Yield of degradation products, wt%			Liquid products: Cnp[b] Density [g/cm³]	Total halogen in oil: Br [ppm] Cl [ppm]
	Liquid [L]	Residue [R]	Gas [G][a]		
Thermal	48	31	21	10.4 0.92	55,000 4300
Cat. 1 g	44	44	12	9.3 0.87	910 2000
Cat. 2 g	37	51	12	9.3 0.84	0 840
Cat. 4 g	38	48	14	9.4 0.83	0 0
Cat. 8 g	34	53	13	9.0 0.78	0 0

[a] G = 100 − (L + R).
[b] Average carbon number of liquid products.

higher than with the thermal degradation (Table 7). The bromine content was 55,000 ppm, and the chlorine content was 4300 ppm, with an average carbon number (Cnp) of 10.4 for thermal degradation of Br-HIPS/PVC. The degradation of HIPS-Br/PVC was carried out with varying amount of catalyst. With 1 g of TR-00301, the bromine content in the oil decreased to 90% and the chlorine content to 50%, in comparison with the noncatalytic (thermal) degradation. The average carbon number of the liquid product decreased due to the cracking of higher-molecular-weight compounds. As a result, the density of the fuel oil also decreased, due to the catalytic degradation. When the catalyst amount was increased from 1 g to 2 g, the bromine was completely removed, but 840 ppm of chlorine was still present in the oil. Further increase in the amount of catalyst (4 g and 8 g) resulted in the complete removal of both chlorinated and brominated compounds from the oil. With 8 g catalyst, the average carbon number was decreased from 10.4 to 9.0 and the density from 0.92 to 0.78 g/cm³.

Figure 16 illustrates the carbon number distribution of the liquid products (C-NP gram) obtained by analyzing their gas chromatogram. The higher hydrocarbons decreased during the catalytic degradation of HIPS-Br/PVC with TR-00301 at 430°C. The hydrocarbons (C_6–C_{11}) during the thermal degradation were 45 wt%, but they increased to about 60 wt% for catalytic degradation. As a result, the hydrocarbons in the range of C_{16-20} decreased. As with the C-NP gram, the carbon number distribution of chlorinated and brominated hydrocarbons was

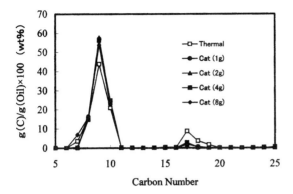

FIG. 16 C-NP gram of liquid product from thermal and catalytic degradation of HIPS-Br/PVC (4/1) at 430°C.

prepared from the gas chromatogram obtained using a gas chromatograph with an atomic emission detector. The chlorine compounds are presented as a Cl-NP gram and brominated compounds as a Br-NP gram in Figures 17 and 18, respectively. The hydrocarbons containing chlorine atoms are distributed in the range of C_9–C_{14}, with a peak at C_{11}, and brominated compounds were observed at C_{11} only. The liquid products were analyzed by GC-MS to identify the compounds. The presence of brominated compounds during the thermal degradation and catalytic degradation with 1 g of catalyst (TR-00301) was observed, and one of the main compounds was identified as 1-(bromoethyl-4-methyl benzene. The main chlorinated compound in the liquid products of thermal degradation and

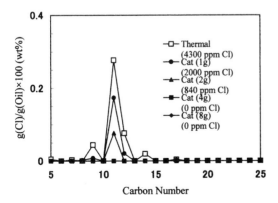

FIG. 17 Cl-NP gram of liquid product from thermal and catalytic degradation of HIPS-Br/PVC (4/1) at 430°C.

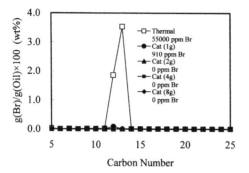

FIG. 18 Br-NP gram of liquid product from thermal and catalytic degradation of HIPS-Br/PVC (4/1) at 430°C.

catalytic degradation (1 g and 2 g of TR-00301) was identified as 1-chloroethyl benzene. The amount of Cl and Br in the oil decreased with the increase of catalyst amount and decreased to an undetectable level (zero) with more than 4 g catalyst (Figs. 17 and 18). We have suggested in our previous report that these halogenated organic compounds are formed via the reaction of styrene or styrene derivatives produced from polystyrene degradation and the hydrogen halides (HCl and HBr) derived from PVC and polybrominated flame retardant degradation. Richard et al. reported that high-temperature degradation of polybrominated flame-retardant materials produced Br-benzenes, Br-phenols, polybrominated dibenzodioxins (PBDD), and polybrominated dibenzofurans (PBDFs) [40]. However, they were subsequently destroyed at high temperature (800°C). Dioxin (PHDD) and dibenzofuran formation (PHDF) during the thermal treatment in the presence of air and plastics containing polybrominated diphenyl ether as a flame retardant has been reported [41]. In the present study we have not found any such compounds in the degradation products.

IV. CONCLUDING REMARKS

When PVC-containing (model) mixed plastics and municipal plastic wastes are thermally degraded to obtain fuel oil, various chloro-organic compounds, such as 2-chloro 2-methyl propane, 2-chloro 2-methyl pentane, chloroethyl benzene and 2-chloro 2-phenyl propane are formed in the liquid products in addition to hydrocarbons. Iron oxide (Fe_3O_4)–carbon composite catalysts are effective for dehydrochlorination of these compounds to their corresponding alkenes and hydrogen chloride selectively in a fixed-bed flow-type reactor, and catalytic activity remains stable for a long time (24 h). A similar catalytic effect is confirmed in the dehydrochlorination of model chloro alkanes, such as chlorocyclohexane; iron oxide (F_3O_4) selectively converts the chlorocyclohexane to cyclohexene and hydrogen chloride with a selectivity of more than 98%. Other iron oxides, α-Fe_2O_3 and γ-Fe_2O_3, are also effective in this dehydrochlorination reaction. Other metal oxides, such as ZnO and MgO, also reveal catalytic activity at the initial stages of the reaction but deactivate drastically with time. The iron oxides are converted to $FeCl_2$ during reaction, which suggests that iron oxides also act as sorbents. However, the iron oxide catalyst's activity remains constant with time, even after iron oxide converted to iron chloride, which implies that iron chloride ($FeCl_2$) also acts as a dehydrochlorination catalyst. Iron oxide–carbon composite catalyst (TR-00301) is very effective for degradation of the HIPS-Br and PVC mixed plastics and simultaneous removal of chlorinated and brominated organic compounds from the liquid products. In this review, we demonstrated that our catalytic process can provide the possibility of the secondary use of waste materials (plastic wastes). It is an important concept in life cycle assessment of plastics and the promotion of the recycling movement.

REFERENCES

1. Monthly report of the Plastic Waste Management Institute, Japan, June 1998.
2. J Aguado and D Serrano. Feedstock Recycling of Plastic Waste. Series editor: James H. Clark, RSC Clean Technology Monographs: The Royal Society of Chemistry, UK, 1999, p 1–29.
3. C-H Wu, C-Y Chang, J-L Hor, S-H Shih, L-W Chen, and F-W Chang. Canadian Journal of Chemical Engineering 72:644–650, 1994.
4. D Braun. Pure Applied Chemistry 26:173–192, 1971.
5. G Ayrey, BC Head, and RC Poller. Journal of Polymer Science, Macromolecular Reviews 8:1–6, 1974.
6. D Braun, B Bohringer, B Ivan, T Klen, and F Tudos. European Polymer Journal 22:299–304, 1986.
7. R Knumann and H Bockhorn. Combustion Science and Technology 101:285–299, 1994.
8. M Blaszo. Rapid Communication of Mass Spectrometry 12:1–4, 1998.
9. MJP Slapah, JMN Van Kasteren, and AAH Drinkenburg. Waste Management 20: 463–467, 2000.
10. A Jimenez, V Berenguer, J Lopez, and A Sanchez. Journal of Applied Polymer Science 50:1665–1674, 1993.
11. R Miranda, H Pakdel, C Roy, H Darmstadt, and C Vasile. Polymer Degradation and Stability 66:247–255, 1999.
12. IC McNeill, L Memetea, and WJ Cole. Polymer Degradation and Stability 47:33–57, 1995.
13. MA Uddin, Y Sakata, Y Shiraga, A Muto, and K Murata. Industrial & Engineering Chemistry 38:1406–1410, 1999.
14. Y Shiraga, MA Uddin, A Muto, M Narazaki, Y Sakata, and K Murata. Energy and Fuels 13:428–432, 1999.
15. Y Sakata, MA Uddin, A Muto, M Narazaki K Murata, and M Kaji. Polymer Recycling 2:309–314, 1996.

16. AJ Chandler, TT Eighmy, J Hartlen, O Hjelmar, DS Kosson, SE Sawell, HA van der Sloot, and J Vehlow. Municipal Solid Waste Incinerator Residues. Elsevier, Amsterdam, 1997.
17. T Gamse, F Steinkellner, R Marr, P Alessi, and I Kikic. Industrial and Engineering Chemistry Research 39:4888–4890, 2000.
18. F Bertini, G Audisio, and K Kiji. Journal of Analytic and Applied Pyrolosis 33: 213–230, 1995.
19. MA Uddin, K Ikeuchi, N Lingaiah, H. Tanikawa, A Muto, and Y Sakata. Proceeding of 2nd International Symposium on Zeolite and Microporous Crystals (ZMPC2000), Sendai, 2000, p 283.
20. S Horikawa, Y Takai, H Ukei, N Azuma, and A Ueno. Journal of Analytical and Applied Pyrolysis 52:167–179, 1999.
21. W Kaminsky, B Schlesselmann, and C Simon. Polymer Degradation and Stability 51: 151–158, 996.
22. K Saito. Annual Report of the Agency of Industrial Science and Technology, Ministry of International Trade and Industry, Japan, 1990, p 37.
23. N Lingaiah, MA Uddin, A Muto, Y Sakata, and T Imai. Proceeding of 1st International Symposium on Feedstock Recycling of Plastics, Sendai, 1999, pp 119–120.
24. EN Balco, E Zybylski, and F VanTrentini. Applied Catalysis B Environment 2:1–8, 1993.
25. A Gampine, and DP Eyman. Journal of Catalysis 179:315–325, 1998.
26. N Lingaiah, MA Uddin, A Muto, and Y Sakata. Chemical Communications, 657–658, 1999.
27. T Imai, T Matsui, and Y Fujii. Proceeding of the International Chemical Congress of Pacific Basin Societies (PACIFICHEM 2000), Hawaii, 2000, ENVR273.
28. M Brebu, MA Uddin, A Muto, Y Sakata, and C Vasile. Energy and Fuels 15:559–564, 2001.
29. M Brebu, MA Uddin, A Muto, Y Sakata, and C Vasile. Energy and Fuels 15:565–570, 2001.
30. N Lingaiah, MA Uddin, A Muto, Y Sakata, T Imai, and K Murata. Applied Catalysis A: General 207: 79–84, 2001.
31. N Lingaiah, MA Uddin, A Muto, T Imai, and Y Sakata. Fuel 80:1901–1905, 2001.
32. Annual Report of the New Energy and Industrial Development Organization, Japan, Project No. 8G-010, 1998.
33. K Murata and T Makino. Nippon Kagaku Kaishi, 192–2000, 1975.
34. N Lingaiah, MA Uddin, Y Shiraga, H Tanikawa, A Muto, Y Sakata, and T Imai. Chemical Letters, 1321–1322, 1999.
35. B Coq, G Ferrato, and F Figueras. Journal of Catalysis 101:434–445, 1986.
36. N Blazso and E Jakab. Journal of Analytic and Applied Pyrolosis 49:125–143, 1999.
37. N Lingaiah, MA Uddin, K Morikawa, A Muto, K Murata, and Y Sakata. Green Chemistry 3: 74–75, 2001.
38. G Tavoularis and MA Keane. Applied Catalysis A 182:309–316, 1999.
39. G Tavoularis and MA Keane. Journal of Molecular Catalysis A 142:187–199, 1999.
40. CS Richard, AR Wayne, AT Debra, and B Dellinger. Chemosphere 23:1197–1204, 1991.
41. J Vehlow, B Bergfeldt, K Jay, H Seifert, T Wanke, and FE Mark. Waste Management Research 18:131–140, 2000.

Index